Engineering Materials

This series provides topical information on innovative, structural and functional materials and composites with applications in optical, electrical, mechanical, civil, aeronautical, medical, bio- and nano-engineering. The individual volumes are complete, comprehensive monographs covering the structure, properties, manufacturing process and applications of these materials. This multidisciplinary series is devoted to professionals, students and all those interested in the latest developments in the Materials Science field, that look for a carefully selected collection of high quality review articles on their respective field of expertise.

Indexed at Compendex (2021)

More information about this series at https://link.springer.com/bookseries/4288

Onoyivwe Monday Ama · Suprakas Sinha Ray ·
Peter Ogbemudia Osifo
Editors

Modified Nanomaterials for Environmental Applications

Electrochemical Synthesis, Characterization, and Properties

Springer

Editors
Onoyivwe Monday Ama
Department of Chemical Sciences
University of Johannesburg
Johannesburg, South Africa

Peter Ogbemudia Osifo
Department of Chemical Engineering
Faculty of Engineering
Vaal University of Technology
Vanderbijlpark, South Africa

Suprakas Sinha Ray
Department of Chemical Sciences
University of Johannesburg
Johannesburg, South Africa

Centre for Nanostructures and Advanced Materials, DSI-CSIR Nanotechnology Innovation Centre
Council for Scientific and Industrial Research
Pretoria, South Africa

ISSN 1612-1317 ISSN 1868-1212 (electronic)
Engineering Materials
ISBN 978-3-030-85554-3 ISBN 978-3-030-85555-0 (eBook)
https://doi.org/10.1007/978-3-030-85555-0

This Springer imprint is published by the registered company Springer Nature Switzerland AG
The registered company address is: Gewerbestrasse 11, 6330 Cham, Switzerland

Preface

Through the last three decades, nanomaterials offer unique properties that can be exploited in novel environmental applications. The electrochemical, magnetic, and optical properties of these materials are used to detect the presence of environmental pollutants or other undesirable heavy metals or the presence of organic ions in water. Therefore, the earth's and physical/biological components immeasurable contaminations lead to environmental issues such as diseases and death in the severe case detox environmental issues such as diseases and death in the severe case due to human and anthropogenic activities. Recently, globally awareness, sterner laws, and their implementation in most developed nations leading to a huge level of environmental protection are severe long-term effects with the advent of nanotechnology which depends on the ability to design, control and create materials at the nanoscale, environmental related issues have been increasingly undertaken through prevention, monitored using sensors and the use of nano-adsorbents for contaminants treatment by using novel nanomaterials developed using innovative and cutting edge synthesis techniques. In this book, there are nine chapters, the various electrochemical techniques employed in the synthesis of different nanomaterials are presented to the readers. Due diligence is also given how various nanomaterials created using these electrochemical techniques are employed to determine the surface morphologies, their various surface sizes and distribution, active functional groups on their surfaces, their elementary composition, magnetic and electrical properties that impact their use as sensors and nano-adsorbents. The ensuing chapters also deal with heavy metals ions detection via electrochemical processes using electrochemical synthesized sensors, the use of synthesized nanomaterials in the prevention of corrosion, the utilization of surface modified materials in the treatment of industrial wastewater containing heavy metals, their photochemical use in the detection of pharmaceutical waste and the application of these electrode materials as biosensors for environmental utilization. This book is expected to be a research guide for students and scientists to go on-board with courses development on material and environmental sciences and engineering, giving insight to readers into the captivating world of nanotechnology and how their advent has currently transformed the world and the environment at large. Therefore, this book will play a good row in our community that will cut across electrochemistry, water science, materials science, and nanomaterials. Hence, its application successfully highlights the nanomaterials and environmental.

Short description, this book addresses fundamental issues that persist concerning nanomaterials for environmental application, which cuts across electrochemistry, materials science, water science, and nanotechnology. This is an ideal book for water scientists, material scientists, researchers, engineers, and chemicals; therefore, it also includes undergraduate and post-graduate students interested in this exciting field of research. Furthermore, this book will also help industrial researchers and R&D managers who want to bring advanced nanomaterials for environmental applications into the marketplace.

All authors are sincerely appreciated for their valuable contribution as well as reviewers of proposals and chapter manuscripts. Our thanks to Dr. Zachary Evenson, The Associate Editor at Springer Nature, for suggestions and guidance during every various stage of the book preparation and production. Our thanks to the University of Johannesburg, Department of Chemical Science, the University of Vaal, the Department of Chemical Engineering, the Council for Scientific and Industrial Research (CSiR), the Department of Science and Innovation is highly appreciated.

Johannesburg, South Africa — Dr. Onoyivwe Monday Ama
Johannesburg/Pretoria, South Africa — Prof. Suprakas Sinha Ray
Vanderbijlpark, South Africa — Prof. Peter Ogbemudia Osifo

Contents

1 Electrochemical Synthesis of Nanomaterials 1
D. M. Jeroh
1.1 Introduction 1
1.2 Nanomaterials 2
1.2.1 Quantum Dots 2
1.2.2 Carbon Nanotubes 3
1.2.3 Graphenes and Fullerenes 3
1.3 Nanomaterials Synthesis 3
1.3.1 Electrochemical Synthesis 4
1.4 Parameters Influencing Electrodeposition of Nanomaterials 6
1.4.1 Temperature 6
1.4.2 Current Density 6
1.4.3 Nature of Substrate 7
1.4.4 Electrolyte pH 7
1.4.5 Voltage 7
1.5 Conclusion 8
References 8

2 Electrochemical Characterization of Nanomaterials 11
Tladi Gideon Mofokeng, Mpho Phillip Motloung,
Onoyivwe Monday Ama, and Suprakas Sinha Ray
2.1 Nanostructured Materials 11
2.2 Electrochemical Techniques for Nanomaterial Characterization 12
2.3 Static Techniques 13
2.3.1 Potentiometry 13
2.4 Dynamic Techniques 14
2.4.1 Electrochemical Impedance Spectroscopy (EIS) 14
2.4.2 Galvanostatic Techniques 17
2.4.3 Cyclic Voltammetry 18
2.4.4 Differential Pulse Voltammetry 20

2.5 Conclusion and Future Perspective 21
References 22

3 Electrochemical Detection of Heavy Metals 25
Uyiosa Osagie Aigbe, Robert Birundu Onyancha, Kingsley Eghonghon Ukhurebor, Otolorin Adelaja Osibote, Onoyivwe Monday Ama, Harrison Ifeanyichukwu Atagana, Peter Osifo Ogbemudia, and Seyi Philemon Akanji
3.1 Introduction 26
3.2 Voltammetry Used for Heavy Metals Detection 29
3.2.1 Cyclic Voltammetry (CV) 31
3.2.2 Differential Pulse Voltammetry (DPV) 33
3.2.3 Square Wave Voltammetry (SWV) 33
3.2.4 Linear Sweep Voltammetry (LSV) 34
3.3 Electroanalytical Techniques for Heavy Metals Detection 35
3.3.1 Stripping Voltammetry Methods (SVM) 35
3.4 Types of Substrate Electrodes Used for Heavy Metals Detection 40
3.4.1 Glassy Carbon Electrode (GCE) 40
3.4.2 Screen-Printed Electrodes (SPE) 41
3.4.3 Carbon Paste Electrode (CPE) 41
3.5 Detection of Heavy Metals Using Nanomaterials 42
3.5.1 Carbon Nanomaterials for Heavy Metals Detection 43
3.5.2 Metal Nanomaterial for Detection of Heavy Metals 50
3.5.3 Metal Oxide Nanomaterials for Heavy Metals Ions Detection 52
3.5.4 Polymers Modified Electrode for Heavy Metal Detection 54
3.6 Conclusion 56
References 57

4 Sensing the Presence of Inorganic Ions in Water: The Use of Electrochemical Sensors 65
Kabir Opeyemi Otun, Idris Olayiwola Azeez, Onoyivwe Monday Ama, William Wilson Anku, Uyiosa Osagie Aigbe, Kingsley Eghonghon Ukhurebor, and Robert Birundu Onyancha
4.1 Introduction 66
4.2 Principle of an Electrochemical Sensor 68
4.3 Materials Used for Electrochemical Sensing of Inorganic Ions in Water 69
4.3.1 Metal Nanoparticles 69
4.3.2 Metal Compounds 70
4.3.3 Graphene 70
4.3.4 Polymers 72
4.3.5 Quantum Dots 72
4.3.6 Carbon Nanotubes 73

4.4 Electrochemical Determination of Inorganic Ions in Water 75
4.4.1 Inorganic Anions 75
4.4.2 Inorganic Cations 80
4.5 Conclusion and Outlook 83
References 84

5 Corrosion Prevention: The Use of Nanomaterials 91
A. Momoh, F. V. Adams, O. Samuel, O. P. Bolade, and P. A. Olubambi
5.1 Introduction 92
5.2 Nanostructured Materials 92
5.3 Nanomaterials in Corrosion Mitigation 93
5.4 Application of Metal/Metal Oxide Nanoparticles as Corrosion Inhibitor 94
5.5 Nanocomposites Corrosion Inhibitors 94
5.6 Nanocontainers as Corrosion Storage Inhibitors 96
5.7 Metal Oxidation Control at Varied Temperature Using Nanomaterials 97
5.7.1 Moderate-Temperature 97
5.7.2 High Temperature 98
5.8 Tribocorrosion Control Using Nanomaterials 99
5.9 Corrosion Protection of Aerospace Alloys Using Nanomaterials 100
5.10 Challenges of the Use of Nanomaterials as Corrosion Mitigation Substances 100
5.11 Conclusions 101
References 101

6 Application of Surface-Modified Electrode Materials in Wastewater Treatment 107
Onoyivwe Monday Ama, Khotso Khoele, Penny Poomani Govender, and Suprakas Sinha Ray
6.1 Introduction 108
6.2 Wastewater and Organic Pollutants 109
6.2.1 Wastewater 109
6.2.2 Organic Pollutants and Their Impact 109
6.3 Photocatalysists and Their Applications 110
6.3.1 Photocatalysts 110
6.3.2 Nanostructured Materials and Photoanodes 110
6.4 Photocatalysis and Advance Oxidation Measurements 111
6.5 Factors that Affect Degradation Processes on Wastewater Pollutants Using a Photoelectrochemical Technique 112
6.5.1 Light Captivation Properties 112
6.5.2 The Reduction and Oxidation Rates on the Surface by the Photo-generated Electrons and Holes 113
6.5.3 Recombination Rate of Electron Charges 113
6.6 Surface-Modified Photoanodes 113

6.7 Morphological Adjustments ... 114
6.8 Synthetic Technique on Novel Materials ... 114
6.9 Variation of Synthetic Process Parameters ... 114
6.9.1 Temperature ... 114
6.9.2 pH Range ... 115
6.10 Summary ... 115
6.10.1 Wastewater Pollutants and Their Impacts ... 115
6.10.2 Photocatalysis and Advanced Oxidation Methods ... 115
6.10.3 Factors that Affect Degradation Processes on Wastewater Pollutants Using a Photoelectrochemical Technique ... 116
6.10.4 Methods Modifying Surface Properties of Photoanodes and Their Applications ... 116
6.11 Conclusions ... 116
References ... 117

7 Photoelectrochemical Application of Nanomaterials ... 121
Seyi Philemon Akanji, Onoyivwe Monday Ama, Omotayo A. Arotiba, Duduzile Nkosi, Idris Azeez Olayiwola, Uyiosa Osagie Aigbe, Robert Birundu Onyancha, and Kingsley Eghonghon Ukhurebor
7.1 Introduction ... 126
7.2 Photoelectrochemical Applications of Nanomaterials ... 127
7.2.1 Applications in Biosensors ... 127
7.2.2 Applications in Photoelectrochemical Water Splitting ... 134
7.2.3 Applications in Photovoltaics ... 140
7.3 Conclusions and Future Scope ... 148
References ... 148

8 Electrode Materials for Pharmaceuticals Determination ... 155
Azeez Olayiwola Idris, Onoyivwe Monday Ama, Kabir Opeyemi Otun, Seyi Philemon Akanji, Usisipho Feleni, Bhekie Mamba, Robert Birundu Onyancha, Uyiosa Osagie Aigbe, and Kingsley Eghonghon Ukhurebor
8.1 Introduction ... 157
8.1.1 Carbon Paste Electrode (CPE) ... 158
8.1.2 Boron-Doped Diamond Electrode (BDDE) ... 163
8.1.3 Glassy Carbon Electrode (GCE) ... 167
8.1.4 Pencil Graphite Electrode (PGE) ... 174
8.1.5 Screen-Printed Carbon Electrode ... 177
8.1.6 Conclusion ... 179
References ... 180

9 Biosensing Applications of Electrode Materials 187
Kingsley Eghonghon Ukhurebor, Uyiosa Osagie Aigbe, Robert Birundu Onyancha, Onoyivwe Monday Ama, Can-voro Osemwengie Amadasun, Joseph Onyeka Emegha, Otolorin Adelaja Osibote, Samuel Ogochukwu Azi, Azeez Olayiwola Idris, and Kabir Opeyemi Otun
9.1 Introduction 188
9.2 Biosensors Operational Principles 190
9.3 Electronic Principles of Biosensor 192
9.4 Biosensors Mode of Operations 193
9.5 Advances in Biosensors Machineries 194
9.5.1 Electrochemical Biosensors 194
9.5.2 Biosensors Used in Environmental Monitoring and Management 196
9.5.3 Optical/Visual Biosensors 197
9.5.4 Silica, Quartz/Crystal and Glass Biosensors 198
9.5.5 Nanomaterials-Built Biosensors 199
9.5.6 Genetically Programmed/Synthetic Fluorescent Biosensors 199
9.5.7 Microbial-Built Biosensors 200
9.6 Benefits and Technological Comparison of Some Biosensors 201
9.7 Research Trends, Future Challenges and Limitations of Biosensor Machineries 204
9.8 Applications of Nanostructured Conducting Polymers for Electrochemical Biosensors 207
9.8.1 Nanostructured Conducting Polymers-Integrated Electrochemical Biosensors 208
9.9 Graphene and Its Applications for Biosensors and Biosensing-Based Electrode Systems 214
9.9.1 Graphene-Built Enzymatic Electrodes 215
9.9.2 Graphene-Built Electrochemical DNA Sensors 218
9.9.3 Graphene-Built Electrochemical Immunosensors 219
9.9.4 Some Commercial Advances in Graphene Biosensors and Biosensing-Based Electrode Systems 220
9.10 Conclusion and Future Contribution to Knowledge 221
References 222

Index 233

Editors and Contributors

About the Editors

Dr. Onoyivwe Monday Ama received a Ph.D. degree in Chemistry at the University of Johannesburg, South Africa 2017. He is the author of thirty-two international journals in high impact ten book chapters and one book. Dr. Ama multi-disciplinary work cuts across electrochemistry, water science, materials science, and nanotechnology. It successfully highlights the applicability of exfoliated graphite nanocomposites in electrochemical technology. His contribution paves the way for the development of electrochemical reactors, powered by solar light in our society, for water treatment shortly.

Prof. Suprakas Sinha Ray is a chief researcher in nanostructured materials at the Council for Scientific and Industrial Research (CSIR) with a Ph.D. in physical chemistry from the University of Calcutta in 2001 and Manager of the Centre for Nanostructures and Advanced Materials, DSI-CSIR Nanotechnology Innovation Centre. Ray's current research focuses on advanced nanostructured materials and their applications. He is one of the most active and highly cited authors in the field of polymer nanocomposite materials, and he has recently, been rated by Thomson Reuters as being one of the Top 1% most impactful and influential scientists and Top50 high impact chemists (out of 2 mil Chemists worldwide).

Professor Ray is the author of five books, co-author of three edited books, 32 book chapters on various aspects of polymer-based nanostructured materials & their applications, and author and co-author of 370 articles in high-impact international journals, 30 articles in national and international conference proceedings. He also has six patents and seven new demonstrated technologies shared with colleagues, collaborators, and industrial partners. So far, his team commercialized 19 different products. His honors and awards include South Africa's most **Prestigious 2016 National Science and Technology Award (NSTF); Prestigious 2014 CSIR-wide Leadership award; Prestigious 2014 CSIR Human Capital development award; Prestigious 2013 Morand Lambla Awardee** (top award in the field of polymer processing worldwide), International Polymer Processing Society, USA. He is also appointed as

Extraordinary Professor, University of Pretoria and Distinguished Visiting Professor of Chemistry, University of Johannesburg.

Prof. Peter Ogbemudia Osifo currently hold the position of an Acting Executive Dean in the Faculty of Engineering and Technology at the Vaal University of Technology (VUT), Vanderbijlpark, South Africa. In 1991, I received a Bachelor of Engineering (B.Eng.) Degree in Chemical Engineering from the University of Benin, Nigeria. In 1992, I joined the Oil Test Services, a petroleum servicing company, in Port-Harcourt, Nigeria. After several years of working, I started research programmed in the department of Biotechnology for Master of Technology degree at Durban University of Technology in South Africa, where I graduated in 2001 with Master's in Biotechnology. After which I continued with my doctoral programmed at the North-West University, South Africa, where I graduated in 2007 with Ph.D. in Chemical Engineering. I joined the Vaal University of Technology as a lecturer since 2003 and, in January 2008 to August 2015, and July 2018 to December 2018, I was the Head of Department of Chemical Engineering, VUT. My teaching areas are primarily on Reactor Technology, Process Optimization and Control and Fluid Mechanics. My current research focus areas include clean technology of wastewater treatment through biodegradation of domestic and industrial wastewater, including membrane bioreactor design and application for wastewater treatment and reuse. Other research focus areas include the use of adsorbents for heavy metals removal from contaminated water, and membrane development from bioresources materials for applications in fuel cells to produce electricity. I have published over 21 papers in accredited journals, eight books chapters and an author of a book; enhanced chitosan material for water treatment. I have attended several local and international conferences. I have examined more than 10 dissertations from various universities and graduated six masters' and one Ph.D. students. I am NRF rated researcher with C2 and have received several research funding from NRF, CHIETA, WRC, SASOL, and TIA for research projects.

Contributors

F. V. Adams Department of Petroleum Chemistry, American University of Nigeria, Yola, Nigeria;
School of Mining, Metallurgy and Chemical Engineering, Centre for Nanomechanics and Tribocorrosion, University of Johannesburg, Doornfontein, Johannesburg, South Africa

Uyiosa Osagie Aigbe Department of Mathematics and Physics, Faculty of Applied Sciences, Cape Peninsula University of Technology, Cape Town, South Africa

Seyi Philemon Akanji Department of Chemical Science, University of Johannesburg, Doornfontein, Johannesburg, South Africa

Onoyivwe Monday Ama Department of Chemical Science, University of Johannesburg, Doornfontein, Johannesburg, South Africa

Can-voro Osemwengie Amadasun Department of Chemical Sciences, University of Johannesburg, Johannesburg, South Africa

William Wilson Anku CSIR-Water Research Institute, Accra, Ghana

Omotayo A. Arotiba Department of Chemical Science, University of Johannesburg, Doornfontein, South Africa

Harrison Ifeanyichukwu Atagana Department of Life and Consumer Sciences, College of Agriculture and Environmental Sciences, University of South Africa, Pretoria, South Africa

Idris Olayiwola Azeez Nanotechnology for Water Sustainability (NaNOWS) Research Unit, College of Science, Engineering and Technology, University of South Africa, Johannesburg, South Africa

Samuel Ogochukwu Azi Department of Physics, Faculty of Physical Sciences, Ambrose Alli University, Ekpoma, Edo State, Nigeria

O. P. Bolade Department of Petroleum Chemistry, American University of Nigeria, Yola, Nigeria

Joseph Onyeka Emegha Department of Physics, Faculty of Physical Sciences, Ambrose Alli University, Ekpoma, Edo State, Nigeria

Usisipho Feleni Institute for Nanotechnology for Water Sustainability (iNanoWS), Florida Campus, College of Science, Engineering and Technology, University of South Africa, Johannesburg, South Africa

Penny Poomani Govender Department of Chemical Science, University of Johannesburg, DoornfonteinJohannesburg, South Africa

Azeez Olayiwola Idris Institute for Nanotechnology and Water Sustainability (iNanoWS), Florida Campus, College of Science Engineering and Technology, University of South Africa, Johannesburg, South Africa;
Department of Physics, Faculty of Physical Sciences, University of Benin, Benin City, Edo State, Nigeria

D. M. Jeroh Department of Physics and Industrial Physics, Nnamdi Azikiwe University, Anambra State, Awka, Nigeria

Khotso Khoele Department of Chemical, Metallurgical and Materials Engineering, The Tshwane University of Technology, Pretoria, South Africa

Bhekie Mamba Institute for Nanotechnology for Water Sustainability (iNanoWS), Florida Campus, College of Science, Engineering and Technology, University of South Africa, Johannesburg, South Africa

Tladi Gideon Mofokeng DST/CSIR National Centre for Nanostructured Materials Council for Scientific and Industrial Research, Pretoria, South Africa

A. Momoh Department of Petroleum Chemistry, American University of Nigeria, Yola, Nigeria

Mpho Phillip Motloung DST/CSIR National Centre for Nanostructured Materials Council for Scientific and Industrial Research, Pretoria, South Africa

Duduzile Nkosi Department of Chemical Science, University of Johannesburg, Doornfontein, South Africa

Peter Osifo Ogbemudia Department of Chemical Engineering, Vaal University of Technology, Vanderbijlpark, South Africa

Idris Azeez Olayiwola Nanotechnology for Water Sustainability (NaNOWS) Research Unit, College of Science, Engineering and Technology, University of South Africa, Johannesburg, South Africa

P. A. Olubambi School of Mining, Metallurgy and Chemical Engineering, Centre for Nanomechanics and Tribocorrosion, University of Johannesburg, Doornfontein, Johannesburg, South Africa

Robert Birundu Onyancha Department of Physics and Space Science, School of Physical Sciences and Technology, Technical University of Kenya, Nairobi, Kenya

Otolorin Adelaja Osibote Department of Mathematics and Physics, Faculty of Applied Sciences, Cape Peninsula University of Technology, Cape Town, South Africa

Kabir Opeyemi Otun Institute for the Development of Energy for African Sustainability, University of South Africa, Johannesburg, South Africa;
Department of Chemical, Geological and Physical Sciences, College of Pure and Applied Sciences, Kwara State University, MaleteIlorin, Nigeria

Suprakas Sinha Ray Department of Chemical Science, University of Johannesburg, Doornfontein, Johannesburg, South Africa;
DST-CSIR National Center for Nanostructured Materials Council for Scientific and Industrial Research, Pretoria, South Africa

O. Samuel Department of Chemical Engineering, Federal Polytechnic, Mubi, Adamawa State, Nigeria

Kingsley Eghonghon Ukhurebor Department of Physics, Faculty of Science, Edo State University Uzairue, Auchi, Edo State, Nigeria;
Climatic/Environmental/Telecommunication Unit, Department of Physics, Edo University Iyamho, Okpella, Edo, Nigeria

Chapter 1
Electrochemical Synthesis of Nanomaterials

D. M. Jeroh

Abstract This chapter discusses the process involved in synthesizing nanomaterials using electrochemical reactions. Nanomaterials and the various classes are introduced. Electrochemical synthesis is reported, some factors which influence the properties exhibited by nanomaterials are also discussed briefly. These factors which influence the properties of nanomaterials are current density, electrolyte temperature, pH of electrolyte, nature of substrate and voltage.

Highlights

This chapter addresses the following topics:

- Nanomaterials and the different categories
- Nanomaterials synthesis with emphasis on electrodeposition
- Parameters influencing electrodeposition.

1.1 Introduction

Nanomaterial's research is a vast and very significant field of study in the advancement of science and technology due to their abundant and essential applications in vast research fields including Physics, Chemistry, Material Science, Medicine, Engineering, etc. Nanomaterials are gaining attention and importance as a result of their many distinct properties which contradicts those of the traditional bulk counterparts [1]. Nanomaterials have at least a dimension in the nanometer scale range. This range is defined as 10^{-9} m.

To effectively understand the properties exhibited by nanomaterials, it is crucial to grow them using either physical or chemical means. A number of physical methods require very expensive equipment's to work with, hence the need to consider methods that will be less expensive and capable of bulk assembly of nanomaterials with good quality. Electrochemical preparation technique is among the possible

D. M. Jeroh (✉)
Department of Physics and Industrial Physics, Nnamdi Azikiwe University, Anambra State Awka, Nigeria

O. M. Ama et al. (eds.), *Modified Nanomaterials for Environmental Applications*, Engineering Materials, https://doi.org/10.1007/978-3-030-85555-0_1

options for nanomaterials fabrication with the advantage of having the possibility to deposit a large variety of materials in varying forms from different solutions and the opportunity to obtain nanomaterials in bulky amounts [2].

The major goal of this chapter is to report how nanomaterials are synthesized from electrochemical processes and also highlight and discuss the important parameters upon which nanomaterials synthesis from electrochemical reactions are relied upon.

1.2 Nanomaterials

Nanomaterials are considered the major building blocks of nanoscience and nanotechnology. Nanomaterials can be obtained from various dimensions: zero, one, two and three dimensions [3]. Nanomaterials garner interest because, at this scale, some distinctive magnetic, optical, electrical and other characteristics materialize and these evolving characteristics have the capability for immense impacts in medicine, electronics and other disciplines [4]. More recently, nanomaterials are considered more suitable as a result of their fascinating physicochemical properties, which are different from those displayed by their bulk equivalent [5].

Nano-sized materials are currently used in many industries. For example, carbon-black particles are exploited for manufacturing rubber tyres which render the rubber tyres wear resistant. Other examples include using nanofibers for insulation and reinforcement of composites, using iron oxides in fabricating the magnetic material employed in the manufacture of videotapes and disk drives (DDs). Some products currently relying on nanotechnology are: magnetic recording tapes; computer hard drives; bumpers on cars; solid-state compasses; automobile catalytic converters; metal-cutting tools; dental bonding agents; longer-lasting tennis ball; burn and wound dressing; ink; etc. [6]. Some examples of nanomaterials include quantum dots, carbon nanotubes, graphene and fullerenes.

1.2.1 Quantum Dots

A quantum dot (QD) is an extremely small semiconductor structure whose diameter is within the nanoscale range, say between 2 and 20 nm. However, their dimensions rely mainly on their preparative materials [7]. According to Pokropivny et al. [8], QDs are crystals that emit a single wavelength of light during electrons excitation. QDs could also be seen as a class of semiconductors which are composed of periodic groups of II-VI, III-V, or IV-VI materials (such as ZnO, CdSe, CdS, PbSe, InP) with size spanning between 1.5 and 20 nm (10–50 atoms) in diameter [9]. Quantum dots conform to quantum mechanical principle (quantum confinement) and they exhibit band gap that ascertains the necessary wavelength pertaining to the absorption and emission spectra [10]. It is believed that in the future QDs could serve as quantum bits and may possibly constitute the foundation of quantum computers [8].

1.2.2 Carbon Nanotubes

Carbon nanotubes are materials possessing a tube-like shape with diameters in the nanoscale range. CNTs are synonymous with cylinders made of sheets of graphites, having enclosures at both ends, with carbon atoms on the top of the hexagon [11]. CNTs are also known as "bulkytubes" and are depicted by a cylindrical nanostructure in the appearance of a tube [12]. CNTs exist as a single wall or multi-wall carbon nanotubes.

1.2.3 Graphenes and Fullerenes

Graphene consists of one atom thick planar sheet of carbon atoms closely spaced in a honeycomb crystal lattice which is the fundamental structural assembling block of CNTs and fullerenes [12].

Fullerenes are composed of allotropes of carbon wholly consisting of carbon and exist as hollow spheres (or buckyballs), ellipsoid (bucky tubes or carbon nanotubes) [13]. Fullerenes have 60 carbon atoms (C_{60}).

1.3 Nanomaterials Synthesis

Nanomaterials can be synthesized by "top-down" and "bottom-up" techniques.

The "top-down" technique involves getting the nanomaterials by disintegrating the bulk material bit by bit till the desired dimension is achieved. Figure 1.1 demonstrates the "top-down" approach.

Grinding and lithography are types of the "top-down" approach. Lithography involves depositing a computer-generated pattern on a substrate's surface. However, this technique is associated with some disadvantages such as the introduction of crystallographic defects to the generated patterns. Another disadvantage is the elevated cost of equipment and the inability to obtain a single nanostructure through this method. Moreover, despite the defects, the "top-down" approach plays a principal role in synthesizing and engineering nanomaterials in that the present state of nanoscience can be viewed as an amalgamation of bottom-up chemistry and top-down engineering techniques [14].

The "bottom-up" approach involves preparing bulk materials by assembling of a structure atom by atom till the desired thickness of the bulk structure is achieved. Bottom-up approaches include self-assembly, precipitation from chemical solutions and aerosol techniques. Figure 1.2 demonstrates the "bottom-up" approach.

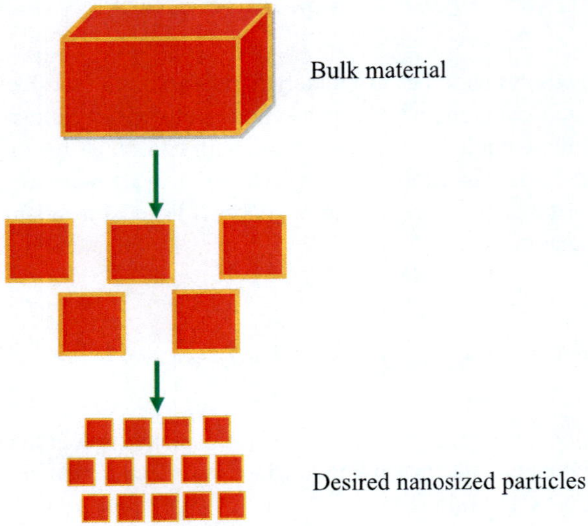

Fig. 1.1 Demonstration of "top-down" approach

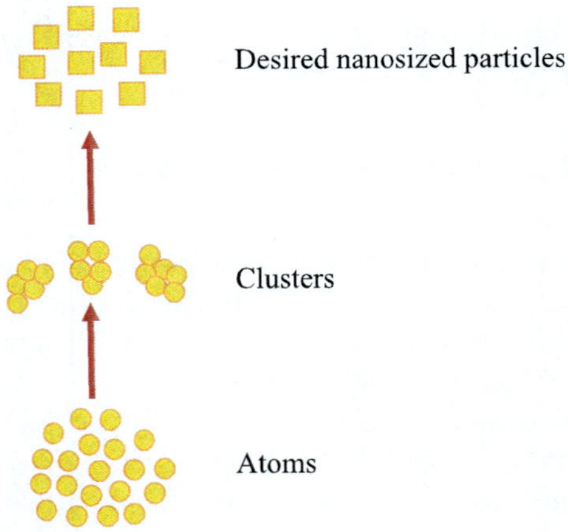

Fig. 1.2 Bottom-up approach of synthesizing nanomaterials

1.3.1 Electrochemical Synthesis

1.3.1.1 Electrodeposition Technique

Electrochemical deposition involves any type of electrochemical reactions which results in the deposition of a material [15]. According to [16], electrochemical deposition is defined as the production of a coating on a surface from an aqueous solution composed of several substances. Electrochemical deposition usually occurs inside an electrochemical cell. In this type of deposition technology, the film deposit can be formed on a substrate by allowing current to pass through an electrochemical cell with the aid of a power source connected externally. Alternatively, it can be grown by making a complex solution where chemical reactions take place continuously without any power source due to the sufficiently high potential between the surface and a solution [16].

A typical electrochemical cell comprises a reaction vessel (a beaker in most cases) and maybe two or three electrodes. The two-electrode cell comprises a working electrode (substrate) and a counter-electrode. The reactions can be controlled in the two-electrode cell by the application of current between the working electrode and the counter-electrode. In the three-electrode cell, deposition of the required film on a substrate can be achieved either by controlling the potential or the current and the corresponding potential or current is measured.

Figure 1.3 illustrates a simple electrochemical cell for synthesizing nanomaterials.

In electrochemical synthesis, when a potential different (PD) is applied across the system, a direct current (DC) source supplies current which flows in one direction through the external circuit. The DC source could be either a motor, generator or a rectifier. Electrons are the main carriers of current in the external conductors. Electricity is transferred in the solution through electrically charged particles (ions). When a PD is applied, the positive ions referred to as cations move towards the negative electrode (cathode), while the negative ions known as anions move towards the positive electrode (anode). As this mechanism takes place, deposition of materials

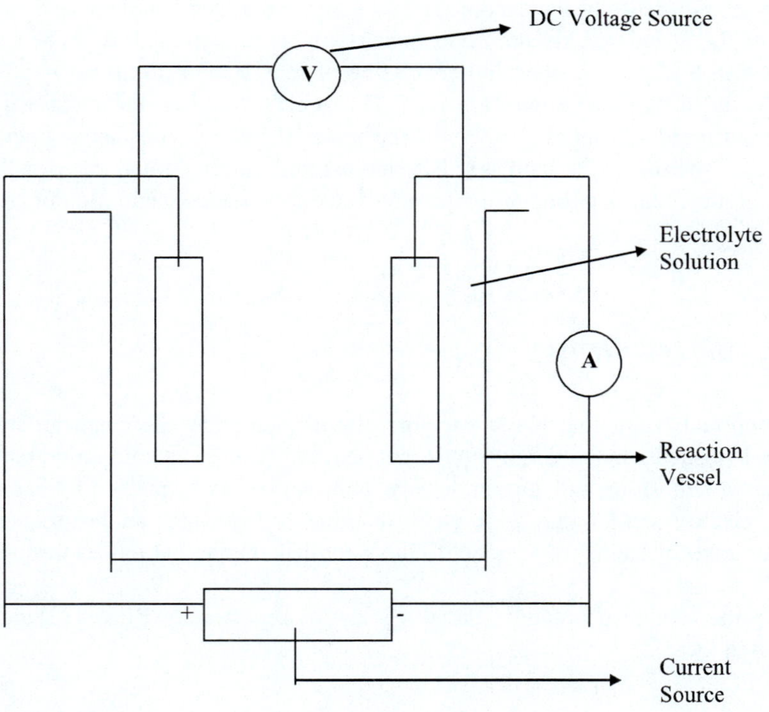

Fig. 1.3 Schematics of a simple electrochemical cell

occurs on the substrate from the aqueous solution. Electrochemical deposition is an interesting method for preparing materials in large quantities because it relies on inexpensive equipment, enables large area deposition and offers eased control of growth factors through applied current, PD, bath temperature and pH [17].

1.4 Parameters Influencing Electrodeposition of Nanomaterials

Some factors have been identified to have some notable influence on electrodeposition of nanomaterials. Some of these include temperature, current density, substrate nature, applied voltage and solution pH.

1.4.1 Temperature

Temperature is vital in influencing the end properties exhibited by nanomaterials. An increase or decrease in the electrolyte temperature may result in significant modification of the properties of the nanomaterials being synthesized. An increased temperature of the electrolyte results in an increase in solubility, catalysis of reactions, increase in transport number and energizing of ions leading to higher deposition density and the rate of deposition [18]. The general trend is that increased bath temperature reduces internal stress of the material [19]. A substantial grain size reduction is witnessed at increased bath temperature implying that the growth rate of the grains reduces at higher temperatures due to lower concentration of oxygen [20].

1.4.2 Current Density

An important factor that plays a major role in influencing the characteristics of electrodeposited nanomaterials is "current density." There is a correlation between the rate of deposition and current density. This can be best explained by Faraday's law of electrolysis. Faraday's law states that the level to which an electrochemical reaction takes place is proportional to the electrical charge that passes through the electrolyte [21].

In a mathematical notation, Faraday's law as reported by Bijoy [22] may be expressed as:

$$m = \left\{ \frac{Q}{F} \right\} \left\{ \frac{M}{Z} \right\} \tag{1.1}$$

In 1.1, m represents the altered mass of the substance at the electrode, Q represents the electric charge in the system, M equals the material's molar mass, F is Faraday's constant given as 96.485 Cmol^{-1} and Z is transferred electrons per ion [22].

At low current density, the speed of nucleation centers creation is much lower compared to crystal growth of the existing seeds, while for high current density, a decrease of ions concentration near the deposited surface is observed, or hydrogen evolution takes place and the obtained layers become non-continuous, spongy and porous [2]. In the electrochemical synthesis of Ni/SiCw nanocomposite coatings, [23] reported that electrodeposits acquired at low current density resulted in smooth surface morphology and fine grains in comparison to electrodeposits acquired at elevated current density. However, in the experiment conducted by Zaki et al. [24], limitation of nucleation and growth of deposits of copper on Ni-zincates Al surface was witnessed at low current density (0.5 mA/cm^2), while improved nucleation was recorded at elevated current density (6.0 mA/cm^2).

1.4.3 Nature of Substrate

Synthesizing nanomaterials using electrochemical synthesis technique requires a conducting substrate. Substrates commonly employed for electrodeposition are Fluorine-doped tin oxide (FTO) and Indium-doped tin oxide (ITO) conducting substrates. Synthesis cannot occur using a non-conducting substrate.

1.4.4 Electrolyte pH

The constituent bath (electrolyte) pH is another important parameter that affects the growth of nanomaterials electrochemically. An electrolyte is said to be acidic having pH value less than seven, while it is considered alkaline with a pH value greater than seven. Increase in electrolyte pH results in increased deposition rate [25], which obviously resulted in improved characteristics displayed by nanomaterials. Low pH results in the evolution of hydrogen, which breaks through the layer, resulting in the deposited material being hard and in turn result in a lot of stress to the layer [2]. Hence, in fabricating nanomaterials from electrochemical processes, a suitable pH is essential to develop a good quality material as the final product.

1.4.5 Voltage

Applied voltage also has a significant effect on deposition of nanomaterials using electrochemical deposition technique. The applied voltage has a linear relationship to the current that is passed within the electrolyte. In electrochemical synthesis, a higher voltage implies the availability of more energy at the cathode which will eventually result in reduction while the anode will record more energy leading to oxidation. In the growth of zinc selenide nanofilms, an improved crystallinity of the films at increased potential (voltage) [26] is reported.

1.5 Conclusion

Electrochemical synthesis of nanomaterials has been discussed. Some parameters which influence nanomaterials properties obtained by electrochemical synthesis were identified and discussed. It is emphasized that temperature, current density, voltage and nature of substrate play vital roles in influencing the end properties of nanomaterials. To obtain good quality nanomaterials using electrochemical process, it is pertinent to optimize all synthesizing conditions.

References

1. R. Singaravelan, S.B.S. Alwar, Electrochemical synthesis, characterization and phytogenic properties of silver nanoparticles. Appl. Nanosci. (2015). https://doi.org/10.1007/s13204-014-0396-0
2. B. Kalska-Szostko, Electrochemical methods in nanomaterials preparation, in *Recent Trend in Electrochemical Science and Technology*, ed by U.K. Sur (InTech, China, Europe, 2012)
3. H. Hoffmann, Advanced nanomaterials. Retrieved online from www.ltp.epfl.ch/files/Introduction to Nanomaterials Lecture Support All.pdf on 3 April 2017 (2009)
4. A. Alagarasi, Introduction to nanomatertials. Retrieved online from https://nccr.iitm.ac.in/2011.pdf on 17 Jan 2020 (2011)
5. D. Tonelli, E. Scavetta, I. Gualandi, Electrochemical deposition of nanomaterials for electrochemical sensing. Sensors **19**, 1–28 (2019)
6. V. Pokropivny, R. Lohmus, I. Hussainova, A. Pokropivny, S. Vlassov, Introduction to nanomaterials and nanotchnology. Retrieved online from www.digar.ee at 05 April 2020 (2007)
7. V. Drbohlavova, R. Adam, J. Kizek, Hubalek, Quantum dots—characterization, preparation and usage in biological systems. Int. J. Mol. Sci. **10**, 656–673 (2009)
8. Zobair, The impact of nanotechnology on computers. Glob. J. Res. Eng. **12,** 35–38 (2012)
9. N.H. Zaidi, Synthesis and characterization of semiconductor nanocrystals: photoluminescence and size tunability. Int. J. Adv. Eng. Nano Technol. **2**, 19–21 (2014)
10. J. McDaniels, Quantum dots: science and applications. Retrieved online from www.ion.chem.usu.edu on 22 Sept 2019 (2015)
11. K.K. Sulabha, *Nanotechnology: Principles and Practice*, 3rd edn. Springer International Switzerland and Capital Publishing Company, India (2015)
12. A.W. Salamon, P. Courtney, I. Shuttler, *Nanotechnology and Engineered Nanomaterials* (PerkinElmer Inc, USA, 2010)
13. P.S. Pradeep, G.S. Nalini, Fullerenes revisited: materials chemistry and applications of C_{60} molecules. Resonance, 123–135 (2015)
14. Ghosh, Introduction to nanomaterials and nanotechnology. Retrieved online from www.nptel.ac.in on 04 Dec 2019 (2009)
15. P.T. Tang, Utilising electrochemical deposition for micro manufacturing. in *Multi-Material Micro Manufacture*, ed. by S. Dimov, W. Menz (Whittles Publishing Ltd., United Kingdom, 2008)

16. A. Mendelson, Identification and control of deposition processes. An Unpublished Doctoral Dissertation Submitted to the Faculty of Electronics, Communications and Automation at the Aalto University School of Science and Technology, Espoo, Finland (2010)
17. M.N. Mammadov, A.S. Aliyev, M. Elrouby, Electrodeposition of cadmium sulfide. Int. J. Thin Film Sci. Technol. **1**, 43–53 (2012)
18. A.A. Ojo, I.M. Dharmadasa, Electroplating of semiconductor materials for applications in large area electronics: a review. Coatings **8**, 1–17 (2018). https://doi.org/10.3390/coatings8080262
19. E.S. Guler, Effects of electroplating characteristics on the coating properties, Retrieved online from www.intechopen.com on 12/04/20 (2016). https://doi.org/10.5772/61745
20. S. Hori, T. Suzuki, .S. Tsukasa, .S. Miura, S. Nonomura, Effect of deposition temperature on the electrochemical deposition of zinc oxide thin film from chloride solution. Mater. Trans. **55**, 728–734 (2014)
21. M.W. Losey, J.J. Kelly, N.D. Badgayan, S.K. Sahu, P.S. Rama-Sreekanth, Electrodeposition. in *Reference Module in Materials Science and Materials Engineering*, vol. 13, ed. by S. Hashmi (Oxford, Elsevier, 2017), pp. 1–20
22. B. Bijoy, Electrochemical machining, in *Electrochemical Machining for Nanofabrication, MEMS and Nanotechnology* (2015)
23. L. Lai, .H. Li, .Y. Sun, .G. Ding, .H. Wang, .Z. Yang, Investigation of electrodeposition external conditions on morphology and texture of Ni/SiCw composite coatings. Appl. Sci. **9**, 1–12 (2019). https://doi.org/10.3390/app9183824
24. M.H.M. Zaki, Y. Mohd, N.N.C. Isa, The effect of current densityon surface properties of electrodeposited copper coatings on modified aluminium. AIP Conf. Proc. **2030** (020112) (2018). https://doi.org/10.1063/1.5066753
25. W. Wu, N. Eliaz, .E. Gileadi, The effects of pH and temperature on electrodeposition of Re-Ir-Ni coatings from aqueous solutions. J. Electro. Soc. **162**, 20–26 (2015)
26. A. Kathalingam, T. Mahalingam, .C. Sanjeeviraja, Optical and structural study of electrodeposited zinc selenide thin films. Mater. Chem. Phy. **106**, 215–221, (2007)

Chapter 2
Electrochemical Characterization of Nanomaterials

Tladi Gideon Mofokeng, Mpho Phillip Motloung, Onoyivwe Monday Ama, and Suprakas Sinha Ray

Abstract This chapter reports the recent advancements in the electrochemical characterization of nanomaterials. Various electrochemical techniques including potentiometry, electrochemical impedance spectroscopy and cyclic voltammetry, etc. are discussed in this chapter. Nanoparticle-modified electrodes revealed superior behaviour when applied in modification of the electrodes. Sensitivity, selectivity and shorter response time were improved. Overall, electrochemical techniques provide an efficient and cost-effective route for characterization of nanomaterials.

2.1 Nanostructured Materials

Nanomaterials have garnered immense interest in the past decades because of their unique chemical and physical properties. The term nanoparticle is self-explanatory, as it refers to a material whose at least one dimension is in a nanometre range. The dimensions of the nanoparticles (NPs) mostly differ according to their origin and their synthetic route. For instance, the structural dimensions of cellulose nanoparticles depend entirely on the cellulose source and treatment. NPs exist in diverse forms including spherical shapes, rod-like, platelets, cylindrical, etc. and they can be distinguished according to their sizes, geometry and properties. The comprehensive reviews on the nature, structure and properties of the NPs have been conducted and reported elsewhere [1–3]. NPs are widely used in many fields of applications, such as in medicine, cosmetics and sensors, among others. It is imperative to realize the properties and behaviour including the toxicity of these materials before their application. As a result, it is essential to employ advanced techniques to gather exhaustive information regarding the NP of interest.

T. G. Mofokeng (✉) · M. P. Motloung · S. S. Ray
DST/CSIR National Centre for Nanostructured Materials Council for Scientific and Industrial Research, Pretoria 0001, South Africa

O. M. Ama · S. S. Ray
Department of Chemical Sciences, University of Johannesburg, Doornforntein, Johannesburg 2028, South Africa

O. M. Ama et al. (eds.), *Modified Nanomaterials for Environmental Applications*, Engineering Materials, https://doi.org/10.1007/978-3-030-85555-0_2

Various techniques can be employed to characterize the NPs. The microscopic techniques including electron microscopy (EM), optical microscopy (OM), atomic force microscopy (AFM), etc. are commonly employed for the morphological analysis, such as size, shape and structure of the NPs. Besides, the scattering techniques including small-angle X-ray scattering (SAS) and light scattering (LS) have also been successfully employed, most especially for structural-dimensional analysis of the NPs. However, the analysis with the SAS techniques is based on modelling through different form factors, unlike the microscopic techniques which are based on direct imaging. Although these techniques provide sufficient information, there are drawbacks that to some extent impede their application. The above-given techniques are complex and are also not cost effective [4]. In addition, most of the NPs are insulating materials; therefore, this necessitates an additional exercise of coating the material with a conducting layer, prior testing in case of the scanning electron microscopy (SEM) and staining in confocal and transmission electron microscopy TEM.

Electrochemical (EC) techniques also form part of the characterizing techniques used for the NPs and a lot of research work has been carried on the EC characterization of different NPs. Compared to the analytical techniques; however, the EC-based techniques are portable and easy to handle, fast, highly sensitive and cost effective [4–6]. Hence, research interest has been diverted into these techniques. Therefore, this chapter explores different electrochemical techniques utilized in characterization of the NPs.

2.2 Electrochemical Techniques for Nanomaterial Characterization

The concept of electrochemistry dates to eighteenth century [7]. Generally, electrochemistry evaluates the chemical changes as a result of electron movement. A typical electrochemical cell comprises three electrodes. A working electrode (WE), where the electrochemical reaction of interest occurs. The potential of the WE are unknown and it is measured against reference electrode (RE) with known potential. An auxiliary electrode (AE) completes the external circuit as it allows the current to flow across the cell. It is noteworthy that in the EC measurements, the potential and current cannot be measured simultaneously; consequently, the analysis is either carried under *zero* current, or at controlled current or potential. Based on this, the EC techniques can be classified into both static and dynamic. In the static techniques, the potential is measured at *zero* current. A typical static technique is a potentiometer. The potentiometric technique is mostly employed to determine the concentrations of ions in the solutions. This technique offers certain advantages such as high selectivity and the response time is short [6, 8, 9]. Contrary, dynamic techniques involve a controlled potential and the resulting current is measured. Different EC techniques are summarized in Table 2.1.

Table 2.1 Summarized EC techniques [8–10]

EC technique	Controlled variable	Specific technique
Static	$i = o$	Potentiometry
Dynamic	Potential	Differential pulse voltammetry Cyclic voltammetry
Galvanostatic	Current	Chronopotentiometry
Impedance measurement	Current	Electrochemical impedance spectroscopy

2.3 Static Techniques

2.3.1 Potentiometry

Potentiometric techniques mostly are applied in quantitative analysis of ions in solutions thanks to their unique properties, which include shorter response time, high selectivity and low detection limits [8]. The potentiometric techniques were employed in wide range of applications including determination of food additives in various food products [11], determination of organic carbon content of soil [12] and in pharmaceutical determination of drugs in biological fluids [13]. The potentiometric cell has a WE that responds to the analyte's activity and a reference electrode with a known or a fixed potential. The WE of the potentiometric cell can be modified to enhance the selectivity and response time. Ion-selective electrodes (ISEs) electrodes have gained enormous interest since they possess interesting properties over traditional electrodes such as reproducibility, accuracy and high selectivity. Additionally, ISEs require no treatment prior application. Further developments have led to a fabrication of ISEs and ISEs-screen-printed electrodes (SPEs) based on the nanoparticles, with the aim of enhancing sensitivity, selectivity, response time and reproducibility of these electrodes [14–16].

Several studies on development of nanoparticle-modified electrodes have been reported [17–20]. In these studies, nanoparticles were effective in enhancing the selectivity, reducing the response time and improving the stability of the electrodes. Akl et al. [17] developed SPE-modified with triazole surfactant assembled on silver (Ag) nanoparticles (SNPs-MSPE) for potentiometric determination of uranium ions in spike water. The Ag-based SPE showed enhanced properties such as short response time, which was about 3 s for nanoparticle-modified SPE compared to the SPE containing no Ag particles, which showed a response time of about 8 s. Moreover, the Ag-based SPE revealed high selectivity for UO_2 (2+) ions, low detection limits and it was stable. Similar behaviour was also reported by Frag et al. [18] who modified carbon paste electrode with CuNiO nanoparticles for potentiometric determination of Cu^{2+} in spiked water. Novotný et al. [20] used silver amalgam electrode (AgAE) for potentiometric determination of silver in silver colloidal nanoparticle solution. In terms of sensitivity, modified AgAE was much more effective than normal AgE.

The behaviour and performance of the NPs depend on a number of factors such as surface area and size. As a result, it is crucial to consider such parameters when designing the WE-electrodes modified with nanoparticles. Moreover, looking at things from a commercialization perspective, it is essential to consider the cost-effective strategies for synthesizing these NPs. This will result in cheaper electrodes, with enhanced properties including high selectivity and stability. Therefore, the studies summarized above revealed the improved electrochemical performance after the inclusion of nanoparticles on the WE electrode.

2.4 Dynamic Techniques

2.4.1 Electrochemical Impedance Spectroscopy (EIS)

The electrochemical impedance spectroscopy (EIS) technique is mostly used to study the interfacial properties between the electrode and electrolyte. With this technique, the intrinsic property that affects the conductivity of the nanoparticle solution interface can be determined. EIS technique has advantages, which make them favourable over other dynamic EC techniques. They include ease of signal quantification and the ability to separate the surface binding events from the solution impedance, among others. The EIS electrolytic cell comprises three electrodes; WE, reference electrode and the counter electrode which is based on platinum or graphite. The electrochemical reaction on the WE are induced by applying the alternating current (AC) at varying frequency and the impedance is measured [8, 21–23]. The analysis of the EIS measurements is mostly based on Nyquist plots and Bode plots. The Nyquist plot is a plot of real impedance against imaginary impedance at different frequencies, while the latter is a plot of absolute impedance against the frequency. The Nyquist plots are characterized by semicircle arcs with varying diameters. The diameters of these arcs are associated with the resistance to the transfer of the electron. The more depressed semicircles suggest high conductivity, while a less depressed semicircle indicates a high resistance to the electron transfer. EIS is used for various purposes such as performance evaluation of modified working electrodes, biosensing, coatings, semiconducting electrodes, batteries, fuel cells, etc.

Numerous researchers [24, 25] developed nanoparticle-based immobilized antibody electrodes for biosensing applications. In these studies, EIS was employed to determine the electrochemical properties of nanoparticle-modified electrodes and sensing performance. Rezaei et al. [24] and Chen et al. [25] prepared gold (Au)-based electrodes as a platform for immobilization of somatotropin and lysozyme proteins, respectively. The modification of electrodes with Au nanoparticles greatly enhanced the sensitivity of the protein-immobilized Au electrodes than bare Au electrodes. The Nyquist plots obtained from the EIS characterization of bare Au electrode, 1, 6-hexanedithiol HDT-modified gold Au electrode, Au nanoparticles (AuNPs) colloid/HDT-modified Au electrode and the human Growth Hormone (hGH) antibody immobilized (AuNPs) colloid/HDT-modified Au electrode are shown in Fig. 2.1. It is apparent that (AuNPs) colloid/HDT-modified Au electrode was conductive as observed from a drastic decrease of the semicircle compared to HDT-modified Au electrode, which showed large semicircle. On the other hand, an increased arc of hGH-immobilized electrode indicates high electron transfer resistance due to insulating layer of protein on the Au nanoparticles surface. This observation on hGH-immobilized electrode insinuates a successful immobilization of (hGH) antibody on the (AuNPs) colloid/HDT-modified Au electrode [24].

Casero et al. [26] characterized graphene oxide (GO) and both electrochemically (EGR) and photolytically reduced graphene oxide (PGR) with EIS. The authors showed that EIS can also be utilized as a complementary technique to other analytical techniques such as XPS, XRD, Raman, etc. in the analysis of graphene-based nanoparticles. In their case, GO, EGR and PGR were coated onto glass carbon electrode (GCE). The Bodes plot for GO/GCE was different from both EGR/GCE and PGR/GCE plots. The GO/GCE was characterized by two times constant at high and low frequencies, while the GR-based electrodes showed only one time constant associated with the kinetic process. The Nyquist plots were also different for the three electrodes studied. The results obtained from this analysis can indicate the state of graphene, whether it oxidized or reduced. EIS has also been applied in the characterization of many other nanoparticles. These nanoparticles are magnetic nanoparticles [27], single Au nanorods [28], ZnO [29], NiO [30] and TiO_2 [31]. Moreover, other researchers employed EIS to investigate the capacitance behaviour of polymer-based nanocomposites used in supercapacitors. In this regard, the EIS is used as a

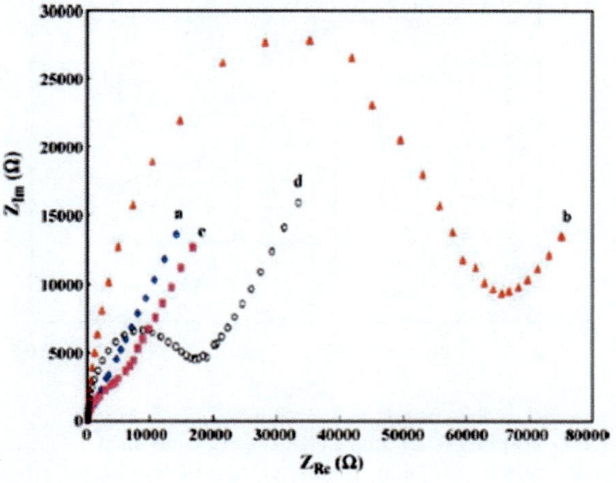

Fig. 2.1 Nyquist plots of (a) bare gold electrode, (b) HDT-modified electrode, (c) HDT/AU-colloid modified electrode, (d) hGH antibody immobilized electrode. "Reprinted from publication title, vol/edition number, author(s), title of article/title of chapter, pages no., copyright (year), with permission from Elsevier [OR APPLICABLE SOCIETY COPYRIGHT OWNER]." Also, Lancet special credit—"Reprinted from The Lancet, vol. number, author(s), title of article, pages no., copyright (year), with permission from Elsevier"

complementary technique to galvanostats (discussed in the next section). Conducting polymers are commonly used in supercapacitors due to their excellent electrical conductivity and high electrochemical activity. However, conducting polymers have poor stability and capacity. Recently, a lot of research work has been conducted on graphene-based nanocomposites to enhance their performance in supercapacitors [32–36]. These studies were based on polypyrrole (PPy) [32, 33] and polyaniline (PANI) [34–36]. In these studies, the incorporation of graphene in both PANI and PPy significantly enhanced their capacitive performance.

Chang et al. [32] reported enhanced electronic properties of PPy-modified gold electrode in the presence of RGO nanocomposites. In particular, the electron charge transfer resistance was low in PPy/RGO-based electrodes compared to PPy-based electrodes and PPy/GO-based electrodes. The Nyquist plots of PPy, PPy/GO and PPy/RGO are depicted in Fig. 2.2. At higher frequencies, it was observed that PPy/RGO and PPy/GO possess low charge transfer resistance compared to PPy-based electrodes. However, at low frequency region, which indicates capacitive behaviour, it was observed that RGO/PPy approached 90°, suggesting better capacitive behaviour. Such observation was also reported in case of PANI-containing graphene-based nanoparticles [34, 36].

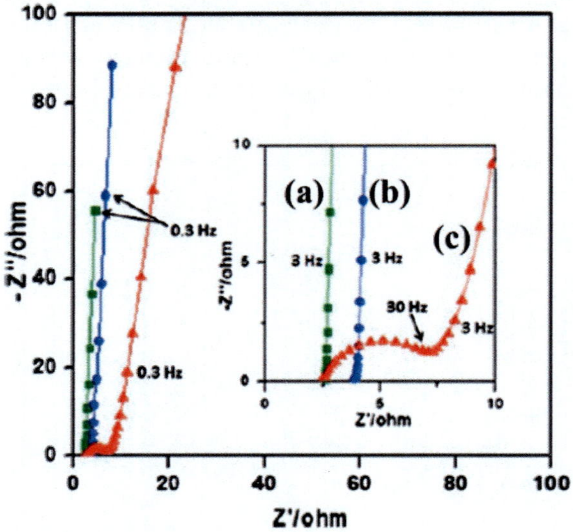

Fig. 2.2 Nyquist plots of (a) PPy/RGO, (b) PPy/GO, (c) PPy-modified gold electrode. "Reprinted from publication title, vol/edition number, author(s), title of article/title of chapter, pages no., copyright (year), with permission from Elsevier [OR APPLICABLE SOCIETY COPYRIGHT OWNER]." Also, Lancet special credit—"Reprinted from The Lancet, vol. number, author(s), title of article, pages no., copyright (year), with permission from Elsevier"

2.4.2 Galvanostatic Techniques

Similar to other EC dynamic techniques, galvanostatic techniques comprise three electrodes. In this technique, the current is applied between the working electrode and counter electrode and the resulting potential is measured across the reference electrode. A common current-controlled technique is chronopotentiometry (CP), which measures current as a function of time. The advantage of current-controlled technique is the constant ohmic drop due to solution resistance. Nevertheless, large double layer charging effect during experiment is a challenge in these techniques. Galvanostatic techniques have been used to characterize the nanomaterials. Numerous studies [37, 38]. have been done on electrodeposition using galvanostatic technique to produce powders with various properties. In galvanostatic electrodeposition, parameters such as current density and pH can be monitored to yield nanoparticles with different morphologies. For instance, Maksimovic et al. [39] obtained copper (Cu) particles with different structures depending on the current density. Similar observation was also noticed in case of nanocrystalline Al-Mg alloy powders [40]. Galvanostatic electrodeposition of nanomaterials onto screen-printed carbon electrode (SPCE) has been performed. P-Fernández et al. modified the screen-printed electrodes by galvanostatic electrodeposition of Cu nanospheres [41] and nickel nanoflowers [42].

The capacitive behaviour of nanoparticle-modified electrodes has also been carried with galvanostatic techniques in conjunction with the EIS technique. The electrochemical capacitance of the materials can be determined by galvanostatic charge/discharge method under controlled current. The capacitance performance of NiO nanoparticles-based has been investigated and reported. Wang et al. [43] demonstrated the effect of particle size on the capacitive behaviour of NiO nanoparticles. The authors observed that NiO with smaller size had longer charging time, suggesting higher capacitance (Fig. 2.3a). Besides, the cycling performance of the electrodes

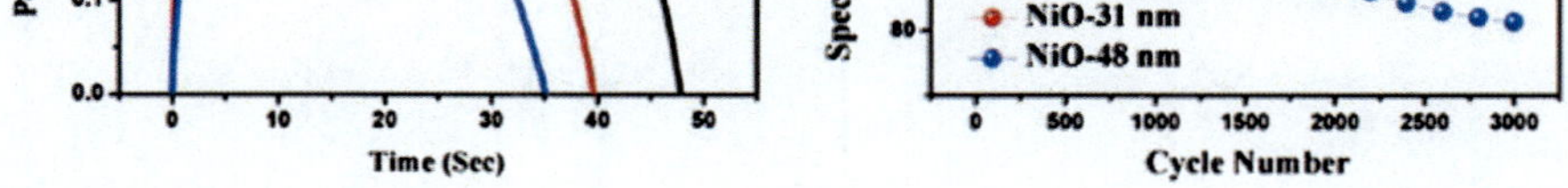

Fig. 2.3 **a** Charge–discharge curves. **b** Cycling test of NiO samples with different sizes. "Reprinted from publication title, vol/edition number, Author(s), title of article/title of chapter, pages no., copyright (year), with permission from Elsevier [OR APPLICABLE SOCIETY COPYRIGHT OWNER]." Also, Lancet special credit—"Reprinted from The Lancet, vol. number, author(s), title of article, pages no., copyright (year), with permission from Elsevier"

as a function of particle was also examined after 3000 charge/discharge cycles and the results are shown in Fig. 2.3b. The capacitance retention for NiO with smallest size was 86.4% and it was the highest among the tested particles with large sizes. On the other hand, Zhu et al. [44] fabricated NiO nanoflakes/graphene nanocomposites-based electrodes for application in pseudo capacitors. The nanocomposite-modified electrode exhibited superior specific capacitance of 240 F g^{-1} at 5 A g^{-1} and 220 F g^{-1} at 10 A g^{-1}, compared to NiO nanoflakes-based electrode, which had capacitance of 100 F g^{-1} at 5 A g^{-1}. Further, the cycling stability of 100–200% retention was noticed in the nanocomposite after 1500 cycles. The electrochemical capacitance of various polymer nanocomposite-based electrodes for supercapacitors were investigated. Li et al. [36] examined the electrochemical properties of PANI/graphene nanosheets electrodes for application in supercapacitors. As expected, PANI/graphene showed highest capacitance compared to neat PANI and graphene electrodes.

2.4.3 Cyclic Voltammetry

Cyclic voltammetry (CV) is a type of technique that is used for measuring the electrochemical properties of materials. In this technique, the working electrode potential is scanned linearly as a function of time resulting in production of current. The produced current is monitored and a cyclic voltammogram curve occurs when current is plotted as a function of potential as shown in Fig. 2.4a. The cyclic voltammogram curve in Fig. 2.4a is divided into two regions, namely, the anodic and cathodic regions. The anodic and cathodic regions are represented by the positive and negative currents, respectively depending on the direction of the scanning potential. In Fig. 2.4a the potential is catholically scanned between point I and IV at a fixed rate. The cathodic

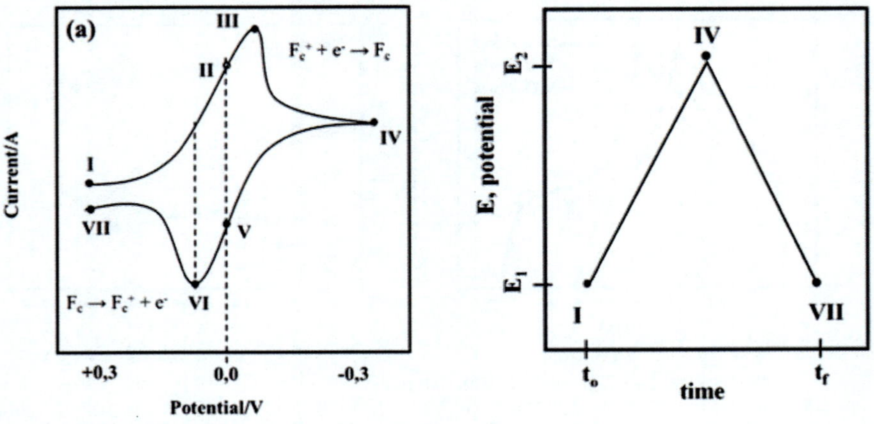

Fig. 2.4 **a** Voltammogram of the reversible reduction of Fc$^+$ solution to Fc. **b** Applied potential as a function of time for CV experiment

scan of potential results in diminishing of the Fc^+ concentration near the electrode as it is reduced to Fc. The scan is reversed when the potential reaches point IV and the potential is anodically scanned back to point I slowly oxidizing the product Fc back to Fc^+. Another information that can be deduced from cyclic voltammogram curves is the electrochemical potential window. Electrochemical potential window of a material is the voltage range along which the material is neither reduced nor oxidized. This electrochemical potential window provides the maximum operating potential range for the cell if appreciable electrolyte degradation is to be avoided [45]. Over the years, a lot of researchers have used CV in their studies for characterization of nanoparticles and polymer nanocomposites [46–48]. Janjan et al. [46] used leaf extracts of guava to develop an eco-friendly and cost-effective method for synthesizing zinc oxide nanoparticles. The cyclic voltammogram obtained by scanning along with the potential range +1.0 to −1.0 V at scan rate 0.1 v/s showed that after addition of guava leaf extract, the cathodic peak shifted to 0.82 V indicating reduction of Zn^{2+} ions. However, an anodic peak occurred at −0.39 V after addition of 1 M NaOH showing reoxidation of the formed Zn. Cea et al. [47] used CV to investigate the blocking effect on a gold electrode, a gold electrode covered tightly by self-assembled monolayers and a gold nanoparticle/organic monolayer/gold sandwich structure. Their experiment was made of a gold electrode placed in an electrochemical cell containing 1 mM and 0.1 M of $[Ru(NH3)_6]Cl_3$ and KCl, respectively. The reference electrode was made of Ag/AgCl, KCl (3 M), whereas a Pt sheet was used as a counter electrode. The cyclic voltammograms of a gold electrode and a gold nanoparticle/organic monolayer/gold sandwich structure showed the presence of Faradaic current, whereas a Faradaic current was reduced significantly in the gold electrode covered by self-assembled monolayers. This suggests that the transfer of electrons between the surface of the electrode and redox probe was blocked effectively. However, low incubation periods (lower than 3 h) used for the preparation of gold electrodes covered by self-assembled monolayers result in no prevention of Faradaic current suggesting poor coverage of the gold substrate. Recently, CV was also used for characterization of polymer composites containing polyvinyldene fluoride (PVDF) and nanofillers such as reduced graphene oxide [48] and a mixture of barium titanate ($BaTiO_3$) and zinc oxide (ZnO) [49]. The cyclic voltammogram curves of Pazhamalai et al. [48] was rectangular shaped and their current increased with increasing scan rates from 1 to 10 V/s suggesting that charge was stored in the PVDF/reduced graphene oxide symmetric supercapacitor. Tabhane et al. [49] prepared PVDF/$BaTiO_3$/ZnO nanocomposites by solution casting method using ZnO nanoparticles (PVDF/$BaTiO_3$/ZnO_s), spin coating method using ZnO nanoparticles (PVDF/$BaTiO_3$/ZnO_p) and flower shaped ZnO particles (PVDF/$BaTiO_3$/ZnO_f). The cyclic voltammogram of the PVDF/$BaTiO_3$/ZnO_f nanocomposite attained electrochemical potential window value of 1.7 V, whereas PVDF/$BaTiO_3$/ZnO_p and PVDF/$BaTiO_3$/ZnO_s nanocomposites attained the electrochemical potential windows of 1.6 and 1 V, respectively. The higher electrochemical potential window of the PVDF/$BaTiO_3$/ZnO_f nanocomposite suggests better capacitive behaviour than both PVDF/$BaTiO_3$/ZnO_p and PVDF/$BaTiO_3$/ZnO_s nanocomposites. In the previous studies [50, 51], it was found that addition of mechanically activated $BaTiO_3$ nanoparticles into PVDF and PVDF/PMMA blend had a profound effect of inducing the ferroelectric β-phase of PVDF. The β-phase gives PVDF its piezoelectric properties. Therefore, the work of Tabhane et al. [49] can be extended by investigating the influence of $BaTiO_3$ nanoparticle sizes on the properties (i.e. in particular electrochemical properties using (CV) of PVDF/$BaTiO_3$/ZnO. Different sizes of $BaTiO_3$ can be achieved through ball milling at different activation times.

2.4.4 Differential Pulse Voltammetry

Differential pulse voltammetry (DPV) is a technique often used for making electrochemical measurements. It is a derivative of staircase or linear sweep voltammetry with a series of regular voltage pulses superimposed on the stairsteps or potential linear sweep. The pulses of increasing amplitude are used for varying the potential and the current is sampled immediately before each potential variation. The current difference is plotted as a function of potential variations. Heydari-Bafrooei et al. [52] developed an electrochemical aptasensor using a nanocomposite composed of titanium dioxide (TiO_2) nanoparticles, chitosan (CHIT), multi-walled carbon nanotubes (MWCNT) and a novel synthesized Schiff base (SB) (TiO_2/MWCNT/CHIT/SB) on the surface of a glassy carbon electrode (GCE) for detecting thrombin. The DPV curves showed that a bare GCE had high peak current than the TiO_2 and TiO_2/MWCNT/CHIT/SB) covered electrodes, whereas the TiO_2/MWCNT covered electrode had the highest peak current. The TiO_2 covered GCE had low peak current due to the relatively low electron transfer aptitude of TiO_2. The higher peak current of the TiO_2/MWCNT covered electrode suggested that MWCNT would enhance electron transfer characteristics of the electrode thus increasing the current response. However, TiO_2/MWCNT/CHIT/SB) covered electrodes showed the lowest peak current and this was linked to the presence of a low conducting CHIT/SB in the composite which ultimately reduced the electron transferability of the electrode. Furthermore, the authors explained that electrochemical properties of the TiO_2/MWCNT/CHIT/SB nanocomposite are greater than those of TiO_2/CHIT/SB nanomodified and that needs further exploration because the results of the TiO_2/CHIT/SB covered electrodes were not shown in the manuscript. In addition, a question that can be asked regarding the TiO_2/MWCNT/CHIT/SB nanocomposite is "which compound between TiO_2, and CHIT/SB is dominant at reducing electron transfer characteristics on the electrode?" Although a novel electrochemical aptasensor was developed with a TiO_2/MWCNT/CHIT/SB nanocomposite but scaling up this process for mass production will be challenging because the production of the TiO_2/MWCNT/CHIT/SB nanocomposite was done using 3 h of sonication. Therefore, one can build from this study by investigating whether lower sonication times have an influence on electronic properties. It is likely that there exists a critical sonication time less than 3 h for optimum morphological and electrochemical properties. Vinay et al. [53] used DPV to investigate the ability of OH functionalized MWCNT (OH/MWCNT) modified electrode to function as an electrochemical sensor for detecting aceclofenac. The concentration of OH/MWCNT on the carbon paste electrode (CPE) ranged between 1 and 190 μM. The DPV curves showed peak current increased with increases in concentrations and the limit of detection for aceclofenac was 0.246 μM, a value high than in other reported studies. Recently, Zhang et al. [54] used differential pulse anodic stripping voltammetry (DPASV) for simultaneous detection of Cd^{2+} and Pb^{2+} ions. The DPASV curves showed two separated and well-defined peaks at potential of −0.56 and −0.81 V corresponding to lead and cadmium ions, respectively. The results proved that the bismuth oxycarbide modified glassy carbon electrode sensor can simultaneously detect two target metal ions. The limit of detection for Cd^{2+} and Pb^{2+} ions were obtained to be 4.30 and 3.97 μg/L, respectively. The research was also done on real water samples where no Cd^{2+} and Pb^{2+} ions were detected. High concentrations of cadmium and lead are the pollutants of water and cause diseases such as kidney poisoning and abdominal cramps in humans. The recommended concentration of lead and cadmium ions in drinking water must not be greater than 10 and 5 μg/L, respectively [54].

2.5 Conclusion and Future Perspective

In the present chapter, electrochemical characterization techniques for nanomaterials (i.e. potentiometry, electrochemical impedance spectroscopy, galvanostatic techniques, cyclic voltammetry and differential pulse voltammetry have been discussed. These techniques supply valuable information about the chemical and physical properties of nanomaterials. It was discernible from the reported studies that the performance of variety of electrochemical techniques can be enhanced by the modification of the working electrodes with the nanoparticles. Inclusion of nanoparticles leads to improved sensitivity, selectivity and shorter response time. Besides, nanoparticle-modified electrodes can be effective in designing batteries and many other applications. However, the process of modifying the electrodes should be well optimized, as shown from the studies that factors such as particle size can have significant influence on the overall performance. Although remarkable advances have been done on electrochemical characterization of nanomaterials, more work still needs to be done for polymer nanocomposites. To be precise, it is known in the polymer nanocomposite field that different sizes, shapes and localization of the same nanofillers influence the mechanical and rheological properties of polymer nanocomposite differently. Therefore, the influence of different sizes, shapes and localization of the same nanofillers in polymer, nanocomposite should be widely explored. Achieving that would enable us to understand the mechanism responsible for different electrochemical behaviour of polymer nanocomposites to further optimize the design of polymer nanocomposites with targeted properties.

References

1. I. Khan, K. Saeed, I. Khan, Nanoparticles: properties, applications and toxicities. Arab. J. Chem. **12**, 908–931 (2019)
2. P. Samyn, A. Barhoum, T. Öhlund, A. Dufresne, Review: nanoparticles and nanostructured materials in papermaking. J. Mater. Sci. **53**, 146–184 (2018)
3. R.J. Moon, A. Martini, J. Nairn, J. Simonsen, J. Youngblood, Cellulose nanomaterials review: structure, properties and nanocomposites. Chem. Soc. Rev. **40**, 3941–3994 (2011)
4. D. Martin-Yerga, Electrochemical detection and characterization of nanoparticles with printed devices. Biosensors **9**, 47 (2019)
5. L.G. Mohtar, P. Aranda, G.A. Messina, M.A. Nazareno, S.V. Pereira, J. Raba, F.A. Bertolino, Amperometric biosensor based on laccase immobilized onto a nanostructured screen-printed electrode for determination of polyphenols in propolis. Microchem. J. **144**, 13–18 (2019)
6. G. Aragay, A. Merkoç, Nanomaterials application in electrochemical detection of heavy metals. Electrochim. Acta **84**, 49–61 (2012)
7. C. Breitkopf, K. Swider-Lyons, Electrochemical science-historical review. *Springer Handbook of Electrochemical Energy*. Springer Handbooks, (Springer, Berlin, Heidelberg, 2017)
8. B. Bansod, T. Kumar, R. Thakur, S. Rana, I. Singh, A review on various electrochemical techniques for heavy metal ions detection with different sensing platforms. Biosens. Bioelectron. **94**, 443–455 (2017)
9. K. Izutsu, Overview of electrochemical techniques, in *Electrochemistry in Nonaqueous Solutions* (2003)
10. É. Lojou, P. Bianco, Application of the electrochemical concepts and techniques to amperometric biosensor devices. J. Electroceram. **16**, 79–91 (2006)
11. Y. Bow, Hairul, I. Hajar, The application of potentiometric methods in determination total organic carbon content of soil. Int. J. Adv. Sci. Eng. Inf. Technol. **4** (2014)
12. N. Dürü, Potentiometric determination of some food additives and their binding to a polycationic species using polyion sensors. Turk. J. Chem. **37**, 308–315 (2013)
13. M. Shamsipur, F. Jalali, S. Ershad, Preparation of a diclofenac potentiometric sensor and its application to pharmaceutical analysis and to drug recovery from biological fluids. J. Pharm. Biomed. Anal. **37**, 943–947 (2005)
14. R.M. Kakhi, Application of nanoparticles in the potentiometric ion selective electrodes. Russ. J. Electrochem. **49**, 515–524 (2013)
15. Z. Taleat, A. Khoshroo, M. Mazloum-Ardakani, Screen-printed electrodes for biosensing: a review (2008–2013). Microchim. Acta **181**, 865–891 (2014)
16. A. Düzgün, G.A. Zelada-Guillén, G.A. Crespo, S. Macho, J. Riu, F.X. Rius, Nanostructured materials in potentiometry. Anal. Bioanal. Chem. **399**, 171–181 (2011)
17. Z.F. Akl, T.A. Ali, A novel modified screen-printed electrode with triazole surfactant assembled on silver nanoparticles for potentiometric determination of uranium. J. Radioanal. Nucl. Chem. **314**, 1865–1875 (2017)
18. E.Y. Frag, R.M.A. Hameed, Preparation, characterization and electrochemical application of CuNiO nanoparticles supported on graphite for potentiometric determination of copper ions in spiked water samples. Microchem. J. **144**, 110–116 (2019)
19. M.M. Khalil, A.A. Farghali, W.M.A. El Rouby, I.H. Abd-Elgawas, Preparation and characterization of novel MWCNTs/Fe-Co doped TNT nanocomposite for potentiometric determination of sulpiride in real water samples
20. L. Novotný, R. Petrankova, Potentiometric determination of silver nanoparticles using silver amalgam electrodes. Anal. Lett. **49**, 161–168 (2016)
21. A.R.R. Bredar, A.L. Chown, A.R. Burton, B.H. Farnum, Electrochemical impedance spectroscopy of metal oxide electrodes for energy applications. ACS Appl. Energy Mater. **3**, 66–98 (2020)
22. F. Lisdat, D. Schäfer, The use of electrochemical impedance spectroscopy for biosensing. Anal. Bioanal. Chem. **391**, 1555 (2008)

23. V. Balasubramani, S. Chandraleka, T.S. Rao, R. Sasikumar, M.R. Kuppusamy, T.M. Sridhar, Review-recent advances in electrochemical impedance spectroscopy based toxic gas sensors using semiconducting metal oxides. J. Electrochem. Soc. **167**, 037572 (2020)
24. B. Rezaei, T. Khayamian, N. Majidi, H. Rahmani, Immobilization of specific monoclonal antibody on Au nanoparticles for hGH detection by electrochemical impedance spectroscopy. Biosens. Bioelectron. **25**, 395–399 (2009)
25. Z. Chen, L. Li, H. Zhao, L. Guo, X. Mu, Electrochemical impedance spectroscopy detection of lysozyme based on electrodeposited gold nanoparticles. Talanta **83**, 1501–1506 (2011)
26. E. Casero, A.M. Parra-Alfambra, M.D. Petit-Domínguez, F. Pariente, E. Lorenzo, C. Alonso, Differentiation between graphene oxide and reduced graphene by electrochemical impedance spectroscopy (EIS). Electrochem. Commun. **20**, 63–66 (2012)
27. A. Rodríguez-López, D. Torres-Torres, J. Mojica-Gomez, C. Estrada-Arteaga, R. Antaño-López, Characterization by electrochemical impedance spectroscopy of magnetite nanoparticles supported on carbon paste electrode. Electrochim. Acta **56**, 8078–8084 (2011)
28. T. Liu, M. Li, Y. Wang, Y. Fang, W. Wang, Electrochemical impedance spectroscopy of single Au nanorods. Chem. Sci. **9**, 4424 (2018)
29. H. Liu, G. Piret, B. Sieber, J. Laureyns, P. Roussel, W. Xu, R. Boukherroub, S. Szunerits, Electrochemical impedance spectroscopy of ZnO nanostructures. Electrochem. Commun. **11**, 945–949 (2009)
30. R. Goel, R. Jha, C. Ravikant, Investigating the structural, electrochemical, and optical properties of p-type spherical nickle oxide (NiO) nanoparticles. J. Phys. Chem. Solids **144**, 109488 (2020)
31. N. Baram, Y. Ein-Eli, Electrochemical impedance spectroscopy of porous TiO_2 for photocatalytic applications. J. Phys. Chem. C **114**, 9781–9790 (2010)
32. H. Chang, C. Chang, Y. Tsai, C. Liao, Electrochemically synthesized graphene/polypyrrole composites and their use in supercapacitor. Carbon **50**, 2331–2336 (2012)
33. D. Zhang, X. Zhang, Y. Chen, P. Yu, C. Wang, Y. Ma, Enhanced capacitance and rate capability of graphene/polypyrrole composite as electrode material for supercapacitors. J. Power Sources **196**, 5990–5996 (2011)
34. D. Gui, C. Liu, F. Chen, J. Liu, Preparation of polyaniline/graphene oxide nanocomposite for the application of supercapacitor. Appl. Surf. Sci. **307**, 172–177 (2014)
35. M. Hassan, K.R. Reddy, E. Haque, S.N. Faisal, S. Ghasemi, A.I. Minett, V.G. Gomes, Hierachical assembly of graphene/polyaniline nanostructures to synthesize free-standing supercapacitor electrode. Compos. Sci. Technol. **98**, 1–8 (2014)
36. J. Li, H. Xie, Y. Li, J. Liu, Z. Li, Electrochemical properties of graphene nanosheets/polyaniline nanofibers composites as electrode for supercapacitors. J. Power Sources **196**, 10775–10781 (2011)
37. A.M. Golsheikh, N.M. Huang, H.N. Lim, R. Zakaria, C. Yin, One-step electrodeposition synthesis of silver-nanoparticle-decorated graphene on indium-tin-oxide for enzymeless hydrogen peroxide detection. Carbon **62**, 405–412 (2013)
38. U.S. Mohanty, Electrodeposition: a versatile and inexpensive tool for the synthesis of nanoparticles, nanorods, nanowires, and nanoclusters of metals. J. Appl. Electrochem. **41**, 257–270 (2011)
39. V.M. Maksimović, L.J. Pavlović, M.G. Pavlović, M.V. Tomić, Characterization of copper powder particles obtained by electrodeposition as function of different current densities. J. Appl. Electrochem. **39**, 2545–2552 (2009)
40. S.S.V. Tatiparti, F. Ebrahimi, Potentiostatic versus galvanostatic electrodeposition of nanocrystalline Al-Mg alloy powders. J. Solid State Electrochem. **16**, 1255–1262 (2012)
41. B. Pérez-Fernández, D. Martín-Yerga, A. Costa-García, Galvanostatic electrodeposition of copper nanoparticles on screen-printed carbon electrodes and their application for reducing sugars determination. Talanta **175**, 108–113 (2017)
42. B. Pérez-Fernández, D. Martín-Yerga, A. Costa-García, Electrodeposition of nickel nanoflowers on screen-printed electrodes and their application to non-enzymatic determination of sugars. RSC Adv. **6**, 83748–83757 (2016)

43. K. Wang, X. Shi, Z. Zhang, X. Ma, Y. Lu, H. Wang, Size-dependent capacitance of NiO nanoparticles synthesized from Ni-based coordination polymer precursors with different crystallinity. J. Alloy. Compd. **632**, 361–367 (2015)
44. Y. Zhu, G. Cao, C. Sun, J. Xie, S. Liu, T. Zhu, X.B. Zhao, H.Y. Yang, Design and synthesis of NiO nanoflakes/graphene nanocomposite as high performance electrodes of pseudocapacitor. RSC Adv. **3**, 19409–19415 (2013)
45. K. Xu, S.P. Ding, T.R. Jow, Toward reliable values of electrochemical stability limits for electrolytes. J. Electrochem. Soc. **146**, 4172–4178 (1999)
46. S.M Janjal, A.A Agale, A.S Rajbhoj, S.T Gaikwad, Synthesis and electrochemical characterization of zinc oxide nanoparticles using green method. Int. J. Appl. Res. 309–312
47. P. Cea, S. Martín, A. González-Orive, H.M. Osorio, P. Quintín, L. Herrer, Nanofabrication and electrochemical characterization of self-assembled monolayers sandwiched between metal nanoparticles and electrode surfaces. J. Chem. Educ. **93**, 1441–1445 (2016)
48. P. Pazhamalai, V.K. Mariappan, S. Sahoo, W.Y. Kim, Y.S. Mok, S.-J. Kim, Free-standing PVDF/reduced graphene oxide film for all-solid-state flexible supercapacitors towards self-powered systems. Micromachines **11**, 1–13 (2020)
49. G.H. Tabhane, S.M. Giripunje, Robust flower-like ZnO assembled β-PVDF/BT hybrid nanocomposite: excellent energy harvester. Polym. Testing **88**, 106564 (2020)
50. T.G. Mofokeng, A.S. Luyt, V.P. Pavlović, V.B. Pavlović, D. Dudić, B. Vlahović, V. Djoković, Ferroelectric nanocomposites of polyvinylidene fluoride/polymethyl methacrylate blend and $BaTiO_3$ particles: Fabrication of β-crystal polymorph rich matrix through mechanical activation of the filler. J. Appl. Phys. **115** (2014)
51. V.P Pavlović, V.B Pavlović, B. Vlahović, D.K Božanić, J.D Pajović, R. Dojčilović, V. Djoković, Structural properties of composites of polyvinylidene fluoride and mechanically activated $BaTiO_3$ particles. Phys. Scr.
52. E. Heydari-Bafrooei, M. Amini, M.H. Ardakani, An electrochemical aptasensor based on TiO_2/MWCNT and a novel synthesized Schiff base nanocomposite for the ultrasensitive detection of thrombin
53. M.M. Vinay, Y.A. Nayaka, H.T. Purushothama, R.O. Yathisha, K.V. Basavarajappa, P. Manjunatha, OH functionalized multi-walled carbon nanotube modified electrode as electrochemical sensor for the detection of aceclofenac
54. Y. Zhang, C. Li, Y. Su, W. Mu, X. Han, Simultaneous detection of trace Cd(II) and Pb(II) by differential pulse anodic stripping voltammetry using a bismuth oxycarbide/nafion electrode. Inorg. Chem. Commun. **111**, 1–5 (2020)

Chapter 3
Electrochemical Detection of Heavy Metals

Uyiosa Osagie Aigbe, Robert Birundu Onyancha, Kingsley Eghonghon Ukhurebor, Otolorin Adelaja Osibote, Onoyivwe Monday Ama, Harrison Ifeanyichukwu Atagana, Peter Osifo Ogbemudia, and Seyi Philemon Akanji

Abstract Heavy metals effluence poses a severe threat to environmental systems and a huge challenge for global sustainability. To monitor and detect heavy metals pollution in the environment, a portable point of care sensing platform is required. The detection of heavy metals ion using electrochemical sensors is considered a unique and perfect class of sensor technology. This chapter focuses on the use of voltammetry techniques for the detection of heavy metals ions using nanomaterials like carbon nanomaterials, polymers, metal oxides nanomaterials and metals nanomaterials. The voltammetry techniques were found to be effective in heavy metal ions detection owing to the ease of fabrication, chemical stability of sensors, low cost, selectivity, and sensitivity of nanomaterials electrochemical sensors. From the observed limits of detection for various metal ions detection from previous research, most electrochemical sensors made from nanomaterials showed selectivity and sensitivity towards heavy metals ions and were below the allowed level for each metal's ions by the World Health Organization.

U. O. Aigbe (✉) · O. A. Osibote
Department of Mathematics and Physics, Faculty of Applied Sciences, Cape Peninsula University of Technology, Cape Town, South Africa

R. B. Onyancha
School of Physical Sciences and Technology, Technical University of Kenya, Nairobi, Kenya

K. E. Ukhurebor
Department of Physics, Faculty of Science, Edo State University Uzairue, Edo, Nigeria

H. I. Atagana
Department of Life and Consumer Sciences, College of Agriculture and Environmental Sciences, University of South Africa, Pretoria, South Africa

P. O. Ogbemudia
Department of Chemical Engineering, Vaal University of Technology, Private Mail Bag X021, Vanderbijlpark 1900, South Africa

O. M. Ama · S. P. Akanji
Department of Chemical Science, University of Johannesburg, Doornfontein, Johannesburg 2028, South Africa

O. M. Ama et al. (eds.), *Modified Nanomaterials for Environmental Applications*,
Engineering Materials, https://doi.org/10.1007/978-3-030-85555-0_3

Graphical Abstract

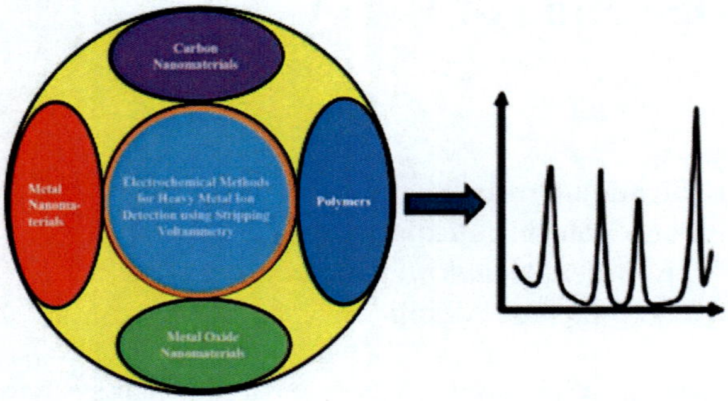

3.1 Introduction

Heavy metals (HMs) contamination of the environment in recent years has gained much attention due to the significant impact on public health [1]. According to their toxicity, metals and metalloid ions are separated into three groups. The first group comprises of metals (metalloids) that are toxic at very low concentrations, like lead (Pb), cadmium (Cd), and mercury (Hg). The second group of metals like arsenic, bismuth, indium, antimony, and thallium are only toxic in higher concentrations. The third group comprises of metals (metalloids) like copper (Cu), zinc (Zn), cobalt (Co), selenium (Se), and iron (Fe), which are essential for different chemical and biochemical processes in the body and are toxic only above a certain concentration. The concentration gap of these HMs is between toxic and maximum permissible limits [2].

Exposure to an abnormal concentration of HMs leads to adverse health effects. Exposure can result from industrial processes like smelting, disposing of factory waste, as well as diet, medication, those for battery outputs, mining activities, electroplating, fossil combustibles, pesticides and paper, tanneries, manure, diverse plastics, and metallurgies [3, 4].

The contamination of groundwater, rivers, and streams, with chemical pollutants, has become one of the utmost serious environmental problems of the century, with human activities increasing the concentration of HMs in the environment. Due to occupational exposure, 35 metals are of concern to us, but 23 of these are HMs[5, 6]. In freshwater, the natural concentration of metals depends on the concentration of the metal in the soil and the underlying geological structures, water acidity, human content, and concentration of particulate matter [7].

Water contamination monitoring is essential for environment preservation and inhibition of the adverse effects that it can have on human health [2]. Similarly, to assess the state of metallic contaminations in air, the examination of atmospheric deposition samples is useful in conducting environmental monitoring of HM concentrations in soil solution and plant leaves [8].

To decrease the pollution of the environment and mitigate the subsequent degradation of water and soil resources, it is essential to precisely determine the concentration of HMs. Various analytical methods have been recognized to detect heavy metals ions (HMI) [9], this includes atomic absorption spectrometry (AAS) [10], inductively coupled plasma mass spectrometry (ICP-MS) and inductively coupled plasma optical emission spectroscopy (ICP-OES) [11]. These methods are highly sensitive and selective, relatively expensive instruments are needed, involves the application of complex operational procedures, and have long detection times [9].

Remarkably, the electrochemical study of HMs is an alternative to conventional techniques and offers both attractive properties and a high degree of sensitivity. Owing to their compact, simple, and portable instrumentation, electrode miniaturization and easy electrode modification, electrochemistry offers exceptional application possibilities in the field of HMs study [11]. The most important electrochemical method can make a trace analysis of HMI correctly by electrochemically depositing the heavy metals to the surface of the working electrode (WE). This outstanding sensitivity is ascribed to the combination of the effective preconcentration step and the following measurement procedures [4].

However, the unique electrochemical technique that has high sensitivity and that can be applied for the identification and detection of HMs contamination is voltammetry. Over the years, substantial developments have occurred in electrochemical methods, including the change of various types of solid electrodes, which are presently applied to analyse different types of samples. Recently, inorganic materials have attracted significant attention owing to their low cost, compatibility, and strong adsorption to HMs. Specifically, in the electrochemical detection of HMs, nanomaterial-based metals, metal oxides, carbonaceous metals, and their composites are the most frequently used materials [9].

In these methods, the surface of the electrode is a powerful tool used for the quantification of an analyte. By controlling the potential, the electrode is used as an adjustable free energy source (sink) of electrons and electron crossing the electron solution interface is determined with great sensitivity by current measurement [12]. For the detection of trace amounts of HMs using the electrochemical method, the WE surface is required to be modified with active electrocatalyst (receptors), which are supposed to have a high collection capacity of target metal ions. This surface change is particularly important for the reduced devices due to the small amount of analyte (typically less than 100 μL) to be used for the analysis. Following this line of reasoning, to modify the working electrode for the electrochemical detection of HMs, an extensive variety of organic and inorganic materials have been used [13].

The performance of voltammetric HMs determination is strongly influenced by the WE properties. The modification of WEs with different materials allows for specific recognition and concentration of metals ions. Non-organic WEs are made of carbonaceous nanomaterials, metal oxides, metal nanoparticles, and their nanocomposites. Nanomaterials have attracted huge consideration as appropriate materials for the modification of the surface of the electrodes through dropping and deposition [9].

The electrode materials synthesis and design is a vital step in the process of HMI detection using the electrochemical analysis technique. Among the electrode materials, metal nanoparticle-modified electrodes are known to improve electrochemical sensitivity due to many active sites, fast transfer speed and their large specific surface area [14]. Modification of surface general result in the transfer of physicochemical properties of the modifier to the electron, improve the electrocatalytic activity owing to the use of materials with a large surface area which permits for a better sensitivity, selectivity towards the analyte owing to immobilized functional groups and dopants, fast diffusion kinetics for some materials, and the extraction and accumulation of an analyte at the electrode surface [12].

For trace metals analysis, the traditional electrodes used are the dropping mercury (DME), hanging mercury drop electrode (HMDE) and the mercury film electrode (MFE). These types of base electrodes (mercury-based electrodes) have been used extensively because of their reproducible electrode surface, decent negative potential gap, ability to dissolve the deposited metals and outstanding performance with stripping voltammetry. But, owing to the toxicity and disposal cost of mercury, these electrodes are no longer the excellent choice for sustainable application and alternative electrodes have been studied severely [3].

In the preconcentration of metal ions, numerous materials have been studied. This makes electrochemistry exceptional and emphasized in the world of analytical chemistry. In the development and arrival of nanomaterials, metal oxides like silver (Ag) or gold (Au) were extensively used owing to their increased surface areas and sites for metals ions to be deposited [15]. The modification of electrodes for electrochemical detection of trace metal ions, nanostructured materials are very attractive, because of their unique electronic, chemical, thermal, and mechanical properties in contrast with conventional materials [16].

Also, as a result of conducting and non-conducting polymers capabilities in supporting metal ions, polymers are used in electrode surface modification, the mixture of various materials has also proved to be effective in the deposition of metals ions [15].

A good candidate for the functionalization of the surface of the electrode is carbon nanotubes (CNTs) and graphene, which show a low detection limit and a wider linear range towards the target analyte. Utmost attention has been given to CNTs by scientists due to their fast electron transfer rate, excellent electrical and thermal conductivity, special texture, and the physicochemical characteristics [4, 17].

The potential of these materials is undisputable in sensing applications, as the novel carbon-derived nanomaterials have properties that are profound in their bulk materials, resulting in their ability to operate with not only a higher selectivity and sensitivity in harsh environments but also over greater temperature and dynamic ranges [18]. Advances in fabricating highly monodisperse samples of CNTs and graphene (Gr) have increase interest in using them for HMs sensing applications [19, 20].

In the detection of trace HMs, bismuth (Bi) is also used in the modification of electrodes due to its high sensitivity and inexpensiveness. When compared to Hg electrodes, it is harmless to use owing to its non-toxicity and having a broader range of potential. Apart from this, it can fuse HMs and these properties allow electrodes modified with Bi to identify a variety of electroactive species instantaneously in a chosen sample [21]. Hence, in meeting the increased demand for high output analysis in the field of environmental detection of HMs, detection of HMs through electrochemical methods in the long-term is the best way to meet the national and international legislative measures employed to reduce the anthropogenic pressure on the environment forms the fundamentals of this chapter. Also discussed in this chapter are the various voltammetry method, types of substrate electrodes used, and nanomaterials modified electrodes used for HMI detection.

3.2 Voltammetry Used for Heavy Metals Detection

The term voltammetry was derived from voltamperometry. The voltammetry method was invented from the polarography by the Czech chemist Jaroslav Heyrovsky in 1922, for which he received a Noble prize in chemistry in 1959. A special type of voltammetry technique is polarography, which describes the obtained current–voltage measurement using the dropping mercury electrode with a constant mercury drop flow. It was meant to show that the graphic electrode polarization was recorded as current–potential curves, with the flow of current meaning depolarization and no flow of current meaning polarization (the potential increases exclusive of or only with a slight current response) [22, 23].

Voltammetry is an electrochemical technique with extreme sensitivity and can be applied for the in-situ detection and identification of HMI contamination [9]. Voltammetry involves the perturbation of the initial zero-current state of an electrochemical cell by applying time-constant or time-varying potential (ramp) to promote a change in the electroactive species concentration on the surface of an electrode by either reduction or oxidation and measuring the resulting current [24, 25]. The different types of voltammetry methods can provide the difference in the variety of chemical information, physics and electrochemistry like high quantitative analysis, rate constants for chemical reaction, electrons involved in redox reactions and diffusion constants. Current signals in voltammetry are produced by two different currents. These are faradic currents based on Faraday's law, which corresponds to the flow resulting from the analyte oxidation or reduction, as its magnitude depends on the analyte concentrations in the solution and all the kinetics steps occurring on the electrode (electron transfer process). The other current is produced by the electrode-solution interface "electric layer double" that create the capacitive current, which contributes only to the background signals in the analytical measurement. The capacitive current must be detached from the faradic current or must be very small for trace analysis, with the latter been achieved with microelectrodes use or using step potential methods like pulse methods [26].

Considering an electrochemical reaction on an electrode with the appropriate application of a potential to it (3.1).

$$M^{n+} + ne^{-} \leftrightarrow R \quad (3.1)$$

The oxidation or reduction of a material at the WE surface, using a suitable potential result in the mass transport of novel material to the surface of the electrode and current generation. For mass transport of novel materials, the three probable procedures are diffusion, migration, and convection [27]. Diffusion is considered when the reactant species move in the electrode surface direction and the molecules of the product leave the interfacial area. The movement of charge ion in the presence of an electric field is termed as migration, while the electroactive species movement through thermal current, by rotating electrode or solution stirring is known as convection. Generally, the initial diffusion of the reactant from the solution to the electrode interface is known as mass transport and the potential applied to the cell induces the electron exchange between the electrode surface and the species in the solution [28].

The reaction rate is determined by measuring the current through the circuit as the current is relative to the surface area of the electrode is given by (3.2).

$$I = \frac{\mathrm{d}Q}{\mathrm{d}t} \quad (3.2)$$

where I is current applied (ampere), q is the charge $= 1.602 \times 10^{-19}$ C (coulombs) and t is the time (s) [29].

In an electrochemical system, the current (I) passing through the metal/solution interface is used to measure the total rate of reactants conversion to products at an electrode. To side-step possible inconsistencies owing to the interface size, the current is typically normalized by the electrode surface area (A) and stated as current density (j). The current density (j) and the rate of reaction (v) are expressed by (3.3) and (3.4):

$$j = \frac{I}{A} \quad (3.3)$$

$$v = \frac{I}{nFA} = \frac{I}{nF} \quad (3.4)$$

where n is the number of electrons exchanged in electrochemical reaction and F is the Faraday constant (96,485 C/mol). The total current passing through the electrochemical cell can be improved by increasing the surface area and the rate of reaction [30].

The current measures how fast species are reduced or oxidized at the surface of the electrode and it is influenced by the redox species concentration, the electrode size, shape and material, the solution resistance, the cell volume, and the number of electrons transferred. In voltammetry, current change owing to redox reactions with potential applied is described by the following laws. In the Nerst and Butler–Volmer equation, the applied potential controls the redox species concentration at the surface of the electrode and the reaction rate, respectively. For a reversible redox reaction as in (3.1), the potential applied forces the oxidized species concentration (C_o) and the reduced species concentration (C_e) at the surface of the electrode. The Nerst equation is given by,

$$E = E^* - \frac{RT}{zF}\ln Q_r \tag{3.5}$$

where E is the cell potential at the interest of temperature, E^* is the standard cell potential, T is the temperature in Kelvin, R is the Universal Gas Constant = 8.314 J/K/mol, z is the number of electrons transferred, F is the Faraday constant, and Q_r is the reaction quotient $= \frac{C_r}{C_o}$.

Q_r changes with the potential applied. When the ratio is larger, M is reduced as the potential applied is more negative and R is oxidized for more positive potential applied as the ratio is smaller. The Butler–Volmer equation is given by (3.6), which gives the relation between current, potential, and concentration.

$$\frac{i}{zFA} = k\left(C_o e^{-\alpha\theta} - C_r e^{(-1-\alpha)\theta}\right) \tag{3.6}$$

where A is the area of the electrode, k is the heterogeneous rate constant, α is the transfer coefficient and $\theta = \frac{xF(E-E^*)}{RT}$ [29].

The voltammetry types consist of the square wave voltammetry (SWV), cyclic voltammetry (CV), differential pulse voltammetry (DPV) and linear sweep voltammetry (LSV). The time waveforms created by their function application is the difference between these methods and they are all effective electroanalytical methods after they are improved to obtain the best electrochemical response. The electrochemical sensing device, electrochemical detecting instrument, and electrolyte are the three key parts of a typical analytical electrochemistry system. Specifically, the electrode morphology and size and the method of fabrication used can also influence the response of the voltammetric system. The electrochemical detection instrument comprising of the WE, reference electrode (RE) and counter electrode (CE) or an auxiliary electrode (AE) as depicted in Fig. 3.1. They can be used to detect specific metal ions after the surfaces of the WE have been modified using different materials [9, 18]. Discussed in the sub-sections are the various electrochemical methods used for HMI detections (voltammetry types-cyclic, linear sweep, square wave, and differential pulse voltammetry) and their working principles.

3.2.1 Cyclic Voltammetry (CV)

The cyclic voltammetry (CV) is an influential and popular electrochemical method generally used to investigate the reduction and oxidation procedures of molecular

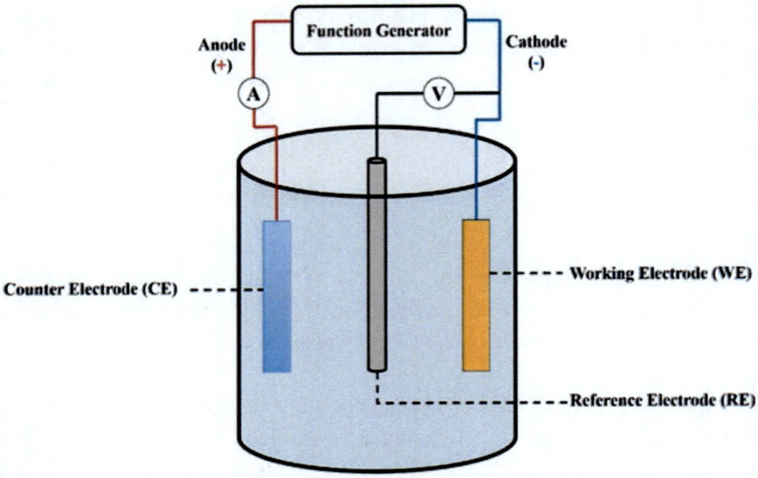

Fig. 3.1 Arrangement of a three-electrode electrochemical measurement

species [31]. This method is based on the step of preconcentration of the analytical species on the surface of the Hg electrode and is used to monitor trace concentrations of metals at a low cost [32].

The CV is used extensively for the redox reaction study and information on how chemical reactions occur is acquired using this technique. The CV is a fast voltage scan method in which the voltage scan direction is reversed. Whereas the directions of the applied potential at the WE are in both forward and reverse, the subsequent current is recorded. The CV can be used in the single or multi-cycles modes. The parameters measured in the CV are the cathodic and anodic peak potentials (E_{pc} and E_{pa}), the cathodic and anodic peak current (I_{pc} and I_{pa}), and the half peak potential ($E_p/2$) at which the anodic and cathodic current reach half of their peak value. Figure 3.2a

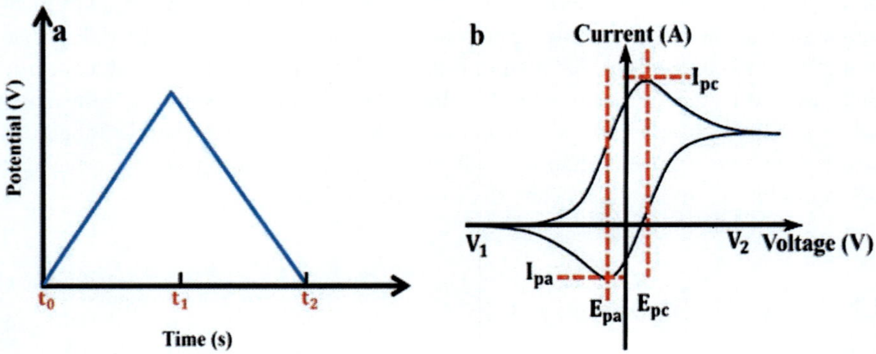

Fig. 3.2 **a** The potential-time waveform and **b** a typical cyclic voltammogram for a reversible redox process

shows the resulting scan of potential against time, the potential of a fixed electrode is scanned linearly, using a triangular potential waveform. The rate at which the potential is cycled (the scan rate) and the number of cycles recorded can be varied respectively to provide a variety of information on the electroactive species which is mostly displayed as a voltammogram as depicted by Fig. 3.2b [23, 33].

For a reversible electrochemical reaction, the well-defined characteristics of the CV recorded is the voltage separation between the current peaks for n electron process is given by (3.7):

$$\Delta E = E_{\mathrm{pa}} - E_{\mathrm{pc}} = \frac{59}{n}\mathrm{mV} \quad (3.7)$$

Secondly, the positions of peak voltage do not alter as a function of the voltage scan rate. Thirdly, the ratio of the peak currents is equal to one as given by (3.8).

$$\left|\frac{I_{\mathrm{pa}}}{I_{\mathrm{pc}}}\right| = 1 \quad (3.8)$$

and finally, the peak currents are proportional to the square root of the scan rate ($I_{\mathrm{p}} \propto v^{1/2}$) [34].

3.2.2 *Differential Pulse Voltammetry (DPV)*

The differential pulse voltammetry (DPV) is an electrochemical method where the cell current is evaluated as a function of the time and the potential between the indicator and RE [35]. The differential pulse voltammetry is carried out using a combination of steps and sweeps that improve the detection limits and speed. It is a technique with a regulated potential, the current is observed after the potential is kept constant for an appropriate time of 40 ms. The charging current discernment is carried out by observing the current right before the end of the pulse (or the final stage in the case of the stairway), where the load current is "capacitive current" in comparison with the faradic component is insignificant. The sensitivity of the DPV is enhanced by deducting the monitored current values at two different times [26].

3.2.3 *Square Wave Voltammetry (SWV)*

In square wave voltammetry, metal ions build-up is primarily due to the adsorption of complexes at the surface of the electrode in open circuit conditions. In the SWV technique, metals detections involve a single step where the metal ions interact with the electrode active sites through adsorption, while the sensitivity in the metal ions quantification is increased significantly by including two-stage SV techniques [33].

The SWV waveform includes a normal square wave, superimposed on the staircase potential base. enables the effective discrimination of faradic processes from charging currents. This is achieved by imposing two opposite square wave pulses of the same height at each staircase potential step and measuring the resultant current at the end of each pulse. During each square wave cycle, the current is doubled, one each at the end of the forward pulse and reverse pulse. Taking the difference between these currents returns a net voltammogram with improved signal-to-noise ratios. The two-measurement difference is plotted against the base potential [23, 36].

The resulting peak current is proportional to the analyte concentration. Exceptional sensitivity occurs when the net current is higher than either the reverse or forward parts. Linked with effective insight against the charging current, very low detection limits can be achieved. The effective scan rate is given by the square wave frequency f (Hz) and the step height ΔE given as $f\Delta E$. The major advantage of this technique is its speed [37].

3.2.4 Linear Sweep Voltammetry (LSV)

In the linear sweep voltammetry (LSV), the current is measured while passing through the WE, and the potential between the WE and RE is swept linearly in time between two pre-determined or fixed values. The linear sweep voltammogram characteristics depend on the rate of the electron transfer reaction(s), the chemical reactivity of the electroactive species and the potential sweep rate. In LSV measurements, the response of the current is plotted as a function of voltage instead of time, with the scan beginning from a fixed potential (V_1) where the current flows are negligible. Overall, as the voltage is swept further to more reductive values, the current begins to flow and goes through a maximum before dropping [30].

In the LSV, a fixed range of potential is employed, and the voltage is imaged from a lower limit to an upper limit as depicted in Fig. 3.3. The voltage scan rate (v) is determined from the slope of the line. The scan starts from the left-hand side of the plot of current against voltage, where no flow of current (V_1). With voltage further swept to the right, the current begins to flow and finally reaches a peak before dropping. With the voltage sweep from V_1 to V_2, the equilibrium position shifts from no change at V_1 to full change at V_2 of the reactant at the surface of the electrode. The precise voltammogram form can be improved by considering the voltage and mass transport effects. The individual curve has the same form, but the total current is observed to increase with an increasing scan rate. This is justified by the size of the diffusion layer and the time taken to record the scan. As the scan rate decreases, the linear sweep voltammogram will take longer to record. Hence, the diffusion layer size above the surface of the electrode will be changed depending on the voltage scan rate used. The flux to the electrode surface is significantly small at a slow scan rate than at a fast scan rate [27].

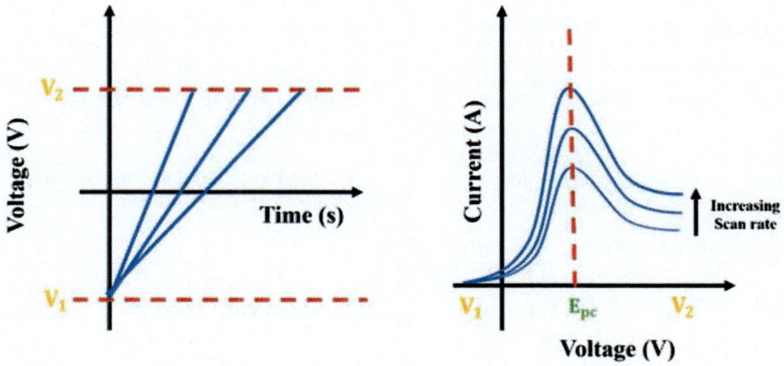

Fig. 3.3 Linear increase of the potential against time

3.3 Electroanalytical Techniques for Heavy Metals Detection

3.3.1 Stripping Voltammetry Methods (SVM)

Stripping voltammetry is a multipurpose electrochemical method used for a wide range of environmental and clinical applications owing to its numerous advantages [3]. The first electrode that received attention for electrochemical detection of HMs owing to their high sensitivity, high reproductivity, and a wide cathodic potential range for heavy metals detection was the mercury-based electrode. The WE of all voltammetric techniques apply mercury electrodes. But owing to the high toxicity of mercury, it is imperative to replace or develop electrodes that are environmentally friendly and exhibit similar characteristics for stripping voltammetric determination [38, 39].

The various stripping voltammetry methods are the most sensitive electroanalytical method making it possible to determine the very low concentration of the analyte (subnanograms). It has received substantial attention and has evolved over the past decade into a very powerful and versatile analytical method [40, 41]. They are capable of routinely identifying, quantifying, and detecting trace components of heavy metals or their complexes from 10^{-5} to 10^{-9} M with excellent sensitivity and selectivity. SV is very similar to polarography, with a small but substantial difference in technique [26, 42].

The preconcentration and dissolution are the two stripping voltammetry stages, with the preconcentration stage involving the metal cations accumulation on the working electrode surface using Faraday's reaction or adsorption [9]. Stripping voltammetry is a multipurpose electrochemical method used for a wide range of applications. This method comprises of two key stages, accumulation (preconcentration) and stripping. The analyte is placed at the working electrode surface during the accumulation stage. The analyte accumulation is either through an oxidation

Table 3.1 List of HMs detectable using stripping voltammetry methods [43]

Heavy metals detectable using the anodic stripping voltammetry method		
Antimony Arsenic Bismuth Cadmium Copper	Gallium Germanium Manganese Indium Lead	Mercury Silver Thallium Tin Zinc
Species detectable using the cathodic stripping voltammetry		
Arsenic Chloride Bromide	Iodide Selenium Sulphide	Mercaptans Thiocyanate Thio compounds
Metals detectable using the adsorptive stripping voltammetry		
Aluminium Cobalt	Nickel Chromium	Uranium Iron

potential (cathode stripping voltammetry) or the reduction potential (anodic stripping voltammetry) or the adsorption to the electrode surface (adsorptive stripping voltammetry) is followed by the current measurement during an opposite potential scan that reverses the redox process used or deposition [3, 26].

The electroanalytical procedures have a universal characteristic primary step, where the analyte of interest is collected on a WE controlled by potential electrolysis. After a while, the preconcentration stage is followed by the stripping stage, which involves the deposit dissolution when a linear lamp is applied to the electrode and a measurable current is created at the surface of the electrode following the reduction or oxidation of the analyte at the characteristic potential. During the stripping stage, the careful interpretation of the resulting peak shape of the current–potential voltammogram is recorded, a piece of anticipated analytical information is obtained, and the characteristic of the given substance used for qualitative identification is the peak potential (E_p). The corresponding analyte in the test solution is proportional to the peak current (I_p) and this analytical quantitative information can be gotten from the area or height of the stripping voltammetric peak [40].

Modern variations of conventional stripping emerge in 1947–1960, with the linear sweep voltammetry being the first followed by the development of square wave and pulse polarography [23]. Table 3.1 shows the list of HMI detectable using the anodic stripping voltammetry, cathodic stripping voltammetry and adsorptive stripping voltammetry. The various stripping voltammetry methods used for the analytical purpose are discussed in the subsection below.

3.3.1.1 Anodic Stripping Voltammetry (ASV)

The anodic stripping voltammetry (ASV) is the most common technique used for HMs detection and these involve the deposition and stripping [44]. The anodic stripping voltammetry technique is increasingly used owing to its wide linear dynamic range, low detection limit, multi-element analysis capability, low cost instrumentation [45, 46], and ultrahigh sensitivity (determination ranges from 10^{-6} to 10^{-11} mol/L and a detection limit of 10^{-10}–10^{-12} mol/L) [41].

The ASV was the first developed stripping method that used a preconcentration stage and was applied mainly to the trace analysis of heavy metal ions using a hanging mercury drop electrode [40]. The two steps of the ASV are the preconcentration step, where the analyte is preconcentrated by electrodeposition into the small-volume mercury electrode (Fig. 2.4). This reaction is represented by (3.9), where M is the analyte metal, n is the charge of the metal in the oxidized, ionic state, and e^- is the electron added by the metal during the reduction reaction.

$$M^{n+} + ne^- + Hg \rightarrow M(Hg) \tag{3.9}$$

As observed from (3.9), the analyte is reduced at the Hg electrode forming a mixture at the negative potential (cathodic). In the next step, the combined analyte is reoxidized and stripped off the electrode by the application of positive potential (anodic). This reaction result in the production and recording of peak current and the reaction is represented by (3.10).

$$M(Hg) \rightarrow M^{n+} + ne^- + Hg \tag{3.10}$$

The analytical information of interest is provided by the resulting voltammogram. The peak current (I_p) is proportional to the concentration of the analyte in the sample of the reproducible deposition condition is expected. The peak potential (E) helps to classify the analyte in the sample (Fig. 3.4) [23].

During electrochemical sensing using ASV, the preconcentration process is important for the selectivity and sensitivity of HMI [47]. Before voltammetric determination, the improved sensitivity and lower detection limits of the ASV, are achieved by the improvement of the analyte, which includes the metal ions reduction through the

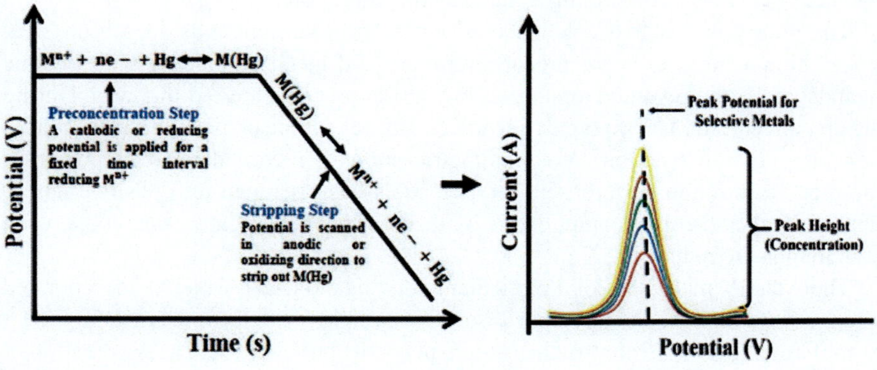

Fig. 3.4 Outline of the potential–time function and the corresponding current versus potential curve for the anodic stripping voltammetry

formation of films or amalgams of metals, reductive or oxidative formation of frugally soluble salts with or without the addition of reagent and the adsorptive build-up of chelates or complexes to the surface of the electrode [48].

The linear potential sweep is most used for anodic stripping, where the peak potentials are distinctive of each analyte, and various metals can be analysed in a single measurement. The observed stripping peak current at a hanging mercury drop electrode (HMDE) is given (at 25 °C) by (3.11).

$$I_{\mathrm{p}} = 2.72 \times 10^{5} n^{3/2} A D^{1/2} v^{1/2} C_{\mathrm{m}} \tag{3.11}$$

where n is the number of electrons involved in the oxidation, A is the electrode area in cm^2, D is the diffusion coefficient for the metal within the mercury drop expressed in cm^2/s, v is the scan rate in V/s, and C_m is the concentration of the metal within the mercury drop given as mol cm^{-3}. For a mercury film deposited on the surface of an inert substrate (like C or noble metal), the stripping peak current is given by (3.12).

$$I_p = \frac{n^2 F^2 v^{1/2} A I C_m}{2.7RT} \tag{3.12}$$

where F is the Faraday (9.65×10^4 C), I is the mercury film thickness in cm, and other symbols are as defined previously [49].

3.3.1.2 Adsorptive Stripping Cathodic Voltammetry (AdCSV)

The adsorptive cathodic stripping voltammetry (AdCSV) is another promising technique for trace metals analysis owing to their low instrumentation, good selectivity, high sensitivity, and good performance with various matrices. This method is based upon the adsorptive build-up of the complexes of the metal ion with an appropriate ligand at the electrode scanning in the negative direction [39, 50].

The reality is that hydrophobic (and amphiphilic) compounds adsorb on electrodes, they can be used for preconcentration, and metal complexes with organic ligands and some organic compounds have been preconcentrated this way. During the preconcentration, compounds adsorbed on the electrode do not experience oxidation or reduction reaction. A voltammetric curve is recorded subsequently after the preconcentration period, with the potentials been scanned into positive potentials when the adsorbed compound is oxidized and to the negative potentials when compounds are reduced [22].

The AdCSV makes use of a particular added ligand (AL), which is incorporated into the water sample and forms an adsorptive complex with the trace metal(s) under examination (3.13). To control the sample pH, a pH buffer is used, as the creation of the AdCSV ligand–metal complex is pH-dependent.

$${}_{y}M^{n+} + {}_{z}AL^{m-} \leftrightarrow M_y(AL)_z^{(yn-zm)} \tag{3.13}$$

A potential scan is carried out when a minute fraction of the ligand–metal complex is adsorbed onto the surface of the mercury drop (3.14). To determine the adsorption efficiency, the adsorption step is conducted in a carefully controlled potential, and in most cases, the adsorption potential selected is slightly more positive (0.1 V or more) than the reduction potential of the ligand–metal complex. The direction of the scan is towards more negative potentials and the ensuing current is measured.

$$M_y(AL)_z^{(yn-zm)} \leftrightarrow M_y(AL)_z^{(yn-zm)}(\text{adsorbed}) \tag{3.14}$$

The reduction of a reducible group on the metal or the ligand in the adsorbed complex is a result of the current produced (3.15). The types of scans applied during AdCSV include linear sweep, but the fast pulse-voltammetric waveforms are also used if the reduction of the ligand–metal complex is electrochemically reversible [51].

$$M_y(AL)_z^{(yn-zm)}(\text{adsorbed}) + e^- \leftrightarrow_y M^{(n-1)+} +_z AL^{m-} \tag{3.15}$$

3.3.1.3 Cathodic Stripping Voltammetry (CSV)

The cathodic stripping voltammetry (CSV) is the reverse of ASV, as the analyte is accumulated as an oxidized species, and is stripped by a potential sweep in the negative direction [52].

In determining organic and inorganic compounds that form insoluble salts with an electrode material, the cathodic stripping voltammetry method is used. The CSV involves the positive deposition of the analyte with the application of the anodic potential to the working electrode. This is followed by the stripping in a negative-going potential scan. Ions like chloride, bromide, iodide can be preconcentrated in the form of economically soluble salts, such as Hg_2Cl_2, Hg_2Br_2, and Hg_2I_2, where the mercury electrode potential is kept sufficiently positive to oxidize mercury to Hg_2^{2+} ions which precipitate these salts. The reaction in this step is given by (3.16).

$$A^{n-} + Hg \leftrightarrow HgA + ne^- \tag{3.16}$$

The CSV sensitivity depends on the amount of the analyte that can be plated in a given time, the density of the insoluble salt firm formed and the separation rate of the insoluble mercury compound during the step of stripping [22, 23].

3.4 Types of Substrate Electrodes Used for Heavy Metals Detection

In analytical chemistry, electrode materials are of great importance. The quest for the replacement of mercury electrodes with comparable or improved electrochemical behaviour for the development of miniaturized sensors is shown by the contribution of important research [53].

In the past, electrodes were any electroconductive materials that were started by metals like platinum (Pt) or Au. This was later substituted by the glassy carbon owing to their ease to use and wide potential range. After this, carbon paste was used due to its easy preparation. To increase the sensitivity and attract analytes like metals ions to the surface of the electrode, and with nanomaterials introduction and conducting polymers, the surface areas for the pre-conducting metals ions are considerably increased, making the technique ideal for trace metals analysis by its simplicity and low cost of the electrochemical techniques [15].

Getting exceptional performance in terms of specificity and sensitivity, the modification of a surface is significant [54]. In the design of electrode materials for electrochemical heavy metals sensing, most reported materials for electrode are hybrid materials mixed with multiple materials with different properties, comprising of materials with improved conductivity to promote the transfer of an electron, improved porosity for higher surface area with more active sites for active component loading or metal ion attachment and surface functional groups for selectivity towards a specified target metals ion and improved metal attachment [20].

In the electroanalytical method, the surface of the electrode is an influential tool for the analyte quantification, by controlling the potential, the electrode can be used as a sink of electrons. Electrons crossing the interface of the electron solution are determined with great sensitivity by measuring current. The electrochemical sensors for multianalyte determination are sub-divided into four classes, which are the glassy carbon electrodes (GCE), screen-printed electrodes (SPE), carbon paste electrodes (CPE) and other types of electrodes [12].

3.4.1 Glassy Carbon Electrode (GCE)

For electrochemical studies of several analytes, the glassy carbon electrodes (GCEs) have widely been used, with bare glassy carbon electrode employed for multianalyte detection [12]. The GCEs are produced by employing a cautiously controlled heating programme of a pre-modelled polymeric resin body in a passive atmosphere. Unlike various no graphitizing carbons, it is resistant to both gases and acid attacks. The glassy carbon structure comprises of graphite plane that is randomly arranged in a complex topology. It also has isotropic properties and in electrode devices, a specific orientation is not required [15].

The different pre-treatment for the synthesis and the activating of the surface of the glassy carbon electrode for electrochemical measurements is through mechanical treatment, which involves the abrasion with emery paper and polishing with alumina. Other techniques used are laser treatment, treatment of glassy carbon with ultrasound or the activation of the glassy carbon using carbon arc and the electrochemical treatment of the glassy carbon electrode [55]. For the pre-treatment of the surface of the electrode through electrochemical or non-electrochemical techniques, a comprehensive modification in the diverse electron transfer rate constant of an array of redox systems, frequently by various orders of magnitude is caused by the GCE. This is true for redox systems that proceed via adsorption and a sequential electron transfer stage [56].

3.4.2 Screen-Printed Electrodes (SPE)

Normally, SPE is designed for disposable applications with single or numerous times use. Gold, carbon, platinum, or silver ink are typically printed on plastic or ceramic substrate for electrochemical analysis in environmental, medical and food applications. The most common material used for the screen-printed electrode modification is carbon and they are particularly suitable for onsite HMs detection [15, 20].

Screen-printed electrodes (SPE) are devices produced by different ink printing on different types of plastic or ceramic substrate [12]. The fundamental principle for the SPE fabrication process is that the ink passes through parts of a mesh on a screen plate to the substrate on the other side, whereas the mesh remaining parts are blocked to form blank spaces on the substrate [9]. The various ink composition used for printing on electrodes defines the sensitivity and selectivity required for every analysis. Screen printing is an attractive technology for electrochemical sensors production as it allows for large-scale production at low cost, reproducible and disposable electrodes with extended capacity for modification [12].

3.4.3 Carbon Paste Electrode (CPE)

Carbon paste electrode (CPE) was invented and introduced in 1958 by Adams [57]. Some of the advantages of CPE over other electrodes are the ease and speed of preparation, the making of a novel producible surface and the minimal cost of carbon paste. The CPE is also a suitable conductive material for chemically modified electrodes (CMEs) preparation and the modified CPEs have good electrocatalytic activity, selectivity, sensitivity and show a low detection limit when compared to traditional CPEs [12].

With ingenious design, extra benefits like reproducibility, stability, and fast response time can be gotten. Carbon-based electrodes like carbon nanotubes, biomass-derived, graphene, porous-coordinated metal frameworks and related materials, and other carbon materials have been extensively used for the determination of both HMs and compounds [15, 20]. The carbon paste modified electrodes are prepared through the incorporation method, where a certain amount of modifier, binder and graphite powder of suitable concentration is mixed and ground to prepare the chemically modified electrode (paste), which is then filled in a cavity of the glass tube, telfon, etc. and copper wire is inserted into the cavity to establish electrical contact [9, 12].

3.5 Detection of Heavy Metals Using Nanomaterials

Electrochemical methods for HMI detection in an aqueous solution are grouped according to the unique electrical indicators produced by the HMI presence in the solution. HMI presence can cause variation in different electrical parameters such as voltage, current, electrochemical impedance, charge, and electroluminescence [58].

Electrode's sensitivity can be improved after modification of the electrode surface and these modifications are accomplished through the application of other materials to the surface of the electrode. The preparation techniques needed for modified electrodes construction are categorized into four types according to the different reactions that ensue inside or on the WE surface. This includes electrochemical polymerization, adsorption, electrochemical deposition, and covalent bond formation [9].

After the electrode modification, electrodes feature like wettability are defined clearly, with parameters like electrolyte and electrolyte concentration, pH, and buffer to use, deposition potential, modifying agents and involving materials concentration, deposition time, interferences and scan rate optimized. This method optimization is then applied to standards to get systematic performance followed by techniques validations [15]. The modification of electrodes by different materials is applied in stripping voltammetry, with optimization of parameters and subsequent use of methods with real samples. Results are normally compared with standard methods, or the standard materials are used for confirmation [15, 25].

The performance of voltametric heavy metals determination is strongly influenced by the WE properties. The modification of WEs with different materials allows for specific recognition and concentration of metals ions. Non-organic WEs are made of carbonaceous nanomaterials, metal oxides, metal nanoparticles, and their nanocomposites. Nanomaterials have attracted huge consideration as appropriate materials for the modification of the surface of the electrodes through dropping and deposition [9].

Nanomaterials are nanostructured materials generally describe as nanocomposites assembled in a three-dimension structure with one of the dimensions in the nanometer size range (0.1–100 nm). They generally have a large specific surface area, controlled surface activity, special quantum size effect, surface/interface effect, macroscopic quantum tunnelling effect, photoelectric effect, volume effect and catalytic effect.

These exceptional properties give nanostructured materials outstanding electrical, optical, magnetic, electrical, and other attributes [59].

Nanomaterials supported sensors provide superb detection of environmental contaminants on the nanomolar to the sub-picomolar level. Interest in these NM supported sensors stems from their potential for simple in-field detection of contaminants without the need for exclusive laboratory equipment. The NM supported sensors comprise of three parts, a nanomaterial(s), a recognition element that provides specificity and a signal transduction method that provides a way of transmitting the analyte presence. The high surface area to volume ratio and the superficial surface functionalization makes nanomaterials highly sensitive to surface chemistry changes, hence enabling nanosensors to accomplish low detection limits. The enhanced sensitivity of nanomaterials supported sensors is because nanomaterials are of comparable size as the analyte of interest (HMI) and are capable of interrogating hitherto isolated matrices [60]. The following materials used in the modification of electrodes from carbon nanomaterials to metallic nanoparticles and other natural nanostructure materials for HMI detections are discussed in the subsection.

3.5.1 Carbon Nanomaterials for Heavy Metals Detection

3.5.1.1 Carbon Nanotubes (CNTs) for Heavy Metals Detection in Water

Carbon nanomaterials are composed entirely of sp^2 bonded graphitic carbon found in all reduced dimensions like zero-dimensional fullerenes, 1-D carbon nanotubes (CNTs), and 2-D graphene. Within the nanometer scale dimensions, carbon nanomaterials properties are strongly dependent on their interactions with other materials and atomic structures [19]. The most used nanocarbon derivatives in electrochemical sensing are CNTs and Gr. The thin gap between the electronic shell ensures carbon hybridizes into three states which are sp, sp^2, and sp^3. For the formation of different organic compounds and nanocarbon materials, these states are exploited [33].

CNTs were first detected and described in 1952 by Radushkevish and Lukyanovich. SWCNTs were later observed by Oberin in 1976. The CNTs discovery in recent time was ascribed to Iijima, who described multi-walled carbon nanotubes (MWCNTs) in carbon soot obtained during C60 carbon molecule fabrication in an arc evaporation test [61].

Generally, CNTs comprised of two well-defined structural groups, which are the single-walled carbon nanotubes (SWCNTs) and multi-walled carbon nanotubes (MWCNTs). The SWCNTs consists of simple geometry with a rolled single graphene sheet with a diameter of 0.4–2 nm and up to 20 nm in length, while the MWCNTs comprise of numerous concentric tubes fitted into each other with a diameter up to 100 nm as depicted in Fig. 3.5 [3].

The uncommon importance given to CNTs, and graphene is due to the vital role they play in current advances based on nanomaterials including conductive and high-strength composites, artificial implants, drug delivery systems, sensors,

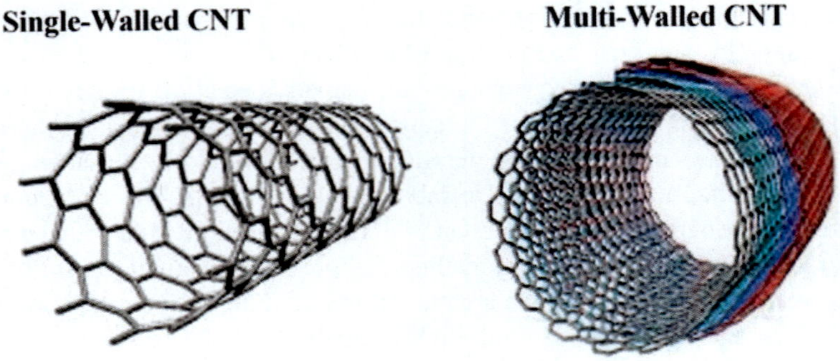

Fig. 3.5 Structures of single-walled carbon nanotubes and multi-walled carbon nanotubes [62]

energy conversion and storage devices, radiation sources and field emission displays, hydrogen storage media and nanometer-sized semiconductor devices, probes, and interconnects. CNTs is a beehive-shaped tube, having a thickness of approximately 1/50,000th of human hair. CNTs exist in three unique geometries, which are armchair, zigzag, and chiral. Their mechanical, electrical, optical, and other properties are determined by chirality [61].

In the development of the electroanalytical procedure, CNTs are fascinating materials for electrode modification and are extremely used for electrochemical detection of HMI due to their unique characteristics like good conductivity, high rate of electron transfer, large surface area and the modulation behaviour of CNTs near biomaterials, providing lower detection limits [1, 35, 63]. CNTs display greater electron transfer capabilities when used as electrodes in electrochemical reactions [18].

Owing to bare CNTs been hydrophobic, they are characteristically modified with functional groups to make them hydrophilic [38]. CNTs are attached to electrode surface either by functionalization or by incorporation on polymeric film and exhibit low background current, chemical inertness with wide potential window, low background current, and suitability for different types of analysis. This electrode surface functionalization changes the attraction for HMs [18, 33].

Lately, several studies have shown that CNTs modified electrodes could be used as a practical platform for electrochemical detection of HMIs [44]. Three electrodes in an electrochemical cell using CNT thread was developed for the detection of trace levels of Hg^{2+}, Pb^{2+} and Cu^{2+} using the ASV [17]. The WE was fabricated from CNTs threaded partly insulated with a coating of polystyrene solution and aspirated into a glassy capillary. The limit of detection with a 120 s deposition time were 1.05, 0.57 and 0.53 nM for Hg^{2+}, Pb^{2+} and Cu^{2+} respectively. The limit of detection of Hg^{2+}, Pb^{2+} and Cu^{2+} is well below the allowed limits by the United States Environmental Protection Agency (EPA) and the World Health Organization (WHO). Graphenated multi-walled carbon nanotubes (G-MWCNT) modified by indium tin oxide electrode (ITO) was also developed for the electrochemical detection of Pb (II) and Cu (II) in water [64]. The square wave anodic stripping voltammetry was used to assess the modified electrode performance and it was observed that Pb (II) and Cu (II) had a wide detection range from 0.05 to 2.5 μM with a limit of detection of 6 nM and 0.05–2.5 μM with a limit of detection of 12 nM using G-MWCNT modified ITO. The limit of detection (LOD) of Pb (II) and Cu (II) was based on signal-to-noise ratio of 3.

Also, CNTs modified with Bi oxide (Bi_2O_3) electrode were fabricated for the detection of Cd (II), and Pb (II) in water using the differential pulse anodic stripping voltammetry (DPASV) method. The linear response range for Cd (II), Pb (II) and In (III) using the DPASV method were 1.5–20 and 2.5–20 μg/L. The limit of detection for Cd (II), Pb (II) and In (III) was 0.22, 0.26 and 0.65 μg/L at a deposition time of 120 s [10]. A modest and easy technique was also employed for Cd (II) and Pb (II) ions detection using 1,3,6,8-pyrenetetrasulfonic acid sodium salt-functionalized carbon nanotubes nanocomposites (PyTS–CNTs) loaded on the pyrolytic graphite electrode [4]. The synthesized PyTS–CNTs showed good sensitivity and selectivity for HMI sensing due to the sulfonic acid group addition and different parameters were optimized using the DPASV. The stripping peak current of PyTS–CNTs/Nafion/PGE was observed to vary linearly with the HMI concentration ranging from 1.0–90 and 1.0–110 μg/L for both Cd (II) and Pb (II) and the limit of detection was assessed to be about 0.8 and 0.02 μg/L for Cd (II) and Pb (II), respectively (Fig. 3.6).

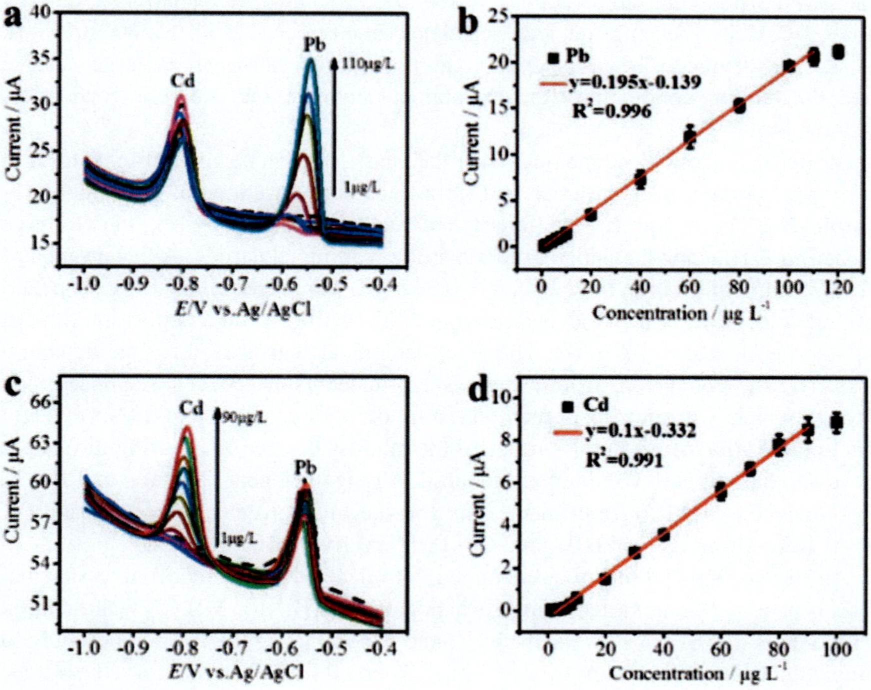

Fig. 3.6 The DPASV curves of different concentrations of lead and cadmium with corresponding calibration plot (**a**, **b**) and the anodic peak current as a concentration function in the presence of Cd (II) and Pb (II) (50 μg/L) (**c**, **d**) [4]

The CNTs covalently were modified with cysteine before deposition on the electrode surface, which was subjected to DPASV analysis was used for Pb (II) and Cu (II) ions detection in water. The limit of detection was found to 1 and 15 ppb for Pb^{2+} and Cu^{2+} respectively. This method was also used for lead and copper ions detection in spike lake water, with average recoveries of 96.2% and 94.5% for both Pb (II) and Cu (II) [65]. A summary of CNTs modified electrodes for the electrochemical HMI detection from previous studies is given in Table 3.2.

3.5.1.2 Graphene for Heavy Metals Sensing

Gr is one of the thinnest materials in the world. It is flexible but rigid than diamond and at room temperature conducts electricity more effectively than any other material. The basic structure of all graphite materials is Gr and is a one atom thick planar sheet of sp^2 bonded carbon atoms in a honeycomb crystal lattice (Fig. 3.7). Gr promotes an exceptional electron transfer for some species and excellent catalytic behaviour towards small biomolecules [12].

Gr unusual nanostructure show outstanding electrical conductivity, electron transfer rate, mechanical strength and large surface area makes it promising for several potential applications in the field of flexible electronics, energy storage devices, sensing devices, catalysts, membranes, and nanocomposites. In sensing devices, Gr has shown great potential because of its large surface area, outstanding conductivity, electrochemical stability, and wide electrochemical window [44, 77, 78].

Stripping voltammetric methods over the years have been shown from studies to have great potential for the trace and ultra-trace concentrations of HMIs determination [40]. The stripping voltammetric techniques were used by [79] to determine chromium (III) using the pencil graphite as the working electrode. A linear range of 12.5–75 mg/L and a limit of detection of 0.31 mg/L were determined using the pencil graphite electrode, which was measured at −0.8 to 1.0 V with a deposition time of 60 s and scan rate of 50 mV/s. The electrochemical sensor used for the detection of Cd (II) and Pb (II) ions from water was fabricated using reduced graphene oxide (RGO), which was micro-patterned with a lithography method, and the subsequent electrodeposition of Bi on the surface of the micro-patterned RGO using the square wave anodic stripping voltammetry method. A range of linear detection of 1.0–120 $\mu g/L$ was observed for both metal ions and the limit of detection of 0.4 and 0.1 $\mu g/L$ were observed for Pb (II) and Cd (II) respectively (Fig. 3.8) [80].

An electrochemical sensor based on graphite electrode extracted from waste zinc-carbon battery was modified using Bi nanoparticles (BiNP), MWCNT and Nafion for the detection of trace levels of Pb (II) and Cd (II) using the ASV in herbal food supplement samples. A limit of detection 1.06 and 0.72 ppb was observed for trace levels of Cd (II) and Pb (II). Also, the range of linear concentration was 5–1000 ppb for both Cd (II) and Pb (II) [81].

The square wave anodic stripping voltammetry method was used to determine traces of Pb (II) in the Hatyai city tap water using a boron-doped diamond (BDD)

Table 3.2. Applications of stripping voltammetry in the determination of toxic metals using CNTs

Nanomaterials	Detection method	Analyte	LOD	References
Metal catalyst free carbon nanotube (MCFCNT)	Square wave stripping voltammetry (SWSV)	Mn^{2+}	120 and 93 nM	[52]
Bismuth-modified carbon nanotube electrode (Bi-CNT electrode)	Square wave anodic stripping voltammetry (SWASV)	Pb^{2+} Cd^{2+}, and Zn^{2+}	1.3, 0.7, and 12 μg/L	[66]
C60 activated carbon and MWCNT modified glassy carbon electrode	–	Hg^{2+}	0.03 mM	[67]
Carbon nanoparticles (CNPs)	SWASV	Pb^{2+}, and Cu^{2+}	0.3 and 0.5 ppm	[68]
Carbon nanotube tower electrode	SWSV	Pb^{2+}, Cd^{2+}, Cu^{2+} and Zn^{2+}	2.5, 2.8, 2.8 and 4.4 ppb	[69]
L-cysteine self-assembled monolayers (SAM) functionalized gold nanoparticles/singlewalled carbon nanotubes/glassy carbon electrode (L-cys/AuNPs/SWCNTs/GCE)	DPASV	Cu^{2+}	0.02 nM	[70]
Triphenylphosphine-modified carbon nanotube composite	SWASV	Pb^{2+}, Hg^{2+} and Cd^{2+}	6.0×10^{-5}, 9.2×10^{-5} and 7.4×10^{-5} μM	[71]
Bismuth-oxychloride particle-multiwalled carbon nanotube composite modified glassy carbon	SWASV	Pb^{2+} and Cd^{2+}	1.9 and 4.0 μg/L	[72]
Carbon nanofibers	Anodic stripping voltammetry	Pb^{2+}	1.73 nM	[73]
Carbon nanotube based electrodes	Square-wave anodic stripping voltammetry (SWASV)	Cu^{2+}	3ppb	[74]
Single-walled carbon nanotube-poly (m-amino benzene sulfonic acid) modified gold electrode (SWCNT-PABS)	Square wave anodic stripping voltammetry (SWASV)	Hg^{2+}	0.06 μM	[75]

Fig. 3.7 Structure of graphene (2-D) [76]

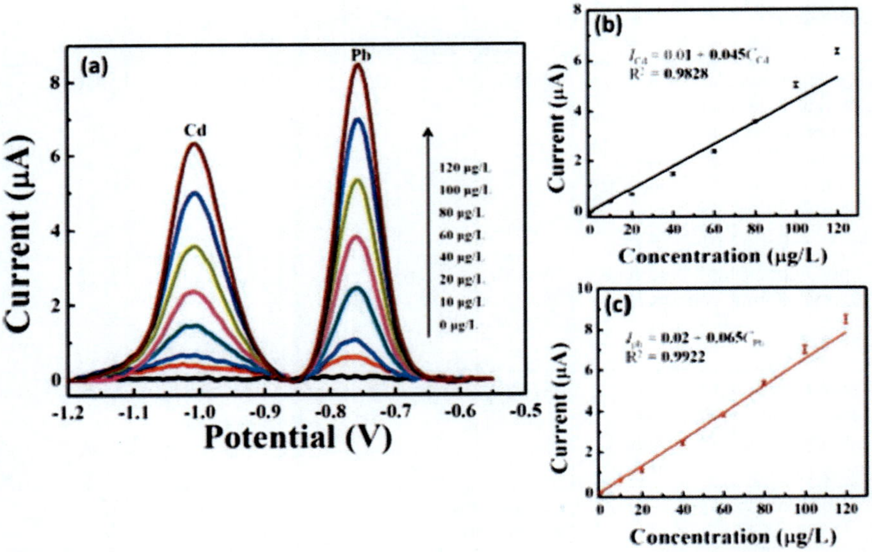

Fig. 3.8 **a** SWASV for different cadmium and lead ions concentrations and the corresponding calibration curves of Cd (II) and Pb (II) ions respectively (**b**, **c**) [80]

electrode. The preconcentration stage parameters studied included pH, concentration, electrolyte influences and volume, with effective preconcentration of Pb (II) traces attained at pH 1.26. The stripping peak of -0.460 V was gotten by scanning potential in the negative direction and the limit of detection was determined to be 0.3 $\mu g/L$. The various concentrations of Pb (II) from water samples taken from regions in Hatyai city were found to be in the range of 0.0–0.8 $\mu g/L$, which were lower than the drinking water contamination standard ($< 10.0 \mu g/L$) by the World Health Organization [45].

A highly sensitive, strong, and low cost microfluid electrochemical carbon-based sensor (µCS) was fabricated for the detection of trace levels of Pb^{2+} and Cd^{2+} [13]. Distinct anodic stripping peaks were observed for Cd (II) and Pb (II) at −0.78 and −0.53 V, and the peak currents for both metals were also observed to increase linearly with increased concentration as shown in Fig. 3.9b, d. Peak broadening was also observed at analyte concentrations below 10 µg/L for both cadmium and lead, which was due to the heterogeneity of the active sites on the graphite foil surface and the varied interactions associated with the metal's particles deposited. The SWASV technique was applied in the detection of Cd (II) and Pb (II), and the limit of detection was 1.2 and 1.8 µg/L for Cd (II) and Pb (II), respectively. The fabricated electrochemical sensor was found to be extremely robust, highly reproducible measurements using a single microfluid electrochemical carbon-based sensor.

Graphene oxide (GO) modified with L-Cysteine in presence of polyvinylpyrrolidone (PVP) was used for electrochemical sensing of Hg^{2+} using the square wave anodic stripping [82], Also, polyglycine-modified graphene paste electrode (PGMGPE) was used for the electrochemical detection of Hg^{2+} and Pb^{2+}in water and

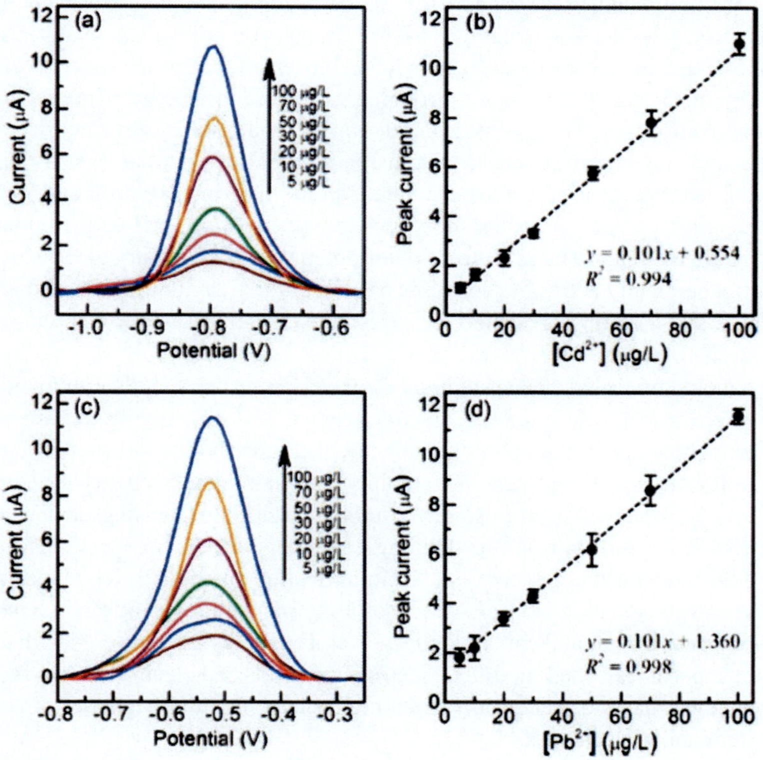

Fig. 3.9 Square wave voltammograms of **a** Cd (II), and **c** Pb (II) in 0.1 M acetate buffer on the microfluidic electrochemical carbon-based sensor, calibration curves of **b** Cd (II) and **d** Pb (II) [13]

biological samples. The cyclic and differential pulse voltammetry techniques were used to study metal ions behaviour and the limit of detection was 6.6 and 0.8 μM for Hg^{2+} and Pb^{2+} respectively [83]. Simple, inexpensive, portable, and effective electrochemical sensors (PIGE, PIGE/MWCNT/HNO_3 and PIGE/MWCNT/EDTA/HNO_3) were developed for evaluating four metal ions, which are Cd (II), Pb (II), Cu (II), and Hg (II). The limit of detection using PIGE/MWCNT/HNO_3 was 2.98 mol $\times 10^{-7}$/L for Cd (II), 3.81 mol $\times 10^{-7}$/L for Cu (II), 4.83 mol $\times 10^{-7}$/L for Pb (II) and 6.79 mol $\times 10^{-7}$/L for Hg (II) [84].

3.5.2 Metal Nanomaterial for Detection of Heavy Metals

Metal nanomaterials have exceptional chemical and physical properties owing to their size, shape, and structure. A range of metal nanomaterials have been utilized in electrode modification and effectively applied to HMIs detection due to their exceptionally large surface area and electrical conductivity. These metal nanomaterials enabled electrochemical sensors to have outstanding performance in terms of sensitivity, selectivity, and linear range for HMI detection. Metal nanomaterials mostly used as electrochemical sensors for HMI are Bi, Hg, Ag and Au nanomaterials [44].

In the design of electrochemical sensor and biosensor platforms, various key roles played by metal nanoparticles are through a simplistic synthesis, effortless surface functionalization, stress-free sensor production method, catalysis of electrochemical reaction and the improved electron transfer process [85]. Initial electrochemical analytical techniques used electrodes based on the hanging Hg drop and Hg film, and these were considered as the most sensitive devices for HMI detection besides the stripping methods. The HMI to be determined with the mercury electrodes must be electrochemically active but concerns of mercury toxicity to the environment and its storage and handling prompted the gradual exclusion of mercury application in electrodes [9].

Au nanomaterials have been studied extensively and successfully employed as an effective electrocatalyst in several electrochemical reactions owing to their exceptional properties like biocompatibility, electrical conductivity, uniform particle size, easy method of synthesis, superior stability, and complete recovery in chemical redox processes. The opportunity of the miniaturization of sensing devices to the nanometer scale and with an outstanding chemical sensing prospect was opened using gold nanoparticles. Micro and nanofabrication methods offer a spectacular approach for the development of Au sensor array platforms for the electrochemical detection of environmental contaminants. Several advantages are offered with the use of Au nanoparticles-based modified electrode or nanosized electrodes, this includes enhanced electroactive species diffusion, high selectivity, high signal-to-noise ratio and better catalytic activity [44, 86].

Au nanoparticles modified screen-printed carbon electrode was developed and used for the real-time Hg (II) and Pb (II) ions in water detection using the square wave anodic stripping voltammetry. The cyclic voltammetry was used to determine the electrochemical properties of the developed electrodes and the limit of detection of 1.8 and 4.62 μg/L for Hg (II) and Pb (II) respectively [87]. Also, Au nanoparticles Gr-modified glassy carbon electrode was used for the detection of Pb (II). The limit of detection was 0.34 μg/L and Pb (II) ions detected in actual water samples had a recovery range of 92.94–101.47% using nano-gold-Gr/glassy carbon electrodes. This showed that the fabricated electrode possessed substantial detection performance for actual water samples [88].

Ag nanomaterials have attracted widespread attention owing to their extraordinary properties such as electrical conductivity, easy preparation, large surface area, amplified electrochemical signal and excellent biocompatibility. Significant impact has been made by the development of the Ag nanoparticles sensors platform for environmental applications. Over the last twenty years, a lot of effort has been made towards the design of novel analytical techniques using Ag nanoparticles and their composites for environmental monitoring and food safety. The prospect of generating novel analytical platform for emerging environmental inorganic, organic and biological contaminants has been solved using Ag nanoparticles-based electrodes due to their sensitivity and specific nature. Ag nanoparticles combination with several matrixes like silicate network, metal oxides, polymers, fibres, dendrimers, Gr, etc. provides improved sensing performance with high stability due to the prolonged usage of the materials. The stability and sensitivity of the sensor platform relate to the distribution and the inhibition of Ag nanoparticles aggregation in the matrices. The steric repulsion and electrostatic repulsion are the two ways of stabilizing Ag nanoparticles using stabilizing agents like polymers [44, 86].

PEG functionalized silver nanoparticles synthesized through the chemical reduction method, was used for the detection of arsenic (III), cobalt (II), Cd (II), iron (III), Cu (II), Hg (II), nickel (II) and magnesium (II). A noticeable alteration from yellow to blue colour owing to PEG functionalized silver nanoparticles aggregation was observed for arsenic samples. The limit of detection was 1 ppb with exceptional linearity of approximately 0.99 using the PEG functionalized silver nanoparticles for arsenic detection [86].

The interfacial adsorption of cadmium (Cd (II)) on a Nafion-coated Bi film electrode and was stripped from the electrode using the square wave cathodic voltammetric. The reduction of Cd(II) ions was observed using 0.1 M acetate buffer electrolyte containing 0.04 mM Cd-Cupferron and 0.4 mg/L of Bi (III) of an accumulation potential of 300 mV against the Ag/AgCl at an accumulation time of 60 s. The linear concentration and limit of detection under optimum conditions were determined to be in the range of 3.0–40.0 μg/L and 0.38 μg/L respectively [39].

Bi electrode was first introduced in the year 2000 as Bi film electrodes on glassy carbon (GC) and carbon fibre substrate. In trace metal analysis, Bi has become a major alternative for mercury-based electrodes. With different metals, Bi can form alloys and offers the advantage of excellent resolution of peaks, high sensitivity, very low toxicity, and wide potential window [3]. In trace analysis, Bi has an advantage in its capability to form binary or multi-component alloys with many other HMs. Also, the conductance problem is avoided as it is generally diamagnetic metal and compare with mercury, Bi is insensitive to dissolved oxygen hence making unnecessary any deaeration step. Most time, Bi film is plated on the electrode by potentiostatic reduction of Bi (III) solution before analysis [89].

Gr)/Bi nanocomposite film-modified glassy carbon electrode (GCE) synthesized through electrodeposition of exfoliated graphene oxide (GO) and the in-situ plating of Bi film was used to determine the trace concentrations of Zn (II) , Cd (II), and Pb (II) ions. The observed linear calibration curves ranged from 1 to 100 µg/L for Zn^{2+}, Cd^{2+}, and Pb^{2+} respectively. The limits of detection were 0.18, 0.11 and 1.80 µg/L forCd (II), Pb (II) and Zn (II). The electrode was also successively used in the analysis of trace metals in the environment [90]. Also, a Bi film glassy carbon electrode (BiFE) prepared in-situ in 0.1 M acetate buffer solution was employed in the trace determination of Zn (II), Cd (II) and Pb (II) concentrations using the square wave anodic stripping voltammetry. The capacitive and resistive behaviour of the BiFe was not influenced by the trace metal concentrations. The limits of detection were 0.4 µg/L for Pb^{2+}, 0.3 µg/L for Cd^{2+} and 1.1 µg/L for Zn^{2+} [91].

3.5.3 Metal Oxide Nanomaterials for Heavy Metals Ions Detection

Metal oxides nanomaterials like Fe_3O_4 [92–94], ZnO [95], Cr_2O_3 [96], CuO [97], SnO_2 [98], ZrO_2 [99], TiO_2 [100], MgO [101], and MnO_2 [9] have been used for HMI detection in water. The metal oxide nanomaterials electrochemical sensors systematic performance is ascribed to their elevated HMI adsorption [44].

The chemical techniques for nanostructured metals oxides include the metal ions reduction and the controlled separation of the formed metals atoms from the bulk solution. Uniform nanoparticles like nanorods, nanofibres, nanobelt, nano comb and nanotubes are obtained, and this family of one-dimensional nanostructures provides a great model system for electrochemical sensing of environmental contaminants [85].

Chromium (III) oxide modified with carbon paste electrode (Cr-CPE) and a screen-printed electrode (SPE) were fabricated by [102] and used for electrochemical detection of Zn (II), Cd (II), Pb (II), Cu (II), and silverAg (I) ions using the SWASV method. The limit of detection for Zn (II) was 25 and 80 $\mu g/L$, Cd (II) was 3 and 10 $\mu g/L$, Pb (II) was 3 and 10 $\mu g/L$, Ag (I) was 3 and 10 $\mu g/L$ and Cu (II) was 3 and 10 $\mu g/L$ respectively. On the screen-printed electrode, the limit of detection for Zn (II) was 350 $\mu g/L$, Pb (II) was 3 $\mu g/L$, Cd (II) was 25 $\mu g/L$, and for Cu (II) was 3 $\mu g/L$.

Also, Cu nanoparticle (CuNP) attached with RGO were prepared through the liquid phase reduction method. The CuNP/RGO nanocomposites modified bare glassy carbon electrode was used for the detection of Hg (II), Cu (II), Cd (II) and Pb (II) using the SWASV. An extraordinary electroanalytical activity and selective

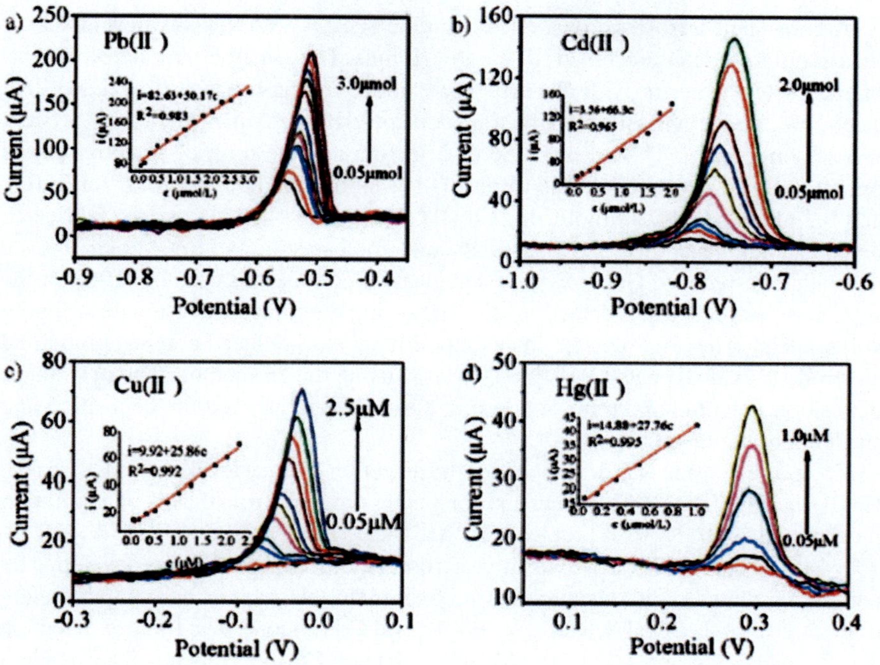

Fig. 3.10 a–d Responses of the SWASV and the corresponding calibration plots of CuNP/RGO nanocomposites modified bare glassy carbon electrode for Pb(II), Cd(II) Cu(II), and Hg(II) at different concentrations [14]

sensitivity to Hg (II), Cu (II), Cd (II) and Pb (II) were displayed by the nanocomposites, with sensitivities of 27.76, 25.86, 66.30 and 50.17 μA/μM displayed by each ion. The limit of detection for each metal ions were 0.186, 0.203, 0.111 and 0.051 μM for Pb (II), Cd (II), Cu (II) and Hg (II) respectively as shown in Fig. 3.10 [14].

A simple, fast, and sensitive sensor built on a glassy carbon electrode modified with tin oxide nanoparticles on MWCNTs was used for the determination of trace levels of thallium (Ti (II)) and Hg (II) using the SWASV. The limit of quantification and detection ranged from 3.3–4.0 μg/L and 0.9–1.2 μg/L respectively under optimization. The synthesized material was also used for the determination of the target trace elements in surface water samples and the accuracy of the analytical result got using the fabricated sensor was equivalent to those gotten by ICP-MS [16].

Fe_3O_4 nanoparticles fabricated through a novel impinging stream rotating packed bed reactor, with a saturation magnetization of 60.5 emu/g was used for the detection of trace levels of Pb (II), Cu (II), Hg (II) and Cd (II) ion using SWASV. The fabricated modified electrode showed selective detection towards Pb (II) with a high sensitivity of 14.9 μA/μM and the response to the other metal ions were negligible. The modified electrode also showed good stability and possible practical application in the electrochemical determination of lead [103].

A core–shell ferroferric oxide@polyaniline (Fe_3O_4@PANI) was developed for the quantitative detection of Cd (II) andPb (II) ions. The sensitivity for detection was increased by ferroferric oxide due to its large surface area and owing to PANI excellent electronic conductivity, the electron transport speed property in the detection process was also improved. The linear range of detection and the limit of detection under optimal experimental conditions were 0.1–10^4 nmol/L and 0.03 nmol/L forPb (II), and 0.9 nmol/L and 0.3 nmol/L for Cd (II). The electrochemical sensor fabricated had outstanding stability, high sensitivity, and good specificity [104].

Magnetite-reduced Gr oxide modified glassy electrode was used by [105] for the selective electrochemical detection of Cd (II) using the square wave anodic stripping voltammetry. The limit of detection using the magnetite-reduced graphene-modified electrode for Cd (II) detection was 0.056 μM using the 3σ method. The pH values, deposition potential and time, supporting electrolyte influenced the deposition and stripping of the metal ions.

Zr/ZrO_2 electrode was developed for the detection of trace level of Cd (II), Pb (II), Cu (II) and Hg (II) ions. The concentration ranges obtained from the linear calibration plots were 8×10^{-8}–10^{-5} M for lead, 5×10^{-8}–2×10^{-6} M for Cu(II) and 4×10^{-8}–10^{-5} M for Hg(II) [106]. An innovative carbon modified electrode was developed by combining titanium dioxide/zirconium dioxide into the graphite carbon paste electrode to electrochemical detecting of Pb (II) and Cd (II) ions. The limit of detection was 7.6×10^{-6} M and 1.1×10^{-5} M for Pb (II) and Cd (II), respectively. The electron transfer coefficient was estimated to be 0.44 and 0.33 with the heterogeneous electron transfer rate constant of 5.64×10^{-3} and 2.42×10^{-3} (cm/s) for Pb (II) and Cd (II), respectively [107].

3.5.4 Polymers Modified Electrode for Heavy Metal Detection

Polymers are the perfect class of electrode modifiers of multi-phase components on the nanoscale dimension, with each phase having its exceptional properties that are different from conventional composite in terms of improved conductivity, enhanced surface area and various electrochemical properties with the synergic effect of polymer and nanomaterial [35]. The electrical conductivity property makes them useful materials plus having low-suspended ionization potential, small energy optical transitions and high electron affinity. Hence, they are appropriate materials for the improvement of sensitivity and performance of several sensors, as they are easy to synthesize, environmentally stable, cheap, and biocompatible [108].

The electroactive surface modification through deposition of polymer represents an extensive field of study and concerning the analysis of trace metals, polymers films permit the immobilization of a huge number of ligands onto the electrode causing the accumulation of complex metal ions. For surface modification, polymers used may be natural or deliberately prepared through chemical synthesis [89]. Polymer-based materials have attracted more research interest in the fabrication of electrochemical sensors for HMI. This section looks at the sensing mechanism and design of electrochemical sensors based on ions imprinted polymers (IIPs), conducting polymers and composites and electroactive polymers. Ion imprinted polymers (IIPs) are extremely selective electrode materials that are widely used for HMIs owing to their outstanding advantage like low cost, relative ease and simple synthesis procedures, strong selectivity, and mechanical and chemical stability [109].

In the field of materials science, conducting polymers (CPs) like polyaniline (PANI), polypyrrole (PPy), and polythiophene (3,4-ethoxythiophene) (PEDOT) are the leading materials [109]. The key reason for interest in conducting polymers (CPs) is that the slight perturbation at the CPs surface or their bulk creates sturdy changes in their electroactivity. Another reason for interest in CPs outside the simplicity they offer in conducting surface modification is that they are easily functionalized with extra chemical functions and the simplest way is to fix CPs with doping ions carrying the desired function [43].

A paper-based microfluid device built using a commercial chromatography paper strip combined with a screen-printed carbon electrode was used for the detection of Pb(II) and Cd (II) in aqueous samples. All measurements were carried out using the SWASV with a frequency of 20 Hz, an amplitude of 25 mV and a potential step of 5 mV. Electrochemical deposition step was carried out at −1.4 V for 120 s and the signal of the SWASV was observed to be greatly improved and showed outstanding analytical performance for Cd(II) and Pb(II) detection from 0 to 100 ppb and with a limit of detection of 2.3 and 2.0 ppb, respectively (Fig. 3.11). The electrochemical device also showed good stability and selectivity for a real sample of groundwater and gas dissolved salty soda water with physical contamination. This device has the prospect of being suitable for point of care in the monitoring of the environment, food safety and public health [110].

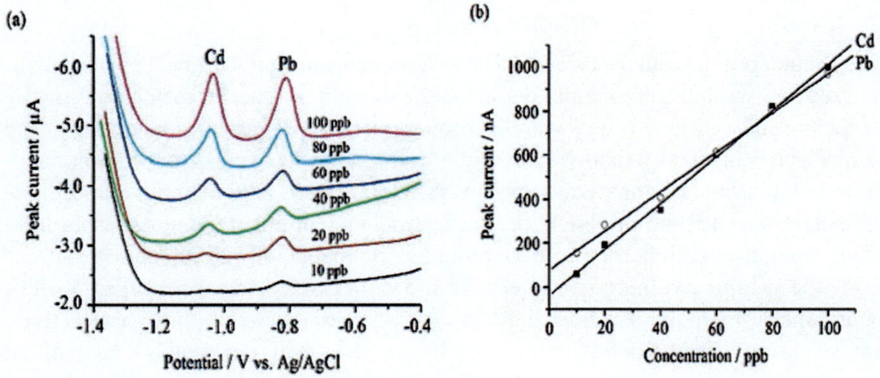

Fig. 3.11 SWASV voltammograms for Pb (II) and Cd(II) trace analysis in the Bi (III) in the paper-based microfluid devices presence. **a** The resulting calibration plots for trace Pb (II) and Cd (II) from 10 to 100 ppb and **b** calibration curves of Pb (II) and Cd (II) ions [110]

Ethylenediaminetetraacetic acid bonded conducting polymer modified electrode (EDTA-CPME) was fabricated through the polymerization of 3',4'-diamino-2,2';5',2''-terthiophene monomer on a glassy carbon electrode. Experimental parameters that affect the real-time analysis of the metal ions were optimized and EDTA-CPME calibration plots using the square wave voltammetry were found in the range of concentration of 5.0×10^{-10} and $1.0\ M \times 10^{-7}$ for copper and between 7.5×10^{-10} and $1.0\ M \times 10^{-7}$ for Pb (II) and Hg (II). The limit of detection for copper, mercury and lead were 2.0×10^{-10}, 5.0×10^{-10} and $6.0\ M \times 10^{-10}$. Coating EDTA-CPME with Nafion film, extraordinary improvement in the stability was observed for more than one month and the analytical availability of the EDTA-CPME was confirmed by the certified standard urine reference material and tap water application [111].

Also, the electrochemical method was used for the synthesis of ethylenediaminetetraacetic acid (EDTA) functionalized PPy and SWNTs nanocomposite on stainless steel electrode (SSE). The electrochemical detection of Pb (II) ions using the nanocomposite showed good sensitivity and selectivity towards HMIs and the limit of detection was 0.07 μM for lead ions [112]. Also, [113] used poly(3,4-ethylenedioxythiophene)/graphitic carbon nitride (PEDOT/g-C_3N_4) composites for the electrochemical detection of Cd^{2+} and Pb^{2+} using the DPV. The widest linear responses for both metals (Cd^{2+} and Pb^{2+}) from the composite ranged from 0.06–12 mM and 0.04–11.6 mM and the limit of detection was 0.0014 mM and 0.00421 mM, respectively. Table 3.3 summarizes recent research on HMI detection based on polymers modified electrodes, including electrode materials, analytes, method of detection, the limit of detection and linear ranges.

3.6 Conclusion

The chapter provides an overview of the development and application of electrochemical sensors created from nanomaterials in the detection of heavy metals ions owing to their improved electronic/catalytic properties, high electrical conductivity, and wide electrochemical window. Recently, electrochemical devices made from nanomaterials (carbon nanomaterials, polymers, metal oxide nanomaterials, and metal nanomaterials) for the modification of electrochemical electrodes or sensors have gained much attention for the determination of metal ions in aqueous solutions using the stripping voltammetry methods due to their size and ease application. The voltammetry methods combine the low cost of maintenance with high selectivity and sensitivity for the detection of trace levels (low levels) of metals ions without prior samples treatment. Under optimization, results from previous studies showed trace level detection of heavy metal ions using nanomaterial modified electrodes with different stripping voltammetry methods showed selectivity and sensitivity towards metals ions at a short time and each metal ions detected were below the allowable levels by the World Health Organization and other environmental organizations.

Table 3.3 Applications of stripping voltammetry in the determination of toxic metals using polymers

Nanomaterials	Detection method	Analyte	LOD	References
Polyaniline methylene blue coated screen printed iron electrode (PANI-IB)	Differential pulse anodic stripping voltammetry (DPASV)	Hg^{2+}	54.27 ± 3.3 (μg/L)	[114]
Phytic acid, polypyrrole and ZIF type metal-organic framework (PA/PPy/ZIF-8@ZIF-67)	DPV	Pb(II) and Cu(II)	2.9 and 14.8 nM	[115]
Conductive polyaniline (ANI) modified with metal-organic framework iO-66-NH 2	DPV	Cd^{2+}	0.3 μg/L	[116]
Liquid crystal polymer J (LCP) based bismuth (Bi) film electrode	SWASV	Zn^{2+}	1.22 nM	[117]
Crosslinked chitosan/carbon	Square wave anodic (SWA)	Pb^{2+}, Cd^{2+} and Cu^{2+}	0.6, 0.8 and 0.1 ppm	[1]
Carbon black–Nafion–glassy carbon electrode	Differential pulse anodic stripping voltammetry	Cd(II) and Pb(II)	0.8 nM (0.9 μg/L) and 5 nM (1.0 μg/L)	[118]
Disodium ethylenediaminetetraacetate dihydrate (EDTA) modified Carbon Paste electrode	Square wave anodic stripping voltammetry	Pb(II)	5.87×10^{-5} mol/L	[119]

References

1. K.H. Wu, H.M. Lo, J.C. Wang, S.Y. Yu, B.D. Yan, Electrochemical detection of heavy metal pollutants using crosslinked chitosan/carbon nanotubes thin film electrodes. Mater. Express **7**(1), 15–24 (2017)
2. A. Odobašić, I. Šestan, S. Begić, Biosensors for determination of heavy metals in waters, in *Environmental Biosensors* (IntechOpen 2019)
3. T. Wang, W. Yue, Carbon nanotubes heavy metal detection with stripping voltammetry: a review paper. Electroanalysis **29**(10), 2178–2189 (2017)
4. R. Jiang, N. Liu, S. Gao, X. Mamat, Y. Su, T. Wagberg, Y. Li, X. Hu, G. Hu, A facile electrochemical sensor based on PyTS–CNTs for simultaneous determination of cadmium and lead ions. Sensors **18**(5), 1–13 (2018)
5. M. Tian, L. Fang, X. Yan, W. Xiao, K.H. Row, Determination of heavy metal ions and organic pollutants in water samples using ionic liquids and ionic liquid-modified sorbents. J. Anal. Methods Chem. (2019)
6. H. Malassa, M. Al-Qutob, M. Al-Khatib, F. Al-Rimawi, Determination of different trace heavy metals in ground water of South West Bank/Palestine by ICP/MS. J. Environ. Prot. **4**, 818–827 (2013)
7. J. Raj, A. Raina, T.D. Dogra, Direct determination of zinc, cadmium, lead, copper metal in tap water of Delhi (India) by anodic stripping voltammetry technique. E3S Web Conf. EDP Sci. **1**, 09009 (2013)

8. Y. Khdaychi, L. Idrissi, S. Souabi, Y. Kadmi, Simultaneous electrochemical analysis of heavy metals in atmospheric deposits. J. Mater Environ Sci **9**(7), 2189–2200 (2018)
9. Y. Lu, X. Liang, C. Niyungeko, J. Zhou, J. Xu, G. Tian, A review of the identification and detection of heavy metal ions in the environment by voltammetry. Talanta **178**, 324–338 (2018)
10. T.T.P. Nguyen, X.G. Trinh, D.T.T. Uyen, Using electrode made of carbon nanotubes and bismuth oxide for the determination of metal concentration by anodic stripping voltammetry. J. Chem. (2019)
11. J. Kudr, H.V. Nguyen, J. Gumulec, L. Nejdl, I. Blazkova, B. Ruttkay-Nedecky, D. Hynek, J. Kynicky, V. Adam, R. Kizek, Simultaneous automatic electrochemical detection of zinc, cadmium, copper, and lead ions in environmental samples using a thin-film mercury electrode and an artificial neural network. Sensors **15**(1), 592–610 (2015)
12. N. Jadon, R. Jain, S. Sharma, K Singh, Recent trends in electrochemical sensors for multianalyte detection—a review. Talanta **161**, 894–916 (2016)
13. L.L. Shen, G.R. Zhang, W. Li, M. Biesalski, B.J. Etzold, Modifier-free microfluidic electrochemical sensor for heavy-metal detection. ACS Omega **2**(8), 4593–4603 (2017)
14. D. Li, C. Wang, H. Zhang, Y. Sun, Q. Duan, J. Ji, W. Zhang, S. Sang, A highly effective copper nanoparticle coupled with RGO for electrochemical detection of heavy metal ions. Int. J. Electrochem. Sci. **12**, 10933–10945 (2017)
15. P. Chooto, Modified electrodes for determining trace metal ions. Application of the voltammetry, M. Stoytcheva, R. Zlatev (eds.) (Intech, Rijeka, Croatia: InTech, 2017), p. 129.
16. S.H. Mnyipika, P.N. Nomngongo, Square wave anodic stripping voltammetry for simultaneous determination of trace Hg (II) and Tl (I) in Surface water samples using SnO2@ MWCNTs modified glassy carbon electrode. Int. J. Electrochem. Sci **12**, 4811–4827 (2017)
17. D. Zhao, D. Siebold, N.T. Alvarez, V.N. Shanov, W.R. Heineman, Carbon nanotube thread electrochemical cell: detection of heavy metals. Anal. Chem. **89**(18), 9654–9663 (2017)
18. A.C. Power, B. Gorey, S. Chandra, J. Chapman, Carbon nanomaterials and their application to electrochemical sensors: a review. Nanotechnol. Rev. **7**(1), 19–41 (2018)
19. D. Jariwala, V.K. Sangwan, L.J. Lauhon, T.J. Marks, M.C. Hersam, Carbon nanomaterials for electronics, optoelectronics, photovoltaics, and sensing. Chem. Soc. Rev. **42**(7), 2824–2860 (2013)
20. H. Hou, K.M. Zeinu, S. Gao, B. Liu, J. Yang, J. Hu, Recent advances and perspective on design and synthesis of electrode materials for electrochemical sensing of heavy metals. Energy Environ. Mater. **1**(3), 113–131 (2018)
21. S. Palisoc, M. Natividad, Y.A. Malabuyo, R.C. Pereja, Determination of heavy metals in root crops using bismuth nanoparticles modified graphene paste electrode. Agron. Res. **17**(1), 245–260 (2019)
22. F. Scholz, Voltammetric techniques of analysis: the essentials. ChemTexts **1**(4), 17 (2015)
23. O.A. Farghaly, R.A. Hameed, A.A.H. Abu-Nawwas, Analytical application using modern electrochemical techniques. Int. J. Electrochem. Sci. **9**(1), 3287–3318 (2014)
24. S. Wilke, H. Wang, M. Muraczewska, H. Müller, Amperometric detection of heavy metal ions in ion pair chromatography at an array of water/nitrobenzene micro interfaces. Fresenius J. Anal. Chem. **356**(3–4), 233–236 (1996)
25. G. Aragay, A. Merkoçi, Nanomaterials application in electrochemical detection of heavy metals. Electrochimica Acta **84**, 49–61 (2012)
26. J. Barón-Jaimez, M.R. Joya, J. Barba-Ortega, Anodic stripping voltammetry—ASV for determination of heavy metals. J. Phys. Conf. Ser. **466**(1), 012023 (2013)
27. D. Andrienko, *Cyclic Voltammetry* (Wiely Pub, New York, 2008), pp. 3–12
28. P. Protti, Introduction to modern voltammetric and polarographic analisys techniques. AMEL Srl 10 (2001)
29. P.S. Joshi, D.S. Sutrave, A brief study of cyclic voltammetry and electrochemical analysis. Int. J. ChemTech Res. **11**, 77–88 (2018)
30. L.A. Pašti, N.M. Gavrilov, S.V. Mentus, Voltammetric techniques in electrocatalytic studies, in *Voltammetry theory, types and applications* (Nova Publisher, New York, 2014), pp. 1–337

31. N. Elgrishi, K.J. Rountree, B.D. McCarthy, E.S. Rountree, T.T. Eisenhart, J.L. Dempsey, A practical beginner's guide to cyclic voltammetry. J. Chem. Educ. **95**(2), 197–206 (2018)
32. F.C. Pereira, N.R. Stradiotto, M.V.B. Zanoni, Electrochemical reduction and cathodic stripping voltammetric determination of clotrimazole. J. Braz. Chem. Soc. **12**(2), 202–207 (2001)
33. S.K. Pandey, P. Singh, J. Singh, S. Sachan, S. Srivastava, S.K. Singh, Nanocarbon-based electrochemical detection of heavy metals. Electroanalysis **28**(10), 2472–2488 (2016)
34. V. CH, A. Srividya, A. Ajitha, U. Rao, An overview on cyclic voltammetry and its application in pharmaceutical analysis. Int. J. Chem. Pharm. Sci. **5**(2) (2014)
35. A.K. Srivastava, S.S. Upadhyay, C.R. Rawool, N.S. Punde, A.S. Rajpurohit, Voltammetric techniques for the analysis of drugs using nanomaterials based chemically modified electrodes. Curr. Anal. Chem. **15**(3), 249–276 (2019)
36. P. Dauphin-Ducharme, N. Arroyo-Currás, M. Kurnik, G. Ortega, H. Li, K.W. Plaxco, Simulation-based approach to determining electron transfer rates using square-wave voltammetry. Langmuir **33**(18), 4407–4413 (2017)
37. N.A. Alarfaj, Adsorptive stripping anodic voltammetric determination of thioctic acid in bulk and pharmaceutical formulations. Int. J. Biomed. Sci. IJBS **5**(1), 54 (2009)
38. M. Li, H. Gou, I. Al-Ogaidi, N. Wu, Nanostructured sensors for detection of heavy metals: a review. ACS Sustain. Chem. (2013)
39. N. Meepun, S. Siriket, S. Dejmanee, Adsorptive stripping voltammetry for determination of cadmium in the presence of cupferron on a nafion-coated bismuth film electrode. Int. J. Electrochem. Sci. **7**, 10582–10591 (2012)
40. A.H. Alghamdi, Applications of stripping voltammetric techniques in food analysis. Arab. J. Chem. **3**(1), 1–7 (2010)
41. L. Eddaif, A. Shaban, J. Telegdi, Sensitive detection of heavy metals ions based on the calixarene derivatives-modified piezoelectric resonators: a review. Int. J. Environ. Anal. Chem. **99**(9), 824–853 (2019)
42. W.M. Peterson, R.V. Wong, *Fundamentals of stripping voltammetry*. American Laboratory (1981)
43. G. March, T.D. Nguyen, B. Piro, Modified electrodes used for electrochemical detection of metal ions in environmental analysis. Biosensors **5**(2), 241–275 (2015)
44. X. Liu, Y. Yao, Y., Ying, J. Ping, Recent advances in nanomaterial-enabled screen-printed electrochemical sensors for heavy metal detection. TrAC Trends Anal Chem 187–202 (2019)
45. P. Chooto, P. Wararatananurak, C. Innuphat, Determination of trace levels of Pb (II) in tap water by anodic stripping voltammetry with boron-doped diamond electrode. Sci. Asia **36**(2), 143–157 (2010)
46. D. Stankovic, G. Roglic, J. Mutic, I. Andjelkovic, M. Markovic, D. Manojlovic, Determination of copper in water by anodic stripping voltammetry using Cu-DPABA–NA/GCE modified electrode. Int. J. Electrochem. Sci **6**, 5617–5625 (2011)
47. W. Ye, Y. Li, J. Wang, B. Li, Y. Cui, Y. Yang, G. Qian, Electrochemical detection of trace heavy metal ions using a Ln-MOF modified glass carbon electrode. J. Solid State Chem. **281**, 121032 (2020)
48. J. Santos, M. Smyth, Mercury-free anodic stripping voltammetry of lead ions using a PVS-doped polyaniline modified glassy carbon electrode. Anal. Commun. **35**(10), 345–348 (1998)
49. R.S. Kelly, *Analytical Electrochemistry*: *The Basic Concepts*. Analytical Sciences Digital Library (2009)
50. T.K. Sari, J. Jin, R. Zein, E. Munaf, Anodic stripping voltammetry for the determination of trace Cr (VI) with Graphite/Styrene-Acrylonitrile copolymer composite electrodes. Anal. Sci. **33**(7), 801–806 (2017)
51. E.P. Achterberg, C. Braungardt, Stripping voltammetry for the determination of trace metal speciation and in-situ measurements of trace metal distributions in marine waters. Anal. Chim. Acta **400**(1–3), 381–397 (1999)
52. W. Yue, A. Bange, B.L. Riehl, B.D. Riehl, J.M. Johnson, I. Papautsky, W.R. Heineman, Manganese detection with a metal catalyst free carbon nanotube electrode: anodic versus cathodic stripping voltammetry. Electroanalysis **24**(10), 1909–1914 (2012)

53. R. Porada, K. Jedlińska, J. Lipińska, B. Baś, Voltammetric sensors with laterally placed working electrodes: a review. J Electrochem Soc **167**(3), 037536 (2020)
54. L. Redivo, M. Stredanský, E. De Angelis, L. Navarini, M. Resmini, L. Švorc, Bare carbon electrodes as simple and efficient sensors for the quantification of caffeine in commercial beverages. R Soc Open Sci **5**(5), 172146 (2018)
55. A. Dekanski, J. Stevanović, R. Stevanović, B.Z. Nikolić, V.M. Jovanović, Glassy carbon electrodes: I characterization and electrochemical activation. Carbon **39**(8), 1195–1205 (2001)
56. G. Ilangovan, K.C. Pillai, Mechanism of activation of glassy carbon electrodes by cathodic pretreatment. J. Solid State Electrochem. **3**(6), 357–360 (1999)
57. Y. Li, L.H. Huang, S.M. Chen, B.S. Lou, X. Liu, One-step fabrication of a new carbon paste electrode for dopamine, ascorbic acid and uric acid determination in serum. Int. J. Electrochem. Sci. **10**, 7671–7683 (2015)
58. B. Bansod, T. Kumar, R. Thakur, S. Rana, I. Singh, 2017. A review on various electrochemical techniques for heavy metal ions detection with different sensing platforms. Biosens. Bioelectron. **94**, 443–455 (2017)
59. M. Pan, J. Yang, K. Liu, Z. Yin, T. Ma, S. Liu, L. Xu, S. Wang, Noble metal nanostructured materials for chemical and biosensing systems. Nanomaterials **10**(2), 209 (2020)
60. M.R. Willner, P.J. Vikesland, Nanomaterial enabled sensors for environmental contaminants. J. Nanobiotechnol. **16**(1), 1–16 (2018)
61. G. Rahman, Z. Najaf, A. Mehmood, S. Bilal, S.A. Mian, G. Ali, An overview of the recent progress in the synthesis and applications of carbon nanotubes. C J. Carbon Res. **5**(1), 1–31 (2019)
62. H. He, L.A. Pham-Huy, P. Dramou, D. Xiao, P. Zuo, C. Pham-Huy, Carbon nanotubes: applications in pharmacy and medicine. BioMed Res. Int. (2013)
63. A.R. Sadrolhosseini, A.S.M. Noor, A. Bahrami, H.N. Lim, Z.A. Talib, M.A. Mahdi, Application of polypyrrole multi-walled carbon nanotube composite layer for detection of mercury, lead, and iron ions using surface plasmon resonance technique. PloS One **9**(4) (2014)
64. E. Guo, Simultaneous electrochemical determination of lead and copper based on graphenated multi-walled carbon nanotubes. Int. J. Electrochem. Sci **10**, 7341–7348 (2015)
65. J. Morton, N. Havens, A. Mugweru, A.K. Wanekaya, Detection of trace heavy metal ions using carbon nanotube-modified electrodes. Electroanal. Int. J. Devoted Fundamental Pract. Aspects Electroanal. **21**(14), 1597–1603 (2009)
66. R. Baby, B. Saifullah, M.Z. Hussein, Carbon nanomaterials for the treatment of heavy metal-contaminated water and environmental remediation. Nanoscale Res. Lett. **14**(1), 341 (2019)
67. O. Zaytseva, G. Neumann, Carbon nanomaterials: production, impact on plant development, agricultural, and environmental applications. Chem. Biol. Technol. Agricul. **3**(1), 17 (2016)
68. K.L.S. Castro, S.M. Oliveira, R.V. Curti, J.R. Araújo, L.M. Sassi, C.M. Almeida, E.H.M. Ferreira, B.S. Archanjo, M.F. Cabral, A. Kuznetsov, L.A. Sena, Electrochemical response of glassy carbon electrodes modified using graphene sheets of different sizes. Int. J. Electrochem. Sci **13**(1), 71–87 (2018)
69. S. Wyantuti, R.A. Hafidza, S. Ishmayana, Y.W. Hartati, Anodic stripping voltammetry with pencil graphite electrode for determination of chromium (III). J. Phys. Conf. Ser. **812**(1), 012006 (2017). IOP Publishing
70. X. Xuan, M.F. Hossain, J.Y. Park, A fully integrated and miniaturized heavy-metal-detection sensor based on micro-patterned reduced graphene oxide. Sci. Rep. **6**, 33125 (2016)
71. S. Palisoc, R.I.M. Vitto, M. Natividad, Determination of heavy metals in herbal food supplements using bismuth/multi-walled carbon nanotubes/nafion modified graphite electrodes sourced from waste batteries. Sci. Reports **9**(1), 1–13 (2019)
72. M. Vanitha, N. Balasubramanian, L.M. Joni, C. Panatarani, C., Detection of mercury ions using L-cysteine modified electrodes by anodic stripping voltammetric method. AIP Conf. Proc. **1927**(1), 030001 (2018). AIP Publishing LLC
73. C. Raril, J.G. Manjunatha, Fabrication of novel polymer-modified graphene-based electrochemical sensor for the determination of mercury and lead ions in water and biological samples. J. Anal. Sci. Technol. **11**(1), 3 (2010)

74. L.U. Pikna, M. Heželová, Z. Kováčová, Optimization of simultaneous electrochemical determination of Cd (II), Pb (II), Cu (II) and Hg (II) at carbon nanotube-modified graphite electrodes. J. Environ. Sci. Health Part A **50**(8), 874–881 (2015)
75. G. Maduraiveeran, W. Jin, Nanomaterials based electrochemical sensor and biosensor platforms for environmental applications. Trends Enviro Anal. Chem. **13**, 10–23 (2017)
76. B.S. Boruah, N.K. Daimari, R. Biswas, Functionalized silver nanoparticles as an effective medium towards trace determination of arsenic (III) in aqueous solution. Results Phys. **12**, 2061–2065 (2019)
77. I.A. Tayeb, K.A. Razak, Development of gold nanoparticles modified electrodes for the detection of heavy metal ions. J. Phys. Conf. Ser. **1083**(1), 012044 (2018). IOP Publishing
78. Y. Cheng, F. Sun, J. Lee, T. Shi, T. Wang, Y. Li, Gold-nanoparticles-graphene modified glassy carbon electrode for trace detection of lead ions. E3S Web Conf. **78**, 03007 (2019). EDP Sciences
79. L. Pujol, D. Evrard, K. Groenen-Serrano, M. Freyssinier, A. Ruffien-Cizsak, P. Gros, Electrochemical sensors and devices for heavy metals assay in water: the French groups' contribution. Front. Chem. **2**, 19 (2014)
80. S. Lee, S.K. Park, E. Choi, Y. Piao, Voltammetric determination of trace heavy metals using an electrochemically deposited graphene/bismuth nanocomposite film-modified glassy carbon electrode. J. Electroanal. Chem. **766**, 120–127 (2016)
81. B. Petovar, K. Xhanari, M. Finšgar, A detailed electrochemical impedance spectroscopy study of a bismuth-film glassy carbon electrode for trace metal analysis. Analytica Chimica Acta **1004**, 10–21 (2018)
82. U.O. Aigbe, R. Das, W.H. Ho, V. Srinivasu, A. Maity, A novel method for removal of Cr (VI) using polypyrrole magnetic nanocomposite in the presence of unsteady magnetic fields. Sep. Purif. Technol. **194**, 377–387 (2018)
83. U.O. Aigbe, W.H. Ho, A. Maity, M. Khenfouch, V. Srinivasu, Removal of hexavalent chromium from wastewater using PPy/Fe_3O_4 magnetic nanocomposite influenced by rotating magnetic field from two pole three-phase induction motor. J. Phys. Conf. Ser. **984**(1), 012008 (2018). IOP Publishing
84. U.O. Aigbe, M.K. Khenfouch, W.H. Ho, A. Maity, V.J. Vallabhapurapu, N.M. Hemmaragala, Congo red dye removal under the influence of rotating magnetic field by polypyrrole magnetic nanocomposite. Desalin. Water Treat. **131**, 328–342 (2018)
85. J. Theerthagiri, S. Salla, R.A. Senthil, P. Nithyadharseni, A. Madankumar, P. Arunachalam, T. Maiyalagan, H.S. Kim, A review on ZnO nanostructured materials: energy, environmental and biological applications. Nanotechnology, **30**(39), 392001 (2019)
86. M.M. Abdullah, F.M. Rajab, S.M. Al-Abbas, Structural and optical characterization of Cr_2O_3 nanostructures: evaluation of its dielectric properties. AIP Adv. **4**(2), 027121 (2014)
87. R. Etefagh, E. Azhir, N. Shahtahmasebi, Synthesis of CuO nanoparticles and fabrication of nanostructural layer biosensors for detecting Aspergillus niger fungi. Scientia Iranica **20**(3), 1055–1058 (2013)
88. K. Henkel, J. Haeberle, K. Müller, C. Janowitz, D. Schmeißer, Preparation, properties, and electronic structure of SnO_2, in *Single Crystals of Electronic Materials* (Woodhead Publishing, 2019), pp. 547–572
89. J. Li, S. Meng, J. Niu, H. Lu, Electronic structures, and optical properties of monoclinic ZrO_2 studied by first-principles local density approximation+ U approach. J. Adv. Ceram. **6**(1), 43–49 (2017)
90. N. Rahimi, R.A. Pax, E.M. Gray, Review of functional titanium oxides. I: TiO_2 and its modifications. Progress Solid State Chem. **44**(3), 86–105 (2016)
91. T. Gonçalves, R.V. Silva, J. de Brito, J.M. Fernández, A.R. Esquinas, Hydration of reactive MgO as partial cement replacement and its influence on the macroperformance of cementitious mortars. Adv. Mater. Sci. Eng. (2019)
92. Z. Koudelkova, T. Syrovy, P. Ambrozova, Z. Moravec, L. Kubac, D. Hynek, L. Richtera, V. Adam, Determination of zinc, cadmium, lead, copper, and silver using a carbon paste electrode and a screen-printed electrode modified with chromium (III) oxide. Sensors **17**(8), 1832 (2017)

93. H.L. Fan, S.F. Zhou, J. Gao, Y.Z. Liu, Continuous preparation of Fe_3O_4 nanoparticles through impinging stream-rotating packed Bed reactor and their electrochemistry detection toward heavy metal ions. J. Alloy. Compd. **671**, 354–359 (2016)
94. Y. Kong, T. Wu, D. Wu, Y. Zhang, Y. Wang, B. Du, Q. Wei, An electrochemical sensor based on Fe_3O_4@ PANI nanocomposites for sensitive detection of Pb $^{2+}$ and Cd $^{2+}$. Anal. Methods **10**(39), 4784–4792 (2018)
95. Y.F. Sun, W.K. Chen, W.J. Li, T.J. Jiang, J.H. Liu, Z.G. Liu, Selective detection toward Cd^{2+} using Fe_3O_4/RGO nanoparticle modified glassy carbon electrode. J. Electroanal. Chem. **714**, 97–102 (2014)
96. G.O. Buica, A.B. Stoian, C. Manole, I. Demetrescu, C. Pirvu, Zr/ZrO_2 nanotube electrode for detection of heavy metal ions. Electrochem. Commun. **110**, 106614 (2020)
97. P.K.Q. Nguyen, S.K. Lunsford, Electrochemical response of carbon paste electrode modified with mixture of titanium dioxide/zirconium dioxide in the detection of heavy metals: lead and cadmium. Talanta **101**, 110–121 (2012)
98. M.A. Deshmukh, M.D. Shirsat, A. Ramanaviciene, A. Ramanavicius, Composites based on conducting polymers and carbon nanomaterials for heavy metal ion sensing. Crit. Rev. Anal. Chem. **48**(4), 293–304 (2018)
99. H. Wang, C. Xu, B. Yuan, Polymer-based electrochemical sensing platform for heavy metal ions detection—a critical review. Int. J. Electrochem. Sci **14**, 8760–8771 (2019)
100. J. Shi, F. Tang, H. Xing, H. Zheng, B. Lianhua, W. Wei, Electrochemical detection of Pb and Cd in paper-based microfluidic devices. J. Braz. Chem. Soc. **23**(6), 1124–1130 (2012)
101. M.A. Rahman, M.S. Won, Y.B. Shim, Characterization of an EDTA bonded conducting polymer modified electrode: its application for the simultaneous determination of heavy metal ions. Anal. Chem. **75**(5), 1123–1129 (2003)
102. M.A. Deshmukh, G.A. Bodkhe, S. Shirsat, A. Ramanavicius, M.D. Shirsat, Nanocomposite platform based on EDTA modified Ppy/SWNTs for the sensing of Pb (II) ions by electrochemical method. Front. Chem. **6**, 451 (2018)
103. W. Wu, A. Ali, R. Jamal, M. Abdulla, T. Bakri, T. Abdiryim, A bromine-catalysis-synthesized poly (3, 4-ethylenedioxythiophene)/graphitic carbon nitride electrochemical sensor for heavy metal ion determination. RSC Adv. **9**(60), 34691–34698 (2019)
104. G.H. Hwang, W.K. Han, J.S. Park, S.G. Kang, Determination of trace metals by anodic stripping voltammetry using a bismuth-modified carbon nanotube electrode. Talanta **76**(2), 301–308 (2008)
105. M.M. Radhi, W.T. Tan, M.Z. Ab Rahman, A.B. Kassim, Voltammetric detection of Hg (II) at C60, activated carbon and MWCNT modified glassy carbon electrode. Res. J. Appl. Sci. **5**, 59–64 (2010)
106. A. Simpson, R.R. Pandey, C.C. Chusuei, K. Ghosh, R. Patel, A.K. Wanekaya, Fabrication characterization and potential applications of carbon nanoparticles in the detection of heavy metal ions in aqueous media. Carbon **127**, 122–130 (2018)
107. X. Guo, Y. Yun, V.N. Shanov, H.B. Halsall, W.R. Heineman, Determination of trace metals by anodic stripping voltammetry using a carbon nanotube tower electrode. Electroanalysis **23**(5), 1252–1259 (2011)
108. X.C. Fu, J. Wu, J. Li, C.G. Xie, Y.S. Liu, Y. Zhong, J.H. Liu, Electrochemical determination of trace copper (II) with enhanced sensitivity and selectivity by gold nanoparticle/single-wall carbon nanotube hybrids containing three-dimensional l-cysteine molecular adapters. Sens. Actuators B Chem. **182**, 382–389 (2013)
109. H. Bagheri, A. Afkhami, H. Khoshsafar, M. Rezaei, A. Shirzadmehr, Simultaneous electrochemical determination of heavy metals using a triphenylphosphine/MWCNTs composite carbon ionic liquid electrode. Sens. Actuators B Chem. **186**, 451–460 (2013)
110. S. Cerovac, V. Guzsvány, Z. Kónya, A.M. Ashrafi, I. Švancara, S. Rončević, Á. Kukovecz, B. Dalmacija, K. Vytřas, Trace level voltammetric determination of lead and cadmium in sediment pore water by a bismuth-oxychloride particle-multiwalled carbon nanotube composite modified glassy carbon electrode. Talanta **134**(2015), 640–649 (2015)

111. J.E. Robinson, W.R. Heineman, L.B. Sagle, M. Meyyappan, J.E. Koehne, Carbon nanofiber electrode array for the detection of lead. Electrochem. Commun. **73**, 89–93 (2016)
112. C. Choi, Y. Jeong, Y. Kwon, Detection of trace copper metal at carbon nanotube based electrodes using square wave anodic stripping voltammetry. Bull. Korean Chem. Soc. **34**(3), 801–809 (2013)
113. G.G. Matlou, D. Nkosi, K. Pillay, O. Arotiba, Electrochemical detection of Hg (II) in water using self-assembled single walled carbon nanotube-poly (m-amino benzene sulfonic acid) on gold electrode. Sens. Bio-Sens. Res. **10**, 27–33 (2016)
114. V. Somerset, J. Leaner, R. Mason, E. Iwuoha, A. Morrin, Determination of inorganic mercury using a polyaniline and polyaniline-methylene blue coated screen-printed carbon electrode. Int. J. Environ. Anal. Chem. **90**(9), 671–685 (2010)
115. W. Zhang, S. Fan, X. Li, S. Liu, D. Duan, L. Leng, C. Cui, Y. Zhang, L. Qu, Electrochemical determination of lead (II) and copper (II) by using phytic acid and polypyrrole functionalized metal-organic frameworks. Microchim. Acta **187**(1), 69 (2020)
116. Y. Wang, L. Wang, W. Huang, T. Zhang, X. Hu, J.A. Perman, S. Ma, A metal-organic framework and conducting polymer based electrochemical sensor for high performance cadmium ion detection. J. Mater. Chem. A **5**(18), 8385–8393 (2017)
117. N. Wang, E. Kanhere, A.G.P. Kottapalli, J. Miao, M.S. Triantafyllou, Flexible liquid crystal polymer-based electrochemical sensor for in-situ detection of zinc (II) in seawater. Microchim. Acta **184**(8), 3007–3015 (2017)
118. R. Xie, L. Zhou, C. Lan, F. Fan, R. Xie, H. Tan, T. Xie, L. Zhao, Nanostructured carbon black for simultaneous electrochemical determination of trace lead and cadmium by differential pulse stripping voltammetry. R. Soc. Open Sci. **5**(7), 180282 (2018)
119. S. Touzara, A. Amlil, H. Saâdane, C. Laghlimi, A. Chtaini, EDTA-modified carbon paste composite for electrochemical determination of Pb(II) ions. J. Mater. Sci. Eng. **8**(6), 1–5 (2019)

Chapter 4
Sensing the Presence of Inorganic Ions in Water: The Use of Electrochemical Sensors

Kabir Opeyemi Otun, Idris Olayiwola Azeez, Onoyivwe Monday Ama, William Wilson Anku, Uyiosa Osagie Aigbe, Kingsley Eghonghon Ukhurebor, and Robert Birundu Onyancha

Abstract Electrochemical method has become a widely used analytical technique for sensing application because of its many remarkable features such as fast response time, good stability, high sensitivity and high selectivity. Determination of inorganic anions and cations in water is an important research field in particular because of the several health risks posed by these analytes when present at abnormal concentrations. This chapter, therefore, focuses on the recent advances in electrochemical method for the detection of inorganic ions in water. The principle of electrochemical sensing was described in detail. Various nanomaterials ranging from inorganic, organic and biomaterials used to detect inorganic ions are also discussed. Finally, we reviewed the application of electrochemical sensors for the determination of inorganic anions

K. O. Otun (✉)
Institute for the Development of Energy for African Sustainability, University of South Africa, Private Bag, Johannesburg 1709, South Africa

Department of Chemical, Geological and Physical Sciences, College of Pure and Applied Sciences, Kwara State University, PMB. 1530, MaleteIlorin, Nigeria

I. O. Azeez
Nanotechnology for Water Sustainability (NaNOWS) Research Unit, College of Science, Engineering and Technology, University of South Africa, Private Bag, Johannesburg 1709, South Africa

O. M. Ama
Department of Chemical Sciences, University of Johannesburg, Doornfontein, Johannesburg 2028, South Africa

W. W. Anku
CSIR-Water Research Institute, P.O. Box M32, Accra, Ghana

U. O. Aigbe
Department of Mathematics and Physics, Faculty of Applied Sciences, Cape Peninsula University of Technology, Cape Town, South Africa

K. E. Ukhurebor
Department of Physics, Edo State University Uzairue, Edo State, Nigeria

R. B. Onyancha
School of Physical Sciences and Technology, Technical University of Kenya, Nairobi, Kenya

O. M. Ama et al. (eds.), *Modified Nanomaterials for Environmental Applications*, Engineering Materials, https://doi.org/10.1007/978-3-030-85555-0_4

and cations in water and gave an outlook of future research direction to improve their performances and applicability.

Graphical Abstract

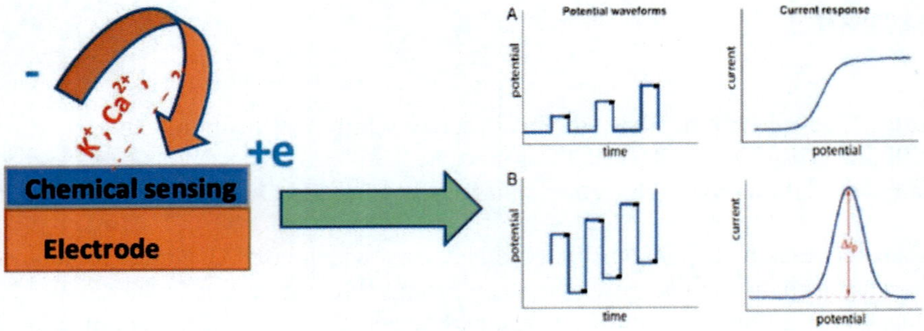

Highlights

1. Detection of inorganic ions in water by electrochemical sensors is discussed.
2. Anions (NO_2^-, SO_4^{2-}, F^-, Cl^- and PO_4^{3-}) and cations (K^+, Na^+, Ca^{2+}, Mg^{2+} and Al^{3+}) are considered.
3. Nanomaterials used for sensing application are grouped into inorganic, organic and biomaterials.
4. An accurate determination of these ions will protect the environment and reduce health risks.
5. Limitations of using ion-selective electrode for sensing inorganic ions are highlighted.

4.1 Introduction

Water is one of the basic substances that sustain life and our natural environment. Water also plays a huge role in industry and serves as one of the basic items consumed by both human and animal [1]. Several measures provide a template to control the quality of water for our domestic needs, as well as the control of effluents and aquatic organisms. Such measures form the basis of analytical measurements. In the same vein, as the demand for clean water resources continue to increase, novel and efficient methods of purifying and detecting contaminants in water are receiving adequate attention from several researchers across the globe. This is very crucial in the sense that the abandoned sources of water such as saltwater may become the only chief source of drinking water for several millions of people during water scarcity. Some of the analytes in water consist of essential inorganic ions, especially those found in drinking water which play a major role in supporting and sustaining health and life of human beings. These include Na^+, K^+, Ca^{2+}, Mg^{2+}, H^+, Cl^-, HCO_3^-, PO_4^{3-} and OH^- among others. The non-essential inorganic ions whose biological functions are not well established but can have adverse effects on both human and animal when present at certain concentrations include but not limited to Pb, Cu, As, Cr, Al, Ag, Sr, Li, Hg, Be, Ba, Au, Cd and Ni [2].

Similarly, the World Health Organization (WHO) and the US Environmental Protection Agency (EPAS) have put in place regulations and permissible exposure limits of different ions and analytes present in water [3]. For instance, the limits for Fe, Zn and Cu are on the order of 1–10 ppm; while the limits are more stringent for Cd, As, Cr, Pb and Hg at the range of 1–10 ppb [4]. The second class are regarded as the heavy metals which are beyond the scope of this chapter. Among the inorganic ions present in water are CO_3^{2-}, HCO_3^-, Cl^-, and SO_4^{2-}, Br^-, F^-, NO_3^-, NO_2- and PO_4^{3-}, which emanated from erosion and minerals dissolution, ores and natural deposits, chemical reactions, agricultural and industrial run-off, manure or decisive addition. When these ions are present more than the maximum contaminant levels, it can lead to different health challenges such as diarrhea, cancer and organ damage to name a few [5–7].

To this end, analytical techniques such as Atomic absorption spectroscopy (AAS), inductively coupled plasma-mass spectrometry (ICP-MS), mass spectroscopy, chromatography, ultraviolet–visible (UV–Vis), capillary electrophoresis, X-ray fluorescence spectroscopy (XFS) [8–14], etc., have been developed to detect or sense inorganic ions in water bodies. Although they are characterized with varying degrees of selectivity, interferences from common ions, and selectivity, their applications are limited by costly instrumentation, lack of technical-know how, complicated sample preparation process, inability to carry out single composition detection and preconcentration procedures which have equally put them at disadvantage in terms of a real-time online and continuous monitoring application. Moreover, the large size of these equipments makes them unsuitable for on-site detection [15]. Consequently, this has called for a way of exploring alternative and cost-effective approaches in sensing inorganic ions in water.

Electrochemical sensing technique has been considered a better technique for sensing inorganic ions in water because it offers extraordinary features such as remarkable sensitivity, good response time, affordability, simple instruments, ease of miniaturization and integration in portable devices [16, 17]. Its simplicity and ease of operation make it more advantageous when compared with spectroscopic techniques. In addition, it is possible to improve the selectivity and sensitivity of the electrochemical sensing technique by using highly efficient electron mediators to chemically modify the bare electrode. This becomes more interesting because, most of the electrochemical sensors for the detection of analytes in water possess large surface area, sites for potential adjustments and exceptional quantum mechanical features which make them suitable for electron modification [18].

4.2 Principle of an Electrochemical Sensor

Electrochemical sensing involves using electrodes to measure the electrochemical changes that occur during the interaction of the chemicals with a sensing electrode. Electrochemical sensors are characterized with good selectivity, quick response time and outstanding sensitivity toward several kinds of analytes in the sample of interest [19]. Electrochemical sensing can also provide a real information about the analyte with the use of a combination of selective detection tool that bind to the sample with transducer. By using the sensor, the interaction with the species to be analyzed is transformed to a signal. Basically, the structure of an electrochemical sensor, which is shown in Fig. 1 contains two essential parts, the transducer and the chemical recognition parts. While the transducer part helps to transform the chemical interaction with analyte into electrical signal, the chemical recognition part is for binding with the analyte [20]. The performance of a sensor depends on the extent of interaction of the recognition specie with the analyte. A good sensor is known by the extent of detection limit, linearity, response duration, stability, selectivity as well as sensitivity. Specifically, electrochemical methods are characterized with very low detection limits, high selectivity and need only a very small amount of analyte to obtain signals. However, based on the quantity being analyzed, electrochemical sensors can be categorized into potentiometric (potential), amperometric (current) or impedimetric (conductivity) as established by Wang [21].

Potentiometric sensors measure the difference in potential between the working electrode and reference electrode when there is little or no flow of current. The measured potential is subsequently used to determine the ion activity in an electrochemical cell. The potential difference is calculated based on Nernst equation where E_{cell} indicates the cell potential when the current equals zero. This value is also what is being referred to as electromotive force (EMF), and ion-selective electrode (ISE) is being used to measure the lowest detection limit of electrochemical devices using this approach [22].

As regards amperometry, a constant potential is introduced to the working electrode and the resulting current is measured relative to time. To study the behaviors of electrochemical sensors for sensing various analytes, different techniques based on voltammetry such as cyclic voltammetry (CV), linear sweep voltammetry, differential pulse voltammetry (DPV), hydrodynamic voltammetry, polarography, and square-wave voltammetry (SWV) are usually employed. These techniques have found useful applications in several fields with different detection limits [23–25].

On the other hand, the voltammetric sensor measures current while applying potential to the working electrode with respect to a reference electrode. The interaction that arises between the surface of electrode and the electrode/electrolyte interface layers generated what is known as the response current [26]. However, the rate of mass transport of the species to the electrode limit electrochemical reaction. Amperometry is different from voltammetry in the sense that it uses a potential step instead of a potential sweep and the obtained current is directly proportional to the concentration of electroactive species in the analyte as illustrated by Cottrell's equation

[27]. In addition, amperometric sensors perform better with respect to selectivity and sensitivity because the redox potential used for sensing is characteristic of the analyte species.

4.3 Materials Used for Electrochemical Sensing of Inorganic Ions in Water

Latest research developments in material science, especially the advancement in nanomaterials has opened new research trends in analytical measurements. These nanomaterials have particularly found useful application in electrochemistry because of their exceptional electronic, physicochemical and mechanical properties [28]. They can be functionalized and assembled on the surface of electrode to construct sensing electrodes for effective detection of inorganic ions. In addition, several other materials with improved electrochemical performance can be fabricated by the assembly of different materials. [29–31]. For instance, polymeric materials are usually assembled on inorganic nanomaterials to synthesize highly sensitive electrochemical materials sensors for the detection of inorganic ions [32]. This chapter therefore is on the recent advances in the use of electrochemical sensors for the detection of inorganic ions in water. The sensors discussed in this chapter are classified based on the materials used in the fabrication of electrodes for sensing the ions, namely inorganic, organic and biomaterials.

Inorganic nanomaterials are often used in the modification of electrode materials because of their intriguing properties such as strong adsorption capability, high surface-to-volume ratio, good thermal and chemical stability [33]. These include metal, oxides of metal, carbon, graphene and silica-based materials as that are frequently used in sensing inorganic ions in water. Examples of organic nanomaterials in sensing inorganic ions in water include small organic molecules and several polymeric organic materials such as polyacetylene, polyethylene glycol, polyacrylonitrile, polyaniline, polypyrrole, polythiophene, poly (3,4 ethylenedioxythiophene) (PDOT), among others. Of the biomaterials used in sensing inorganic ions in water, enzymes (such as oxidase, peroxidase and urease) and functional nucleic acids are the most frequently used [34].

4.3.1 Metal Nanoparticles

These are employed in electrochemical sensing due to their good chemical, optical and electrocatalytic properties [35]. They are categorized into two, namely precious metals such as silver, gold, platinum, etc., and non-precious metals such as iron, copper, cobalt, among others. These metals can be functionalized with small molecules and biomolecules to improve their performance toward sensing inorganic ions in water bodies. For instance, Hameed and co-workers investigated the effects of Co, Cu and Fe for the electrocatalytic activity of nitrite under the same reaction conditions by synthesizing core–shell structure nanocomposites of Cu@Pt/Gr, Ni@Pt/Gr and Co@Pt/Gr. When the potential applied is 0.85 V and pH of 4, their detection ranges were found to be 1 μM–15 mM, 0.01–15 mM, 1 μM–15 mM, while their detection limits were respectively 0.035 μM, 0.49 μM and 0.145 μM. They established that all the three electrochemical sensors possess good stability, reproducibility and selectivity with a satisfactory recovery rate in both tap and river water [36].

4.3.2 Metal Compounds

The class of compounds in this category include metal oxides, metal complexes like metal–organic frameworks (MOFs) and polyoxometalates. These compounds may sometimes be functionalized with organic materials to enhance their performance in the electrochemical sensing of inorganic ions in water. They are relatively cheap and possess not only good specific surface area but also good electrocatalytic properties [37]. For instance, Yantasee et al. designed Fe_3O_4 nanoparticles functionalized with dimercaptosuccinic acid to detect Cu, Cd, Pb and Ag in natural water [38].

Similarly, MOF-525/Gr nanocomposites were synthesized by Kung et al. by thermal assemblage of MOF-525 in dispersed graphene nanoribbons suspension [39]. Metal–organic frameworks (MOFs) are hybrid organic/inorganic crystalline materials constructed from the coordination of metal ion/clusters and organic linkers [40]. The function of MOF-525 in this nanocomposite was to catalyzed nitrite ion while graphene ribbons act as conductive bridges. Furthermore, the prepared MOF-525/GNRs nanocomposites were employed to make an electrode by simply depositing it on glass substrates for the detection of NO_2^- in water system. The linear range of the sensor was found to be 100–2500 μM for the detection of nitrite when the applied potential was 0.85 V. Meanwhile, the detection limit obtained was 0.75 μM, with sensitivity of 93.8 $\mu AmM^{-1}\ cm^{-2}$.

4.3.3 Graphene

Graphene is a 2-D carbon material that consists of sp^2-hybridized carbon atom with a honeycomb hexagonally based lattice structure that extends infinitely [41]. It has been found useful in modifying the electrodes of electrochemical sensors due to its high surface area, good mechanical strength, superb thermal conductivity and good electrochemical activity. Different types of graphene structures, as shown in Fig. 4.1, such as graphene, graphene oxide and reduced graphene oxide do exist which are only different in their molecular structures but also in their electrochemical sensing performance.

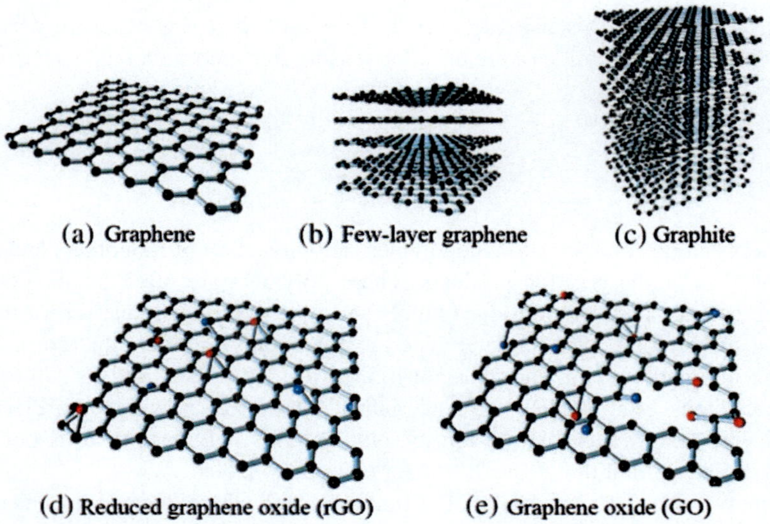

Fig. 4.1 Types of graphene nanomaterials such as **a** graphene, **b** few layer graphite, **c** graphite, **d** reduced graphene oxide and **e** graphene oxide. Reproduced with permission from [45] Copyright 2016 Elsevier

Because graphene performance is limited when it acts alone, it is often functionalized with other materials such as metals, metal oxides, polymers, among other to improve its activity. Interestingly, graphene-based sensors have been used to detect several inorganic ions found in water such as Cl^-, F^-, NO_2^-, K^+, SO_4^{2-}, NO_2^-, etc. For instance, Mani et al. synthesized graphene/multiwalled carbon nanotubes on iron nanoparticles (Gr/MWCNTs/FeNPs) for nitrite determination with the applied potential of 0.77 V and pH of 5 in the modified electrode range of 0.1 μM–1.68 mM and limit of detection is 75.6 nM [42]. This electrode has advantages of good stability, reproducibility, and high resistance to interference from other ions. The recovery of detection in different water bodies was found to be between 96.1 and 103.7%.

In another study by Zhang et al., an electrochemical sensor made of holey graphene (ERHG) characterized with good reproducibility, repeatability and exceptional anti-interference was fabricated to sense nitrite in tap water and drinking water [43]. The potential of 0.92 V and pH of 7.4 was applied to obtain the linear detection range of 0.2 μM–10 mM with sensitivity of 0.054 μM.

Furthermore, Soundappan et al. synthesized crumpled SnO_2-graphene (cGO-SnO_2) nanocomposite sensors using aerosol-based CVD synthesis route and electrospray deposition for sensing of free chlorine [44]. The CGO-SnO_2 working principle is that it directly reduces the hypochlorite ions on the electrode surface which is used to control the concentration of free chlorine. The linear range of the electrode was found to be 0.1–10.08 ppm, with sensitivity of 2.69 μM^{-1}. It was equally concluded that interreferences from other common ions was prohibited by this electrode in drinking water because the electrochemical reduction signals of chlorine were only detected. In addition, a sensitivity of 5.86 μA μ M^{-1} cm^2 was observed when the same CGO-SnO_2 electrodes were used for sensing free chlorine in tap water.

4.3.4 Polymers

Polymers can be constructed by copolymerization reaction of monomers and post-modification of the reaction products. These polymers are often combined with inorganic nanomaterials to produce highly sensitive electrochemical sensor for the sensing of inorganic ions in water. The most common polymers that are used in sensing inorganic ions include but not limited to PPy, PDDA and PyY [46]. For instance, Dong et al. studied the potential of polypyrrole polymer film electrode doped with chloride-to-chloride ion in solution. This polymer film was obtained with a detection limit of 3.5×10^{-5} M for chloride ion [47].

Moreover, Xu et al. prepared PDDA functionalized with reduced graphene oxide (RGO) nanomaterials and was used to modify glassy carbon electrode for electrochemical detection of nitrite ion in water system. The linear concentration range and the detection limit are 0.5 μM–2 mM and 0.2 μM, respectively. It was found out that these electrodes have good selectivity and reproducibility with a recovery of 95.4–16% in drinking water [48].

4.3.5 Quantum Dots

Quantum dots (QDs) are synthetic semiconductor nanoscale crystals that can convey electrons. They possess optical and electronic features that are different from larger particles because of quantum mechanics [49]. Graphene quantum dots (GQD) are one of the most prominent members of this family which are commonly known as fluorescent nanomaterials and have found useful applications in several fields due to their unique optical properties. Quantum encapsulation and edge effects in their structures can also elicit exceptional electrochemical properties [50, 51].

Zhuo et al. investigated graphene-based quantum dots using up-conversion fluorescent sensor for detection of phosphate. The linear range for the sensing of phosphate was found to be 1–12 μM with LOD of 100 nM. This sensor was successfully applied in real river water samples [52].

You et al. also fabricated a novel nitrogen-doped graphene quantum dots decorated N-doped carbon nanofibers (NGQDs@NCFs) for the detection of NO_2^- in water system. The synthesis of NGQDs@NCFs composite material was via electrospinning, pyrolysis combined with a single-step hydrothermal procedure [53]. The excellent flexibility and the free-standing structure of the nanocomposite made it easy to be successfully anchored on the surface of the electrode. The electrocatalytic performance of the electrode was then established by cyclic voltammetry and electrochemical impedance spectroscopy. The results obtained showed that the NGQDs@NCNFs nanocomposite performed better than the previous sensors reported for the sensing of nitrite in terms of linear range (5–300 μM, $R^2 = 0.999$; 400–3000 μM, $R^2 = 0.997$), low detection limit (3 μM), exceptional reproducibility, selectivity and good recoveries in lake and tap water.

In another report, Liu et al. employed hexametaphosphate for the synthesis of CdS QDs and then used in detecting calcium ion and fluoride. The result obtained showed that there was a direct relationship between the fluorescence intensity of hexametaphosphate-doped CdS quantum dots and Ca^{2+} concentration in the range of 10–400 μmol/L with detection limit for Ca^{2+} was found to be 4 μmol/L. In addition, it was possible for fluoride ions to react with Ca^{2+} to produce CaF_2 which will inhibit calcium-induced fluorescence enhancement of CdS quantum dots. This, therefore, suggested that the same system can also be used to successfully detect fluoride ions [54].

4.3.6 Carbon Nanotubes

Carbon nanotubes (CNTs) are fascinating materials that possess different kinds of physical, electrical and mechanical features. They are often employed as electrode materials because of their intriguing properties such as high surface area, ease of post-synthetic modification, small size and superb electrical conductivity [55]. They also possess outstanding sorption for metal ions. CNT-based electrochemical sensors have demonstrated promising for sensing multiple inorganic ions in water system. The modification of CNTs with another molecule also helps to improve the selectivity of the analyte at the interface of the electrode. Analytical sensing of CNTs-modified electrodes often leads to low detection limits, overpotential reductions, high sensitivity and lack of resistance to surface fouling [56] . All these properties made carbon nanotubes a better material to develop good electrochemical sensors and biosensors when compared to their carbon counterparts.

However, CNTs are limited in their applications because it is cumbersome and highly expensive to produce pure and defect-free CNTs. In addition, CNTs can pose serious health risks due to their harmful effects to the lungs that emanated from their agglomeration which can eventually lead to suffocation [57]. Their insolubility in most solvents also put their applications at greater disadvantage. Despite these challenges, an increasing research progress from different researchers across the globe look promising especially through the development of multiwall carbon nanotubes (MWCNTs). MWCNTs are a special type of CNTs that are made of sp^2 carbon and in which multiple singled-walled CNTs are nested inside one another. The electrochemical sensors fabricated from MWCNTs are characterized with enhanced sensitivity, low charge transfer resistance, electrocatalytic activity and minimized fouling [58]. Different sensors based on MWCNTs have been developed for sensing inorganic anions and cations in water.

For instance, Qu et al. developed a novel electrochemical sensor based on Fe_3O_4 nanoparticles-MWCNTs film composite for the determination of nitrite [59]. The

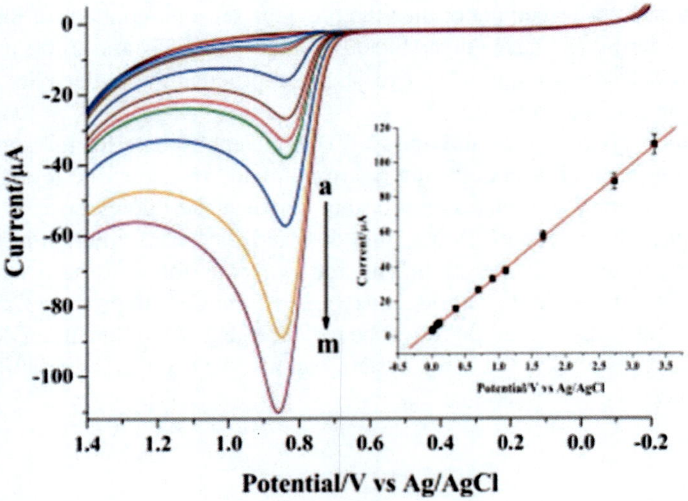

Fig. 4.2 Electrochemical response curves at the PDDA–Fe_3O_4/L-cys/MWCNTs/GC electrode in 0.1 M PBS (pH 5.5) with different concentrations of nitrite

sensor was produced via electropolymerization of L-cystein on the MWCNTs and then immersed on positively charged poly(diallyldimethylammonium chloride) coated with iron oxide nanoparticles by electrostatic interaction on glassy carbon (GC) matrix. At optimized conditions, electrochemical response curves at the PDDA–Fe_3O_4/L-cys/MWCNTs/GC electrode were obtained in 0.1 mol/L PBS (pH 5.5) containing different nitrite concentrations as shown in Fig. 4.2. The linear range was found to be between 7.49×10^{-6} to 3.33×10^{-3} mol/L, with a detection limit of 8.46×10^{-7} mol/L was determined at S/N of 3. The synergistic properties of PDDA–Fe_3O_4, L-cys and MWCNTs contributed significantly to the oxidation of nitrite ion on the electrode surface.

Furthermore, Zhou et al. developed a novel electrochemical sensor based on MWCNTs ferrocene-branched chitosan (CHIT-Fc) composites-covered glassy carbon electrode (GCE) for the sensing of sulfite in boiler water using cyclic voltammetry technique [60]. The sensor exhibited a good performance toward sulfite oxidation. At optimal conditions, the detection limit was found to be 2–8 μM and linear range of 5 μM–1.5 mM at a S/N of 3. A good sensitivity of 13.08 μA/mM was also obtained.

4.4 Electrochemical Determination of Inorganic Ions in Water

The presence of some inorganic ions at high concentrations in water bodies can pose serious health risks. These ions can be cations (such as K^+, Na^+, Ca^{2+}, Mg^{2+} and Al^{3+}) or anion (such as nitrite, sulfite, chloride, fluoride and phosphate). Therefore, an accurate determination of these ions will not only help to protect the environment but also reduce the health risks [61]. Many methods have been developed for the determination of inorganic cations and anions in water such as colorimetry, chromatography, spectrophotometry, chemiluminescence, electrochemical methods and so on. Among them, electrochemical methods are preferred due to the ease of operation, relatively rapid response time and low cost. Consequent to this, this section is on discussion of the determination of inorganic ions in water using electrochemical sensors with a view evaluating their performance in terms of sensitivity, selectivity and stability.

4.4.1 Inorganic Anions

4.4.1.1 *Nitrite (NO_2^-)*

Nitrites are an inorganic anion that has found useful applications in food industry as preservatives and can also serve as fertilizing agents [62]. It exists in water because of chlorination that may trigger the formation of nitrite in the water system. The maximum concentration of nitrite in drinking water according to WHO is 3 mg/L. Elevated levels of nitrite ion in food and water sample may be detrimental to human health and cause diseases such as stomach cancer, Blue Babe Syndrome among other diseases [63–65]. Consequently, it is imperative to develop simple, rapid and sensitive method to detect nitrite ion to protect public health and environment.

There are several techniques used in sensing NO_2^- such as chromatography, potentiometric sensor, capillary electrophoresis, electrochemiluminescence, spectrophotometry and electrochemical techniques [66–68]. Among all, electrochemical method has received adequate attention because of ease of operation, low cost, quick response time, good stability, and high sensitivity. For instance, Mehmeti et al. developed the use of graphene ribbons as an efficient electrochemical sensing material for the determination of nitrite in tap water by using glassy carbon electrode. The summary of some electrochemical techniques used in sensing nitrite in water is shown in Table 4.1.

Table 4.1 Summary of the electrochemical sensing of nitrite in water

Electrode	Technique	Linear range (μM)	LOD (μM)	References
Au-MOF-5	CV	5–65,000	1	[69]
PAOA/GCE	Amperometric-i-t	5–2000	2	[70]
TOSC-MoS_2	Amperometric-i-t	3140–4200	2	[71]
Au/TiO2/CNT	DPV	4–225	3	[72]
CS/MWCNT/CN	LSV	5–1000	0.89	[73]
PEDOT/AuNCs	Amperometry	0.05–2600	0.017	[74]
Cu/CBSA NFNs	DPV	0.5–500	0.1	[75]
Bare glass carbon	SWV	20–6400	0.4	[76]
ZnO@reduced functionalized GO	LSV	10–8000	33	[77]
Renewable copper layer	DPV	100–2500	111	[78]
PANI@GO	Amperometric-i-t	2–44,000	0.5	[79]
3C-SiC	Amperometric-i-t	50–5000	3.5	[80]

Abbreviations
Au-MOF-5 Gold-metal–organic frameworks-5 composite electrode
PAOA/GC Poly(aniline-co–o-aminophenol)-modified glassy carbon electrode
TOSC-MoS_2 TEMPO oxidized straw cellulose/molybdenum sulfide composite
Au/TiO_2/CNT Gold titanium carbon nanotube
CS/MWCNT/CN Chitosan (CS)-grafted multiwalled carbon nanotube carbon nanoparticles
PEDOT/AuNCs Poly(3,4-ethylenedioxythiophene) doped with gold nanoclusters
Cu/CBSA NFNs Copper ions–crosslinked bovine serum albumin nanoflower networks
PANI@GO Polyaniline-Graphene Oxide
3C-SiC Cubic silicon carbide
CV Cyclic voltammetry
DPV Differential pulse voltammetry
LSV Linear sweep voltammetry
SWV: Squarewave voltammetry

4.4.1.2 Fluoride (F^-)

Fluoride plays a major role in human body and participated in different kinds of biochemical reactions with applications in dental and pharmaceutical products. When F^- is taken at appropriate amount, it improves health and helps prevent diseases such as osteoporosis [81]. The acceptable limit of F^- in drinking water as recommended by the USA environmental protection agency (EPA) is 2–4 ppm [82]. However, when F^- is present at higher concentrations, it may trigger several diseases that include nephrotoxic disease, dental fluorosis and even cancer. It is therefore imperative to develop sensitive and selective techniques to detect F^- in both water and biological fluids.

Among the techniques that have been put in place to sense F^-, electrochemical methods, especially ion-selective electrode (ISE) has been documented by the WHO due to its good response time, high sensitivity and selectivity. The major challenge of ISE is the interference from other ions especially when there is hydrogen or hydroxide ion in the analyte. In circumventing this challenge.

Tao et al. developed an electrochemical sensor based on spiropyran skeleton called SPOSi, to exclusively detect F^-. The SPOSi was anchored on single-walled carbon nanotubes (SWCNTs) modified GCE via Π-Π conjugation interaction in order to improve its sensitivity. When the electrode was used to detect F^-, a detection limit of 8.3×10^{-8} M was obtained, which when compared with the normal F^- ISE, a higher sensitivity and stability was obtained [83].

Unlike Tao and co-workers who used non-phenolic spiropyran that complicated the synthesis of the electrode, Appiah-Ntiamoah et al. developed an effective electrochemical sensor based on fluorescein isothiocyanate-MWCNT composite (FITC-OSi-p-MWCNT-GCE) to detect fluoride ion in real water samples [84]. FITC has two phenolic groups which reduces the synthesis of the composites to just three facile stages. The sensor was constructed by modifying GCE with p-MWCNT and FITC-OSi to form FITC-OSi-p-MWCNT-GCE. The sensor displayed superb reproducibility and stability and gave a limit of detection of 0.26 μm, in the linear range of 1–100 μM, which is below the fluoride ion requirements in drinking water recommended by the USEPA and WHO.

4.4.1.3 Sulfite Ion (SO_4^{2-})

Detection of sulfite in aqueous solutions has been the subject of several research studies for several decades, and many methods have been developed in this regard. This became necessary because when sulfite is present at high concentrations, it can produce a change in the organoleptic properties of food and drinks, and can as well induce headache, asthma and dizziness in human [85, 86]. In fact, the FDA has stipulated that caution labels should be affixed on any food or beverage whose concentration of sulfite exceeds 10 ppm with a maximum concentration set at about 50 ppm. Therefore, the development of a selective and sensitive method for sensing sulfite is imperative for analytical applications. Recently, much attention has been given to electrochemical methods like voltammetry and amperometry for the determination sulfite in water samples [87].

However, direct electrochemical oxidation of sulfite usually gives rise to a poor current response, and it is a big challenge when it comes to differentiate the individual analyte current response. To assuage this challenge, scientists have carried out surface modification of the electrodes with different materials to improve their performance.

Ensafi et al. designed a simple, selective, and precise electrochemical sensor based on modified-carbon nanotubes paste to detect sulfite in water and wastewater using cyclic voltammetry [88]. At optimum pH of 7.0, the electrocatalytic peak current revealed linear relationship in the sulfite concentration range of 2.0×10^{-7} to 2.8×10^{-4} mol/L with a detection limit of 9.0×10^{-8} mol/L.

Furthermore, Pandian et al. also carried out a selective electrochemical sensing of sulfite in water samples with the aid of homogeneously dispersed AgNPs assembled on PANI/rGO [89]. This modified electrode performed better in terms of detection limit in nanomolar range when compared with other materials based on Au, Co and graphene which are only capable to detect in micromolar range. In addition, the AgNP@PANI/rGO/GCE shows a remarkable stability, reproducibility, and selectivity in the presence of a host of impurities.

4.4.1.4 Chloride (Cl^-)

Chloride ion forms when a chlorine atom gains an extra electron. It exists generally in form of salts such as NaCl and KCl, which are highly soluble in water. Most natural waters comprise chloride ion and its concentrations are dependent on the mineral content of the neighboring rock or soil. Apart from these sources, chloride may also be present in water because of human activities such as agricultural run-off, road salting, industrial wastewater, and water emanated from gas and oil wells [90]. Chloride is an essential mineral for both aquatic and non-aquatic organisms, as it helps to maintain osmotic pressure, water and acid–base balance in tissues. However, when chloride ion is present at abnormal concentrations in water bodies, it can have adverse effects on the whole aquatic system, and this is the reason why it is important to develop techniques that will help detect chloride in water.

Among all the available methods, electrochemical technique, most especially ion-selective electrode (ISE) appears to be highly promising to detect Cl^- because it makes it easier to monitor Cl^- with respect to its concentration [91]. This is an added advantage for the ISE when compared to other electrochemical methods like amperometry that indirectly detect Cl^-. However, the use of noble metal such as Ag make them unfit for large-scale monitoring due to high cost. Apart from ISEs, other electrodes also depend on noble metal. Thus, fabrication of a cost-effective sensor for Cl^- determination will require no noble metal. Along this line, Zhang et al. constructed a novel carbon paste electrode for high performance electrochemical sensing of Cl^- in water samples using 3-D graphene [100]. By employing DPV, the peak current at 0.05 V was linear with chloride concentration range from 0.5 to 1000 mM, and the LOD at 0.2 mM (S/N = 3) as shown in Fig. 4.3. After 100 tests using the same electrode, the peak current of Cl^- was still 90–110% from the first analysis, which showed that the electrode could be reused.

4.4.1.5 Phosphate Ion (PO_4^{3-})

Phosphorus is an element that is essential for both the living cells and environmental system and plays a major role in the human wellbeing as well as the treatment of wastewater [93]. In the water system, phosphorus is present in form of orthophosphates, which have paved the way for the detection of phosphate in biological and environmental samples. Phosphate is present at a concentration in the range of 10 μM–10 mM in wastewater and biological fluids and this requires the use of analytical techniques for accurate sensing [94]. The common techniques that have

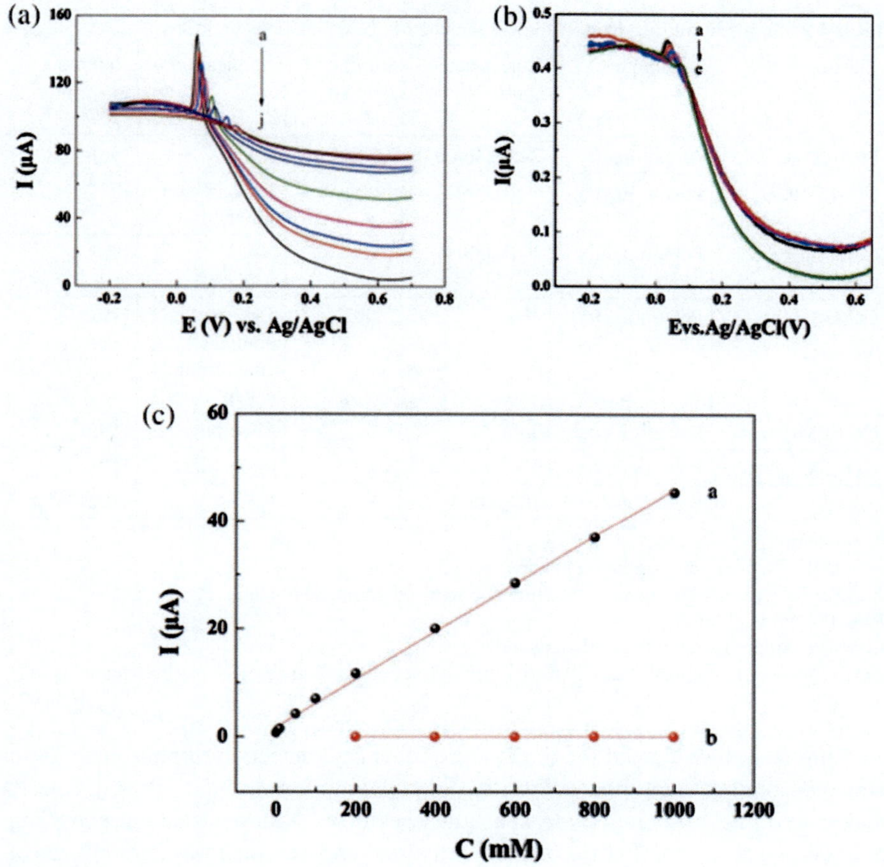

Fig. 4.3 DPV curves of various concentration of Cl^- on **A** 3D-GN/CPE and **B** bare CPE. a → j: 1000, 800, 600, 400, 200, 100, 50, 10, 5 and 0.5 mM, **C** Calibration curves of Cl^- on (a) 3D-GN/CPE and (b) bare CPE. Scan direction: positive

been used include but not limited to colorimetry, spectrophotometry and the use of biomimetic phosphate receptors. However, low sensitivity, intricate analytical procedures, difficult instrumentation limit their practical applications.

Alternatively, electrochemical methods are easy to operate, possess quick response time and can be automated and miniaturized for in situ and online monitoring. Most of the electrochemical methods that have been used to detect phosphate ions in recent times involve interaction of phosphate ion with the receptors, but these methods failed to meet the requirements of the environmental monitoring. However, due to the electrochemically inert nature of phosphate ion, direct sensing of phosphate ion remains a huge challenge. Notwithstanding with this challenge, some successes have been recorded in the direct electrochemical sensing of phosphate ion in water.

Table 4.2 Summary of the electrochemical sensing of phosphate in water

Electrode	Technique	Linear range (ppb)	Limit of detection (μM)	Application	References
Pt/AuNWs	Amperometry	248–14,000	45	NA	[96]
CoNPs-RGO	Potentiometry	1–10,000	NA	Tap and well water	[97]
Chitosan clay/PVC	Potentiometry	1–10,000	0.6	NA	[98]
CuMAPc	Potentiometry	NA	0.002	Tap, river, drainage, underground water	[99]
Cu-BPMP	Potentiometry	3–50	0.5	Mineral water	[100]
UO_2 Salophene	Potentiometry	50–1000	10	River water	[101]
Mb rod	Potentiometry	10–100	1.9	NA	[102]

Abbreviations
Pt/AuNWs Platinum black gold nanowires
CoNPs-RGO Cobalt porous nanostructures on reduced graphene oxide
PVC Polyvinyl chloride
CuMAPc copper monoamino phthalocyanine
Cu-BPMP coper complex of 2,6-bis(bis(2- pyridylmethyl)aminomethyl)-4-methylphenol

Song et al. investigated the direct electrochemical detection of phosphate ion in aqueous solution using phase transition of calcium phosphates, CaP [95]. Specifically, amorphous calcium hydroxyapatite was mixed with octacalcium phosphate and acts as a precursor for the synthesis of hydroxyapatite which was immobilized on Nafion modified glassy carbon electrode (GCE). The novel CaPs/Nafion phosphate sensor developed are based on the influence of phosphate on the peak currents of the CaPs on the electrode surface. The sensor displayed good sensitivity and good selectivity to phosphate ion in aqueous solution. Several other phosphate sensors have been reported as highlighted in Table 4.2.

4.4.2 *Inorganic Cations*

4.4.2.1 Potassium Ion (K^+)

Potassium ion (K^+) is an important mineral found in plant, animal and human nutrition. In human, potassium ions play a central role in regulating the cellular and electrical functions of the body [103]. It also plays a vital role in mediating the osmotic balance of the bodily fluids and regulating the acid-balance of the body by altering the kidney's function to reabsorb bicarbonate-the major extracellular buffer to metabolic acids. Potassium is abundantly present in the environment and natural waters. It can also be present in drinking water because of the potassium permanganate used as an oxidizing agent in the treatment of water.

In some countries, the potassium ion from the chloride salt of potassium (KCl) is commonly being used in exchange for Ca or Mg ion in removing the hardness of water. Although the concentrations of potassium in drinking water are normally low and do not pose serious health risks, the resulting levels of K ion employed as softeners in water treatment can, however, be harmful because of the high solubility of KCl. The abnormal level of K^+ can trigger cardiovascular diseases, kidney disease, hypokalemia among other diseases [104]. Hence, it is important to control the daily intake of potassium present in foods and drinking water. The recommended daily intake of potassium in the diet is between 2000 and 4000 mg [104].

The common method for the determination of potassium ion in water include chromatography, flame photometry, colorimetry, luminescence, and electrochemical based methods. These methods consume lots of time, require sampling, are costly and involve complicated instrumentation which limit their practical application [105]. However, as stated earlier, electrochemical methods have proven to be the best because of their ease of operation, simplicity, low cost and quick response time. The sensitivity, selectivity, reproducibility, and long-term stability of the sensors used in sensing potassium ion in water are outstanding. In the studies involving electrochemical techniques, most of the sensors reported in the literature to determine alkali metal cations are ion-selective electrodes (ISE) [106].

Teixeria et al. developed an electrochemical sensor for K^+ that encompassed carbon paste electrode with a novel $KSr_2Nb_2O_{15}$- an oxide of tetragonal tungsten bronze structure. The sensor exhibited a linear response for K^+ in the concentration range 1.26×10^{-5} at 1.62×10^{-3} mol L^{-1} (E (mV) = 32.7 + 51.1 log $[K^+]$ and LOD of 7.27×10^{-5} mol L^{-1} [107]. It is noteworthy to mention that the lifetime of the sensor is high, which confirmed the sensor stability.

4.4.2.2 Sodium Ion (Na^+)

Sodium is one of the elements that constitute the mineral called NaCl, which is popularly known as common salt. The sodium cation in our body is the major extracellular electrolyte that is responsible for the control of water between intracellular and extracellular [108]. The determination of sodium is very important within both the natural environment and process environment, though its observation in drinking water is much more pronounced as the human intake of sodium at anomalous levels can have adverse effects to human health such as diabetes, renal and hepatic dysfunction, heart and hormonal imbalance, among others. Sodium dissolves highly in water and it is often found in groundwater [109]. All groundwater contains sodium as most rocks and soils comprise salts of sodium which are highly soluble. Although it is generally agreed that sodium is essential to human life, there is no general agreement on the minimum daily intake.

However, it has been estimated that a daily intake of 120–140 mg and 500 mg for the children and adults respectively will meet their daily requirements [110].

The techniques used in determining Na^+ include flame photometry, spectrophotometry, electrophoresis, and chromatography. In studies involving electrochemical techniques, most of the sensors reported in the literature to determine alkali metal cations are ion-selective electrodes (ISE).

Van der Winkel et al. employed Auto Analyzer system to directly determine sodium in river water in the range of 1–1000 ppm using a sodium-selective electrode [111]. EDTA helps to circumvent the interference of other ions such as Mg^{2+} or $SO_4{}^{2-}$, although the sensitivity is restricted by 1 ppm. This system can tolerate a fivefold excess of K and working at high pH can mitigate the interference of H^+. The accuracy and reproducibility of the sensor are good and about 20 samples can be examined in an hour.

4.4.2.3 Magnesium Ion (Mg^{2+})

Magnesium is one of the essential elements for a healthy living. It is found within our cell, and it is a co-factor for more than 400 enzymatic reactions that involve nucleic acid which help to synthesize protein and assist in energy metabolism [112]. An abnormal intake of Mg may have adverse effects on human health such as endothelial dysfunction, pre-menstrual syndrome, decrease in the sensitivity of insulin and psychiatric disorders [113]. It is therefore very crucial to control the levels of magnesium in both soil and water because a healthy soil breeds healthy crops which in turn benefits to keep human life healthy. This has ultimately made the research work in sensing magnesium ion to have received adequate attention in recent times. Several sensors have been used to detect magnesium based on different techniques that include ion-selective electrode, electrochemical conductivity, voltammetry and capillary zone electrophoresis [114].

Chaniotakis et al. used ISE technique to detect Mg^{2+} in water samples by employing Flow Injection Analysis (FIA) method [115]. The sample pretreatment reagents such as pH, buffer solution and calcium complexing agent were used to optimize the detection limits. In addition, this approach employed a combination of flow injection analysis based on a syringe pump as the delivery system together with the reagent that suppresses other interfering ions intensely increased Mg^{2+} selectivity of the electrode. Therefore, it affords new development for rapid and accurate determination of Mg^{2+} in water samples. However, optimum sensing of Mg^{2+} will be obtained if all the experimental parameters are optimized.

4.4.2.4 Calcium Ion (Ca^{2+})

Calcium is one of the most abundant elements in the aqueous environment, biological fluids and biomaterials. It plays a major role in the cell wall and membrane as a counter cation and for organic and inorganic anions. It is very challenging to specifically detect Ca in the presence of other competing ions especially Mg ions. Ion-selective electrode (ISE) is the most common technique to sense inorganic ions including calcium. ISE is based on the relationship between the electrode and the ion activity to detect the ions. They are very appropriate for the detection of Ca ion because of its high selectivity, high sensitivity, low cost, wide variety of tested ions, and small sample size [116].

Li and co-workers investigated a Ca-selective electrode using a copper wire coated with a combination of calcium bis[bis(p-isooctylphenyl) phosophate], dioctyl phenyl phosphate and PVC in cyclohexane [117]. A linear Nernstian response in the range of 2×10^{-5} to 10^{-1} M for Ca^{2+} and a limit of detection of 9×10^{-6} M Ca^{2+} were obtained. The sensor is characterized with low resistance, good stability and high selectivity when applied in natural water.

4.4.2.5 Aluminum Ion (Al^{3+})

Aluminum is the most abundant element and about 8% in the earth's crust. It is present in the environment in form of oxides, silicates, and hydroxides. It can combine with other elements like Na and F and can also form coordination compounds with organic matter. The amount of Al in natural water may vary largely based on the physical, chemical and mineralogical parameters of the water. Dissolved concentration of aluminum in waters usually vary from 0.001–0.05 mg/L with near-neutral pH values but increase to 0.5–1 mg/L in more acidic waters. The source of Al in and the use of coagulants during treatment determine the level of Al in drinking water. However, the toxicity of Al^{3+} is not well established and can have some adverse health risks such as Parkinson's disease, Alzemehier's disease, dialysis, breast cancer among others [118]. Therefore, it is important to develop techniques for the detection of Al^{3+}.

Arvand and co-workers employed ion-selective electrode for the sensing of Al^{3+} in tap and mineral waters by using 6-(4-nitrophenyl)-2-phenyl-4-(thiophen-2-yl)-3,5-diaza-bicyclo[3.1.0]hex2-ene (NTDH) as a novel ionophore [119]. The electrode showed a good response for Al^{3+} over concentration range of 1.0×10^{-6}–1.0×10^{-1} M with a Nernstian slope of 19.6 ± 0.4 mV/decade and low LOD of 6.3×10^{-7} mol L^{-1}. The main merits of this electrode are its simplicity, low cost and precise results that influenced the excellent selectivity toward Al^{3+} over other interference ions.

4.5 Conclusion and Outlook

Electrochemical method has proven to be an emerging technique for the detection of inorganic anions and cations in water with low detection limits and high sensitivity. Substantial progress in both principles and applications has been recorded during the past few decades. When compared with other techniques, electrochemical methods have shown remarkable selectivity, sensitivity, fast response time, low cost, and simplicity. However, there is a challenge associated with the use of ISE which is mainly used in sensing some inorganic ions like chloride. This has to do with the use of noble metal such as Ag which make them unfit for commercial monitoring due to high cost. Apart from ISEs, some other electrodes also depend on noble metal. Therefore, we suggest that fabrication of a cost-effective sensor that will require no noble metal will help in no small amount for industrial application.

References

1. Boyd, C.E., *Water Quality: An Introduction* (Springer Nature, 2019).
2. L.W. Chang, M. László, S. Tsuguyoshi (eds.), *Toxicology of Metals* (Taylor & Francis US, Boca Raton, 1996)
3. United States Environmental Protection Agency, Method 30.0, 1991
4. A.H. Smith, E.O. Lingas, M. Rahman, Contamination of drinking-water by arsenic in Bangladesh: a public health emergency. Bull. World Health Organ. **78**, 1093–1103 (2000)
5. A. Azoulay, P. Garzon, M.J. Eisenberg, Comparison of the mineral content of tap water and bottled waters. J. Gen. Int. Med. **16**(3), 168–175 (2001)
6. X. Zhao, Y. Wu, D. Xing, Z. Ren, L. Ye, Enhanced abatement of organic contaminants by zero-valent copper and sulfite. Environ. Chem. Lett. **18**(1), 237–241 (2020)
7. C. Colombo, A. Crosignani, P.M. Battezzati, *Liver Involvement in Cystic Fibrosis* (1999)
8. F. Barbosa Jr., F.J. Krug, É.C. Lima, On-line coupling of electrochemical preconcentration in tungsten coil electrothermal atomic absorption spectrometry for determination of lead in natural waters. Spectrochim. Acta, Part B **54**(8), 1155–1166 (1999)
9. H.P. Vieira, C.C. Nascentes, A. Germano, C.C. Windmöller, Development of a method for the direct determination of arsenate in honey by hydride generation atomic absorption spectrometry. Anal. Methods **4**(7), 2068–2073 (2012)
10. Q. Zhou, Z. Zheng, J. Xiao, H. Fan, Sensitive determination of As (III) and As (V) by magnetic solid phase extraction with Fe@ polyethyleneimine in combination with hydride generation atomic fluorescence spectrometry. Talanta **156**, 196–203 (2016)
11. S. Burhenn, J. Kratzer, M. Svoboda, F.D. Klute, A. Michels, D. Veža, J. Franzke, Spatially and temporally resolved detection of arsenic in a capillary dielectric barrier discharge by hydride generation high-resolved optical emission spectrometry. Anal. Chem. **90**(5), 3424–3429 (2018)
12. K. Minakata, M. Suzuki, O. Suzuki, Simple and selective determination of arsenite and arsenate by electrospray ionization mass spectrometry. Anal. Chim. Acta **631**(1), 87–90 (2009)
13. Ş Şener, E. Şener, A. Davraz, Evaluation of water quality using water quality index (WQI) method and GIS in Aksu River (SW-Turkey). Sci. Total Environ. **584**, 131–144 (2017)
14. D. Węgrzynek, B. Hołyńska, Simultaneous analysis of trace concentrations of lead and arsenic by energy-dispersive X-ray fluorescence spectrometry. Appl. Radiat. Isot. **44**(8), 1101–1104 (1993)
15. R. Lu, W.W. Li, B. Mizaikoff, A. Katzir, Y. Raichlin, G.P. Sheng, H.Q. Yu, High-sensitivity infrared attenuated total reflectance sensors for in situ multicomponent detection of volatile organic compounds in water. Nat. Protoc. **11**(2), 377 (2016)
16. R. Lu, G. Sheng, W. Li, H. Yu, Y. Raichlin, A. Katzir, B. Mizaikoff, Cover picture: IR-ATR chemical sensors based on planar silver halide waveguides coated with an Ethylene/Propylene copolymer for detection of multiple organic contaminants in water (Angew. Chem. Int. Ed. 8/2013). Angew. Chem. **52**, 2129–2129 (2013)
17. M. Miu, A. Angelescu, I. Kleps, M. Simion, Electrochemical sensors for heavy metals detection in liquid media. Int. J. Environ. Anal. Chem. **85**(9–11), 675–679 (2005)
18. C. Wu, Q. Cheng, K. Wu, Electrochemical functionalization of N-methyl-2-pyrrolidone-exfoliated graphene nanosheets as highly sensitive analytical platform for phenols. Anal. Chem. **87**(6), 3294–3299 (2015)

19. S.A. Alavi-Tabari, M.A. Khalilzadeh, H. Karimi-Maleh, D. Zareyee, An amplified platform nanostructure sensor for the analysis of epirubicin in the presence of topotecan as two important chemotherapy drugs for breast cancer therapy. New J. Chem. **42**(5), 3828–3832 (2018)
20. V.M. Mirsky (ed.), *Ultrathin electrochemical chemo-and biosensors: technology and performance*, vol 2 (Springer Science & Business Media, 2013)
21. J. Wang, Lin, nanotubes del carbonio di Functionalized, del Y. e nanofibers per le applicazioni biosensing. TrAC Tendenza alla Chimica Analitica **27**(7), 619–626 (2008).
22. G. Dorothee, M. Robert, V. Janos, R. Erik, Electrochemical biosensors-sensor principles and architectures. Sensors **8**(3), 1400 (2008)
23. Z. Samec, E. Samcová, H.H. Girault, Ion amperometry at the interface between two immiscible electrolyte solutions in view of realizing the amperometric ion-selective electrode. Talanta **63**(1), 21–32 (2004)
24. I. Palchetti, M. Mascini, Amperometric biosensor for pathogenic bacteria detection, in *Principles of Bacterial Detection: Biosensors, Recognition Receptors and Microsystems*. (Springer, New York, NY, 2008), pp. 299–312.
25. M. Santandreu, S. Alegret, E. Fabregas, Determination of beta-hcg using amperometric immunosensors based on a conducting immunocomposite. Anal. Chim. Acta **396**(2–3), 181–188 (1999)
26. R. Knake, P. Jacquinot, A.W. Hodgson, P.C. Hauser, Amperometric sensing in the gas-phase. Anal. Chim. Acta **549**(1–2), 1–9 (2005)
27. A.J. Bard, L.R. Faulkner, Fundamentals and applications. Electrochem. Methods **2**(482), 580–632 (2001)
28. A. Malinauskas, J. Malinauskiene, A. Ramanavičius, Conducting polymer-based nanostructurized materials: electrochemical aspects. Nanotechnology **16**(10), R51 (2005)
29. J. Wang, Nanomaterial-based electrochemical biosensors. Analyst **130**(4), 421–426 (2005)
30. J.M. Pingarrón, P. Yanez-Sedeno, A. González-Cortés, Gold nanoparticle-based electrochemical biosensors. Electrochim. Acta **53**(19), 5848–5866 (2008)
31. J. Filip, P. Kasák, J. Tkac, Graphene as signal amplifier for preparation of ultrasensitive electrochemical biosensors. Chem. Pap. **69**(1), 112–133 (2015)
32. D. Hernández-Santos, M.B. González-García, A.C. García, Metal-nanoparticles based electroanalysis. Electroanalysis Int. J. Devoted Fundam. Pract. Aspects Electroanalysis **14**(18), 1225–1235 (2002).
33. B.J. Lutz, Z.H. Fan, T. Burgdorf, B. Friedrich, Hydrogen sensing by enzyme-catalyzed electrochemical detection. Anal. Chem. **77**(15), 4969–4975 (2005)
34. R. Xiao, S.I. Cho, R. Liu, S.B. Lee, Controlled electrochemical synthesis of conductive polymer nanotube structures. J. Am. Chem. Soc. **129**(14), 4483–4489 (2007)
35. G. Maduraiveeran, W. Jin, Nanomaterials based electrochemical sensor and biosensor platforms for environmental applications. Trends Environ. Anal. Chem. **13**, 10–23 (2017)
36. R.A. Hameed, S.S. Medany, Evaluation of core-shell structured cobalt@ platinum nanoparticles-decorated graphene for nitrite sensing. Synth. Met. **247**, 67–80 (2019)
37. J.S. Noh, H. Kim, B.S. Kim, E. Lee, H.H. Cho, W. Lee, High-performance vertical hydrogen sensors using Pd-coated rough Si nanowires. J. Mater. Chem. **21**(40), 15935–15939 (2011)
38. W. Yantasee, C.L. Warner, T. Sangvanich, R.S. Addleman, T.G. Carter, R.J. Wiacek, G.E. Fryxell, C. Timchalk, M.G. Warner, Removal of heavy metals from aqueous systems with thiol functionalized superparamagnetic nanoparticles. Environ. Sci. Technol. **41**(14), 5114–5119 (2007)
39. C.W. Kung, Y.S. Li, M.H. Lee, S.Y. Wang, W.H. Chiang, K.C. Ho, In situ growth of porphyrinic metal–organic framework nanocrystals on graphene nanoribbons for the electrocatalytic oxidation of nitrite. J. Mater. Chem. A **4**(27), 10673–10682 (2016)
40. K.O. Otun, Temperature-controlled activation and characterization of iron-based metal-organic frameworks. Inorganica Chim. Acta, p. 119563
41. A.C. Neto, F. Guinea, N.M. Peres, K.S. Novoselov, A.K. Geim, The electronic properties of graphene. Rev. Mod. Phys. **81**(1), 109 (2009)

42. V. Mani, T.Y. Wu, S.M. Chen, Iron nanoparticles decorated graphene-multiwalled carbon nanotubes nanocomposite-modified glassy carbon electrode for the sensitive determination of nitrite. J. Solid State Electrochem. **18**(4), 1015–1023 (2014)
43. J. Zhang, Y. Zhang, J. Zhou, L. Wang, Construction of a highly sensitive non-enzymatic nitrite sensor using electrochemically reduced holey graphene. Anal. Chim. Acta **1043**, 28–34 (2018)
44. T. Soundappan, K. Haddad, S. Kavadiya, R. Raliya, P. Biswas, Crumpled graphene oxide decorated SnO_2 nanocolumns for the electrochemical detection of free chlorine. Appl. Nanosci. **7**(8), 645–653 (2017)
45. S.F. Kiew, L.V. Kiew, H.B. Lee, T. Imae, L.Y. Chung, Assessing biocompatibility of graphene oxide-based nanocarriers: a review. J. Control. Release **226**, 217–228 (2016)
46. F. Zhang, O. Inganäs, Y. Zhou, K. Vandewal, Development of polymer–fullerene solar cells. Natl. Sci. Rev. **3**(2), 222–239 (2016)
47. S. Dong, Z. Sun, Z. Lu, Chloride chemical sensor based on an organic conducting polypyrrole polymer. Analyst **113**(10), 1525–1528 (1988)
48. F. Xu, M. Deng, Y. Liu, X. Ling, X. Deng, L. Wang, Facile preparation of poly (diallyldimethylammonium chloride) modified reduced graphene oxide forsensitive detection of nitrite. Electrochem. Commun. **47**, 33–36 (2014)
49. V.I. Klimov, Spectral and dynamical properties of multiexcitons in semiconductor nanocrystals. Annu. Rev. Phys. Chem. **58**, 635–673 (2007)
50. M.Z. Hu, T. Zhu, Semiconductor nanocrystal quantum dot synthesis approaches towards large-scale industrial production for energy applications. Nanoscale Res. Lett. **10**(1), 1–15 (2015)
51. W. Mao, J. Guo, W. Yang, C. Wang, J. He, J. Chen, Synthesis of high-quality near-infrared-emitting CdTeS alloyed quantum dots via the hydrothermal method. Nanotechnology **18**(48), 485611 (2007)
52. S. Zhuo, L. Chen, Y. Zhang, G. Jin, Luminescent phosphate sensor based on upconverting graphene quantum dots. Spectrosc. Lett. **49**(1), 1–4 (2016)
53. L. Li, D. Liu, K. Wang, H. Mao, T. You, Quantitative detection of nitrite with N-doped graphene quantum dots decorated N-doped carbon nanofibers composite-based electrochemical sensor. Sens. Actuators, B Chem. **252**, 17–23 (2017)
54. S. Liu, H. Wang, Z. Cheng, H. Liu, Hexametaphosphate-capped quantum dots as fluorescent probes for detection of calcium ion and fluoride. Sens. Actuators, B Chem. **232**, 306–312 (2016)
55. M.F. De Volder, S.H. Tawfick, R.H. Baughman, A.J. Hart, Carbon nanotubes: present and future commercial applications. Science **339**(6119), 535–539 (2013)
56. B.L. Hanssen, S. Siraj, D.K. Wong, Recent strategies to minimise fouling in electrochemical detection systems. Rev. Anal. Chem. **35**(1), 1–28 (2016)
57. S. Luanpitpong, L. Wang, Y. Rojanasakul, The effects of carbon nanotubes on lung and dermal cellular behaviors. Nanomedicine **9**(6), 895–912 (2014)
58. Y. Li, Y. Gao, Y. Cao, H. Li, Electrochemical sensor for bisphenol A determination based on MWCNT/melamine complex modified GCE. Sens. Actuators, B Chem. **171**, 726–733 (2012)
59. J. Qu, Y. Dong, Y. Wang, H. Xing, A novel sensor based on Fe_3O_4 nanoparticles–multiwalled carbon nanotubes composite film for determination of nitrite. Sens. Bio-Sens. Res. **3**, 74–78 (2015)
60. H. Zhou, W. Yang, C. Sun, Amperometric sulfite sensor based on multiwalled carbon nanotubes/ferrocene-branched chitosan composites. Talanta **77**(1), 366–371 (2008)
61. P.E. Jackson, Determination of inorganic ions in drinking water by ion chromatography. TrAC, Trends Anal. Chem. **20**(6–7), 320–329 (2001)
62. J. Sadecka, J. Polonský, Determination of inorganic ions in food and beverages by capillary electrophoresis. J. Chromatogr. A **834**(1–2), 401–417 (1999)
63. D. Connolly, B. Paull, Rapid determination of nitrate and nitrite in drinking water samples using ion-interaction liquid chromatography. Anal. Chim. Acta **441**(1), 53–62 (2001)
64. M.J. Moorcroft, J. Davis, R.G. Compton, Detection and determination of nitrate and nitrite: a review. Talanta **54**(5), 785–803 (2001)

65. M.A. Kuiper, J.J. Visser, P.L. Bergmans, P. Scheltens, E.C. Wolters, Decreased cerebrospinal fluid nitrate levels in Parkinson's disease, Alzheimer's disease and multiple system atrophy patients. J. Neurol. Sci. **121**(1), 46–49 (1994)
66. F. Guan, H. Wu, Y. Luo, Sensitive and selective method for direct determination of nitrite and nitrate by high-performance capillary electrophoresis. J. Chromatogr. A **719**(2), 427–433 (1996)
67. J.E. Melanson, C.A. Lucy, Ultra-rapid analysis of nitrate and nitrite by capillary electrophoresis. J. Chromatogr. A **884**(1–2), 311–316 (2000)
68. Z.W. Peng, D. Yuan, Z.W. Jiang, Y.F. Li, Novel metal-organic gels of bis (benzimidazole)-based ligands with copper (II) for electrochemical selectively sensing of nitrite. Electrochim. Acta **238**, 1–8 (2017)
69. D.K. Yadav, V. Ganesan, P.K. Sonkar, R. Gupta, P.K. Rastogi, Electrochemical investigation of gold nanoparticles incorporated zinc based metal-organic framework for selective recognition of nitrite and nitrobenzene. Electrochim. Acta **200**, 276–282 (2016)
70. L. Liu, H. Cui, H. An, J. Zhai, Y. Pan, Electrochemical detection of aqueous nitrite based on poly (aniline-co-o-aminophenol)-modified glassy carbon electrode. Ionics **23**(6), 1517–1523 (2017)
71. H. Wang, F. Wen, Y. Chen, T. Sun, Y. Meng, Y. Zhang, Electrocatalytic determination of nitrite based on straw cellulose/molybdenum sulfide nanocomposite. Biosens. Bioelectron. **85**, 692–697 (2016)
72. L. Sun, H. Li, M. Li, P. Li, C. Li, B. Yang, Simultaneous determination of small biomolecules and nitrite using an Au/TiO_2/Carbon nanotube composite-modified electrode. J. Electrochem. Soc. **163**(13), B567 (2016)
73. S. Li, J. Qu, Y. Wang, J. Qu, H. Wang, A novel electrochemical sensor based on carbon nanoparticle composite films for the determination of nitrite and hydrogen peroxide. Anal. Methods **8**(21), 4204–4210 (2016)
74. X. Fan, P. Lin, S. Liang, N. Hui, R. Zhang, J. Feng, G. Xu, Gold nanoclusters doped poly (3, 4-ethylenedioxythiophene) for highly sensitive electrochemical sensing of nitrite. Ionics **23**(4), 997–1003 (2017)
75. C. Sun, W. Pan, D. Zheng, Y. Zheng, J. Zhu, An electrochemical sensor for nitrite using a glassy carbon electrode modified with Cu/CBSA nanoflower networks. Anal. Methods **11**(39), 4998–5006 (2019)
76. Z.H. Wen, T.F. Kang, Determination of nitrite using sensors based on nickel phthalocyanine polymer modified electrodes. Talanta **62**(2), 351–355 (2004)
77. A. Pandikumar, N. Yusoff, N.M. Huang, H.N. Lim, Electrochemical sensing of nitrite using a glassy carbon electrode modified with reduced functionalized graphene oxide decorated with flower-like zinc oxide. Microchim. Acta **182**(5–6), 1113–1122 (2015)
78. J.C. Gamboa, R.C. Pena, T.R. Paixão, M. Bertotti, A renewable copper electrode as an amperometric flow detector for nitrate determination in mineral water and soft drink samples. Talanta **80**(2), 581–585 (2009)
79. M. Sivakumar, M. Sakthivel, S. Chen, K. Pandi, T. Chen, M. Yu, An electrochemical selective detection of nitrite sensor for polyaniline doped graphene oxide modified electrode. Int. J. Electrochem. Sci. **12**, 4835–4846 (2017)
80. H. Dong, Z. Fang, T. Yang, Y. Yu, D. Wang, K.C. Chou, X. Hou, Single crystalline 3C-SiC whiskers used for electrochemical detection of nitrite under neutral condition. Ionics **22**(8), 1493–1500 (2016)
81. S. Jagtap, M.K. Yenkie, N. Labhsetwar, S. Rayalu, Fluoride in drinking water and defluoridation of water. Chem. Rev. **112**(4), 2454–2466 (2012)
82. US Department of Health and Human Services Federal Panel on Community Water Fluoridation, US Public Health Service recommendation for fluoride concentration in drinking water for the prevention of dental caries. Public Health Rep. **130**(4), 318–331 (2015)
83. J. Tao, P. Zhao, Y. Li, W. Zhao, Y. Xiao, R. Yang, Fabrication of an electrochemical sensor based on spiropyran for sensitive and selective detection of fluoride ion. Anal. Chim. acta **918**, 97–102 (2016)

84. R. Appiah-Ntiamoah, B.T. Gadisa, H. Kim, An effective electrochemical sensing platform for fluoride ions based on fluorescein isothiocyanate–MWCNT composite. New J. Chem. **42**(14), 11341–11350 (2018)
85. T.R. Dadamos, M.F. Teixeira, Electrochemical sensor for sulfite determination based on a nanostructured copper-salen film modified electrode. Electrochim. Acta **54**(19), 4552–4558 (2009)
86. K. Calfumán, M.J. Aguirre, D. Villagra, C. Yañez, C. Arévalo, B. Matsuhiro, L. Mendoza, M. Isaacs, Nafion/tetraruthenated porphyrin glassy carbon-modified electrode: characterization and voltammetric studies of sulfite oxidation in water–ethanol solutions. J. Solid State Electrochem. **14**(6), 1065–1072 (2010)
87. C.S. Pundir, R. Rawal, Determination of sulfite with emphasis on biosensing methods: a review. Anal. Bioanal. Chem. **405**(10), 3049–3062 (2013)
88. A.A. Ensafi, H. Karimi-Maleh, M. Keyvanfard, A new voltammetric sensor for the determination of sulfite in water and wastewater using modified-multiwall carbon nanotubes paste electrode. Int. J. Environ. Anal. Chem. **93**(6), 650–660 (2013)
89. G. Kaladevi, P. Wilson, K. Pandian, Simultaneous and selective electrochemical detection of sulfite and nitrite in water sources using homogeneously dispersed Ag nanoparticles over PANI/rGO nanocomposite. J. Electrochem. Soc. **167**(2), 027514 (2020)
90. J. Bujes-Garrido, M.J. Arcos-Martínez, Development of a wearable electrochemical sensor for voltammetric determination of chloride ions. Sens. Actuators, B Chem. **240**, 224–228 (2017)
91. J.F. Lechner, I. Sekerka, Chloride ion-selective electrode based on HgS/Hg2Cl2. J. Electroanal. Chem. Interfacial Electrochem. **57**(3), 317–323 (1974)
92. M. Zhang, C. Wang, Z. Zhang, J. Ye, P. Fang, A novel carbon paste electrode for sensitive, selective and rapid electrochemical determination of chloride ion based on three-dimensional graphene. Sens. Actuators B Chem **299**, 126951 (2019)
93. C. Warwick, A. Guerreiro, A. Soares, Sensing and analysis of soluble phosphates in environmental samples: a review. Biosens. Bioelectron. **41**, 1–11 (2013)
94. O. Korostynska, A. Mason, A. Al-Shamma'a, Monitoring of nitrates and phosphates in wastewater: current technologies and further challenges. Int. J. Smart Sens. Intell. Syst. **5**(1) (2012)
95. S. Sun, Q. Chen, S. Sheth, G. Ran, Q. Song, Direct electrochemical sensing of phosphate in aqueous solution based on phase transition of calcium phosphate. ACS Sens (2020)
96. C. Forano, H. Farhat, C. Mousty, Recent trends in electrochemical detection of phosphate in actual waters. Curr. Opin. Electrochem. **11**, 55–61 (2018)
97. K. Kargosha, P. Hemmatkhah, S.H. Ahmadi, in *Cobalt-Graphene Nanocomposite Electrode for Phosphate Sensing* (2017)
98. C. Topcu, B. Caglar, A. Onder, F. Coldur, S. Caglar, E.K. Guner, O. Cubuk, A. Tabak, Structural characterization of chitosan-smectite nanocomposite and its application in the development of a novel potentiometric monohydrogen phosphate-selective sensor. Mater. Res. Bull. **98**, 288–299 (2018)
99. M.N. Abbas, A.L.A. Radwan, N.M. Nooredeen, M.A.A. El-Ghaffar, Selective phosphate sensing using copper monoamino-phthalocyanine functionalized acrylate polymer-based solid-state electrode for FIA of environmental waters. J. Solid State Electrochem. **20**(6), 1599–1612 (2016)
100. L. Li, G. Shang, W. Qin, Potentiometric sensing of aqueous phosphate by competition assays using ion-exchanger doped-polymeric membrane electrodes as transducers. Analyst **141**(15), 4573–4577 (2016)
101. N. Pankratova, M. Ghahraman Afshar, D. Yuan, G.A. Crespo, E. Bakker, Local acidification of membrane surfaces for potentiometric sensing of anions in environmental samples. ACS Sens. **1**(1), 48–54 (2016)
102. Y. Li, T. Jiang, X. Yu, H. Yang, Phosphate Sensor Using Molybdenum. J. Electrochem. Soc. **163**(9), B479–B484 (2016)

103. L.D. Talbott, E. Zeiger, Central roles for potassium and sucrose in guard-cell osmoregulation. Plant Physiol. **111**(4), 1051–1057 (1996)
104. H.R. Brunner, L. Baer, J.E. Sealey, J.G. Ledingham, J.H. Laragh, The influence of potassium administration and of potassium deprivation on plasma renin in normal and hypertensive subjects. J. Clin. Investig. **49**(11), 2128–2138 (1970)
105. Z. Chen, J. Guo, S. Zhang, L. Chen, A one-step electrochemical sensor for rapid detection of potassium ion based on structure-switching aptamer. Sens. Actuators, B Chem. **188**, 1155–1157 (2013)
106. J. Ping, Y. Wang, J. Wu, Y. Ying, Development of an all-solid-state potassium ion-selective electrode using graphene as the solid-contact transducer. Electrochem. Commun. **13**(12), 1529–1532 (2011)
107. M.F. Teixeira, B.H. Freitas, P.M. Seraphim, L.O. Salmazo, M.A. Nobre, S. Lanfredi, Development of an electrochemical sensor for potassium ions based on $KSr_2Nb_5O_{15}$ modified electrode. Procedia Chemistry **1**(1), 293–296 (2009)
108. B. Schazmann, D. Morris, C. Slater, S. Beirne, C. Fay, R. Reuveny, N. Moyna, D. Diamond, A wearable electrochemical sensor for the real-time measurement of sweat sodium concentration. Anal. Methods **2**(4), 342–348 (2010)
109. S. Chandra, H. Lang, A new sodium ion selective electrode based on a novel silacrown ether. Sens. Actuators, B Chem. **114**(2), 849–854 (2006)
110. W.D. Basson, J.F. Van Staden, Simultaneous determination of sodium, potassium, magnesium and calcium in surface, ground and domestic water by flow-injection analysis. Fresenius' Zeitschrift für Analytische Chemie **302**(5), 370–374 (1980)
111. P. Van den Winkel, J. Mertens, G.D. Baenst, D.L. Massart, Automatic potentiometric analysis of sodium in river and mineral waters. Anal. Lett. **5**(8), 567–577 (1972)
112. M. Bryce, A. Batsanov, J.K. Howard, L. Goldenberg, Selectiveelectrochemical magnesium and calcium sensors based on non-macrocyclic nitrogen-containing ferrocene ligands. Chem. Commun. **6**, 677–678 (1998)
113. M.H. Asif, S.M.U. Ali, O. Nur, M. Willander, U.H. Englund, F. Elinder, Functionalized ZnO nanorod-based selective magnesium ion sensor for intracellular measurements. Biosens. Bioelectron. **26**(3), 1118–1123 (2010)
114. J. Li, G. Zou, X. Hu, X. Zhang, Electrochemistry of thiol-capped CdTe quantum dots and its sensing application. J. Electroanal. Chem. **625**(1), 88–91 (2009)
115. N.A. Chaniotakis, J.K. Tsagatakis, E.A. Moschou, S.J. West, X. Wen, Magnesium ion-selective electrode: optimization and flow injection analysis application. Anal. Chim. Acta **356**(1), 105–111 (1997)
116. B. Delavaux-Nicot, J. Maynadié, D. Lavabre, S. Fery-Forgues, Ca^{2+} versus Ba^{2+} electrochemical detection by two disubstituted ferrocenyl chalcone chemosensors. Study of the ligand–metal interactions in CH_3CN. J. Organomet. Chem. **692**(4), 874–886 (2007)
117. T.R. Crompton, *Determination of Metals and Anions in Soils, Sediments and Sludges* (CRC Press, 2001)
118. S.M. Ng, R. Narayanaswamy, Fluorescence sensor using a molecularly imprinted polymer as a recognition receptor for the detection of aluminium ions in aqueous media. Anal. Bioanal. Chem. **386**(5), 1235–1244 (2006)
119. M. Arvand, S.A. Asadollahzadeh, Ion-selective electrode for aluminum determination in pharmaceutical substances, tea leaves and water samples. Talanta **75**(4), 1046–1054 (2008)

Chapter 5
Corrosion Prevention: The Use of Nanomaterials

A. Momoh, F. V. Adams, O. Samuel, O. P. Bolade, and P. A. Olubambi

Abstract The devastating economic impact of corrosion necessitates research into suitable corrosion control techniques. An important precursor to the development of such control methods is a comprehensive understanding of the factors that actively contribute to the corrosion of metals. These include environmental parameters such as presence of moisture, the pH and temperature of any surrounding electrolyte solution present, the concentration of oxygen present, etc. However, the level of corrosion observed on the material is only partly due to these factors. It is actually the metallurgical properties of the material that dictate how it will corrode in response to the environmental conditions present. This concept has fueled research into the development of nanomaterials such as nanocrystalline (NC) materials, nanogels, nanocomposites, nanoparticles, carbon dots, and nanocontainers, which possess physical properties that are markedly different from their mainstream microcrystalline (MC) counterparts and other corrosion inhibitors. While the use of nanomaterials in corrosion control undoubtedly has numerous benefits, their use has also been found to pose certain challenges; large scale processing of NC metals can prove difficult as they may have poor thermal stability and are highly susceptible to grain growth. In tribocorrosion control, on the other hand, the use of nanoparticles may result in the deposition of nanosized debris at the region of contact between two surfaces. This could have the undesired effect of increasing friction and damaging the oxide film layer present. This could favor further corrosion.

A. Momoh · F. V. Adams (✉) · O. P. Bolade
Department of Petroleum Chemistry, American University of Nigeria, Yola, Nigeria

F. V. Adams · P. A. Olubambi
School of Mining, Metallurgy and Chemical Engineering, Centre for Nanomechanics and Tribocorrosion, University of Johannesburg, Doornfontein, Johannesburg, South Africa

O. Samuel
Department of Chemical Engineering, Federal Polytechnic, Mubi, Adamawa State, Nigeria

O. M. Ama et al. (eds.), *Modified Nanomaterials for Environmental Applications*, Engineering Materials, https://doi.org/10.1007/978-3-030-85555-0_5

5.1 Introduction

Corrosion is a natural process that converts a refined metal into a more chemically stable form such as oxide, hydroxide, or sulfide. This is the gradual deterioration or destruction of materials (usually a metal) by chemical and/or electrochemical reaction with their environment. Electrochemical corrosion often takes place in electrolytes such as water, soil, or wet gases, while chemical corrosion occurs in nonconductive liquids or dry gases [1]. Both types of corrosion cause a production of rust or mill scale.

In spite of much advancement in the science and technology of corrosion prevention and control, the phenomenon of corrosion (usually of metals and alloys) continues to pose a major concern to many industries around the world [2–5]. The direct and indirect cost of corrosion is huge. A good portion of the loss can be avoided by proper corrosion control and monitoring methods. This means there is the need to utilize cheaper and better methods to prevent corrosion.

Nanoscience and nanotechnology have revolutionized the present-day scientific world, leading to the development of new tools and techniques. Materials at nanometer scale possess many advantages over their micrometer scale counterparts. The key aspect in many applications is the efficient design of nanomaterials in which the desirable pros can be distinguishably emphasized, and the undesirable cons can be properly suppressed. Widespread efforts are underway to promote their effective use in the realm of corrosion control. Employing nanomaterials can enhance a material's performance significantly in certain areas of corrosion control; for example, in an advanced coating [6].

Although in the past, some authors had worked on various applications of nanomaterials for corrosion mitigation [6, 7] and their findings reported this chapter aims at elaborating on already reported work and to also add what is yet to be reported in this area. For instance, Burhagohain and Shrma gave a descriptive study on the use of nanomaterials as corrosion inhibitors with specific applications in the oil and gas industry [6]. These nanomaterials include nanogels, nanocrystals, carbon dots, etc. [8–15]. Nanocomposite materials are applicable in various aspects of corrosion control, some of which will be discussed subsequently.

5.2 Nanostructured Materials

Materials with morphological features on the nanoscale (usually smaller than 100 nm) in at least one dimension are referred to as nanomaterials [16]. They can be classified as nanosized and nanostructured materials. A nanostructured material, which is also referred to as a nanocrystalline or nanophase material can be said to be one with nanoscale morphological features. It is usually constructed using a combination of nanosized (e.g., grain size, particle size, structure size, etc.), materials, and/or nanoscale empty spaces [6]. These materials include secondary structures consisting of primary nanoparticles, nanograined metals/alloys/ceramics, nanoporous structures, nanocomposites, etc.

Nanostructured materials possess unique mechanical, electronic, physical, chemical, and physiochemical properties due to their high surface area which in turn increases the active centers [17]. These properties enable easy coatings of greater wear and tear resistance. The nanomaterials undergo physisorption/chemisorption to the corroded metal surface and inhibit the corrosion of the metal efficiently.

The overall property of a nanostructured material depends essentially on parameters such as the size and distribution of the constituent phases (particle, grain, phase, etc.), nature of grain boundaries, and interaction between the phases [18]. The sizes of the building blocks, which are relatively small, high relative number of atoms in the phase/grain boundaries as well as high surface to volume ratio are responsible for the improved and diversified properties of nanomaterials compared to their microstructured materials. Substantial contribution has been made over the years by various researchers in processing nanostructured materials, this includes nanostructured coatings [19–22].

5.3 Nanomaterials in Corrosion Mitigation

Thin film coatings such as painting have been in existence for many years and have been used to prevent corrosion. However, passive and active coatings, which are benefited appreciably by nanomaterials have recently been used as corrosion inhibitors and are applied in various areas where corrosion is a challenge [23]. A nanocoating is that type of coating that has either nanoscale thickness coating or nanosized grains/phases or has its second phase particles dispersed into the matrix in the nanosize range, etc. [5]. Nanocoatings can either be nanocomposited coatings or nanostructured coatings; these have grain sizes in the range of 100 nm [24, 25]. Vacuum, as well as non-vacuum-based methods, are usually used in the synthesis of nanocoatings. The vacuum based method includes high velocity oxy-fuel thermal spray, electroless plating, electrodeposition, anodization, electro-spark deposition laser beam surface treatment, sol-gel, magnetron sputtering, etc. [26]. Application of nanocoatings on metal surface is basically to improve corrosion resistance of the metal, reduce wear, high temperature oxidation, and corrosion resistance [27].

The subsequent sections show areas in which nanomaterials or nanoparticles can be applied to prevent or mitigate corrosion.

5.4 Application of Metal/Metal Oxide Nanoparticles as Corrosion Inhibitor

Metal/metal oxide nanoparticles have been found to be effective anti-corrosion agents and have found a wide range of applications in aerospace and in various industries. The corrosion inhibition efficiency of metal/metal oxide nanoparticles was reported to result from their self-healing ability and ability to create barrier against corrosive species [12, 28]. Also, an improvement in photodegradation resistance of coatings is achieved using metal/metal oxide nanoparticles.

Obot et al. [29] synthesized silver (Ag) nanoparticles for mild steel inhibition in 0.5 M HCl solution. The authors synthesized the Ag nanoparticles from a bio-derived product-honey, which was used as both capping and reducing agents with sunlight irradiation. Electrochemical corrosion method was used in the corrosion inhibition study. The results obtained showed that Ag nanoparticles synthesized from honey are an effective corrosion inhibitor for mild steel [29]. An inhibition efficiency of 91.5% with a very low inhibition concentration was achieved.

Rivera-Grau et al. [30] studied the effect of Ag nanoparticles in hydroxyethyl imidazoline as mild steel inhibitor against CO_2 corrosion in NaCl and diesel solution. The authors observed that incorporating Ag nanoparticles in hydroxyethyl imidazoline increased inhibition efficiency of the inhibitor by approximately 99% at a very low inhibition concentration. Also, Solomon et al. [31] synthesized green inhibitor using gum Arabic and nanoparticles for corrosion inhibition of mild steel in 15% HCl and 15% H_2SO_4 solutions. Gravimetric, electrochemical, and surface analysis techniques were used to study the corrosion inhibitory effect of gum Arabic/Ag nanoparticles on mild steel in acid solutions. The inhibition efficiencies of the inhibitor were found to increase with an increase in inhibitor concentrations. The inhibitor was found to be more efficient in H_2SO_4 at the lowest concentration, but more efficient in HCl at higher concentrations [31]. According to Solomon et al. [31], the recharging power of the ions in acidic solution (Cl^- and $SO_4{}^{-2}$) could be deciding factor in the efficiency of the inhibitor in these solutions. The Cl^- ions could replenish the steel surface, which further induces greater adsorption of gum Arabic/Ag nanoparticles polycations compared to $SO_4{}^{-2}$ ions hydrated surface. Incorporating Ag nanoparticles to the gum Arabic matrix was seen to increase the inhibition efficiency of the gum Arabic significantly, mostly in HCl solution [31]. In recent publication, Anjum et al. [12] gave a review on the application of metal/metal oxides as corrosion inhibitors.

5.5 Nanocomposites Corrosion Inhibitors

Nanocomposites could be nanoparticles, nanopolymers, or nanogels. Nanopolymers and nanogels are formed due to the high affinity of nanomaterials to polymers, high surface area, and low density. Atta et al. explored gum coated magnetite [32] and acrylamide copolymers/ magnetite [33] nanocomposites as corrosion inhibitors in

1 M HCl solution. The authors chose magnetite nanoparticles because of its tendency to form nanopolymer, magnetic characteristics, low density, and high surface area. Despite these properties exhibited by magnetite nanoparticles, it has the limitation of aggregation, arising from strong inter-particle interactions [33], and poor adherent capacity to the surface of the metals. Thus, myrrh gum was used as a capping agent and was coated on the surface of magnetite nanoparticles to form nanocomposite material. The nanocomposites inhibitor was found to exhibit a mixed inhibitory characteristic and the corrosion resistance of the steel studied increased with increasing inhibitor concentration. An inhibition efficiency of up to 91% was obtained [32].

In another study by Atta et al. [34], hybrid polymer composites based on Ag nanoparticles were synthesized and applied as corrosion protection for line pipe steel. The authors reduced silver nitrate to silver nanoparticles by p-chloroaniline in the presence of polyoxyethylene maleate using 4-nonyl-2-propylene-phenol (NMA) as a stabilizing agent using a semi-batch solution polymerization method [33]. The authors again observed that the nanocomposite inhibitor acted as a mixed type of inhibitor for steel in 1 M HCl.

El-Mahdy et al. [35] on the other hand, synthesized TiO_2 nanoparticles in Poly(Sodium-2-Acrylamido-2-methylpropane) Sulfonate (PNa-AMPS) composite. This resulted in a nanogel which was used as corrosion inhibitor for steel in acidic chloride solution. An inhibition efficiency of 91% was obtained at 250 ppm concentration. The mechanism of inhibition observed for this nanogel was mixed inhibition. Olive/titanium nanocomposite was synthesized using olive leaves and $TiCl_4$ by ethanolic extract [36]. The nanocomposite inhibited mild steel in 1 M HCl solution with an inhibition efficiency that increased from 83.5 to 93.4% at 30 °C. It however decreased to 51.7 and 85.4% at 60 °C after the addition of Ti nanoparticles.

Chitosan based nanocomposites were developed by Srivastava et al. [37] for corrosion inhibition of mild steel in chloride media. Chitosan, a natural biopolymer has green value but cannot be used as inhibitor in its natural form due to its rapid degradation, thus it needs to be converted to a more stable form. The authors converted chitosan to nanocomposites using cobalt and tin sulfide nanoparticles. The nanocomposites were tested on mild steel in 1 M HCl at room temperature. The highest inhibition efficiency (>95%) was obtained from the chitosan/cobalt nanocomposite, while chitosan/tin sulfide exhibited an inhibition efficiency of (>80%). The mechanism was by adsorption and the obtained data fitted in Frumkin isotherm [37].

Chromium (VI) based coating application as anticorrosive coatings were replaced with graphene/polyetherimide (PEI) nanocomposites [38]. The authors synthesized the nanocomposites using a high shear liquid exfoliation route. The resulted nanocomposites showed improved corrosion resistance of mild steel in 3.5% sodium chloride. It was observed that the corrosion potential (E_{corr}) of the graphene-based coating systems was about 400 mV more positive than that of the uncoated mild.

5.6 Nanocontainers as Corrosion Storage Inhibitors

Nanocontainers are nanomaterials with turnable sizes and shapes. These include polymersomes and mesoporous silica particles, which possess adaptable structures with holes or pores which are mostly used for drug delivery and nanospace-confined reactions [39]. They are used to encapsulate inhibitors used in corrosion mitigation.

Raja et al. [10] recently reported on inhibitor-encapsulated smart nanocontainers for the controlled release of corrosion inhibitors. These include titania nanocontainers, polyelectrolyte shells, halloysite nanotubes, and mesoporous silica nanocontainers. Others are nanolayers, porous nanoparticles, nanofibers, and polymer nanocapsules.

Nanocontainers have the capacity to sense any change in pH of surroundings, which makes them effective in corrosion mitigation. Another advantage of nanocontainers is that they are able to encapsulate the corrosion inhibitor without leakage and then release the inhibitor when there is breakage in the coating, which implies self-healing capability.

Noiville et al. [40] studied the release of CE(III) corrosion inhibitors from silica and boehmite nanocontainers. The corrosion inhibitor encapsulation was carried out on aluminium alloys and the authors reported that dense silica nanocontainers were found to exhibit higher Ce(III) loading than the mesoporous ones. On the other hand, Ce(III) release from boehmite nanocontainers was observed to be effective within 2 weeks in the corrosive medium, which was longer than the release from silica (168 h) [40].

Silica nanoparticles were used to encapsulate dodecylamine corrosion inhibitors. The nanocontainers, polyethyleneimine (PEI), poly-styrene sulphonate (PSS), and inhibitor (dodecylamine) layers were alternatively deposited on silica [35]. The released inhibitor from nanocontainers was carried out to determine the corrosion resistance of carbon steel in aerated 0.1 mol/L NaCl solution containing 1 wt.% nanocontainer at varied pH values of 2.0, 6.2, and 9.0 [41]. The electrochemical impedance spectroscopic results indicated that rapid and efficient release of corrosion inhibitors was obtained at pH of 2.

Yeganeh et al. [42] also studied the use of mesoporous silica as the nanocontainer of corrosion inhibitor and various methods of synthesis were discussed. The silica nanocontainers synthesized via in-situ process according to the authors enhances diffusion of corrosion inhibitor in the entire pores of mesoporous materials. Kartsonakis et al. [43] reported release of corrosion inhibitor from titania nanocontainers, which were used in corrosion mitigation of steel in 0.5 M sodium chloride. The results showed that the nanocontainers decreased the corrosion rate almost 100 times, as compared to when only coating was used [42].

5.7 Metal Oxidation Control at Varied Temperature Using Nanomaterials

5.7.1 Moderate-Temperature

Nanocrystalline (NC) materials can offer greater protection against oxidation relative to microcrystalline (MC) materials at relatively moderate temperatures [44, 45]. This is due to contrasting structural features between NC materials and their MC counterparts. It is these structural features that are the defining property of NC materials. More comprehensively, NC materials are those with morphological features on the nanoscale level consisting of a combination of nanosized materials and/or nanoscale empty spaces. These morphological features include grains, grain size, and grain boundaries [45]. The nature of these features explains the corrosion-inhibiting effect of certain NC materials.

Dimtri [46] described a grain as a small region of a metal characterized by a defined and consistent arrangement of atoms in a lattice. Essentially each grain is representative of a small crystal. Thus, the grain structure of a metal is formed during transitions between phases, e.g., from liquid to solid. The region in which two or more grains meet is known as grain boundary. Unlike the actual grains, grain boundaries are not densely packed with atoms in a rigid lattice. This results in higher diffusion rate through the grain boundaries than within the grains.

Corrosion of metals can occur via electrochemical redox reactions involving oxidation at the anode and reduction at the cathode. This process results in the formation of an oxide layer which, depending on the metal in question, may hinder or facilitate further corrosion. Iron, for example, forms a porous hydrated ferric oxide layer which chips off exposing the metal underneath to further corrosion. Other metals like chromium, however, can form a protective oxide layer which will prevent further corrosion. This makes chromium a suitable alloying metal.

Grain size is vital to the formation of protective oxide films like that of chromium. Some steel alloys with high chromium content initially form an iron or nickel oxide layer which is subsequently converted to the protective Cr_2O_3 layer. The kinetics of this conversion is dependent on the diffusion of chromium within the alloy matrix [45]. This diffusion is governed by the grain size. A smaller grain size means higher rate of diffusion since there are fewer atoms within the lattice. Consequently, formation of protective oxide layer occurs at a faster rate. On the other hand, a larger grain size would mean less diffusion of chromium and less chance of formation of the stable Cr_2O_3 layer. As a result, the alloy would be susceptible to further oxidation and corrosion. According to Young [47], quick diffusion of chromium is also necessary to prevent harmful internal oxidation where the oxidant dissolves and diffuses inward forming dispersed metal oxide precipitates.

The rate of formation of oxide layers in metals can be investigated by monitoring the mass gained over time. A simple comparison of NC materials and MC materials shows the enhanced protection offered by the NC materials against oxidation.

Studies have shown that the diffusion coefficient of Cr in Fe can be improved by some orders of magnitude so long the grain size is of nanometer level [48]. A study carried out by Raman and Singh [45] showed that MC Fe-10% Cr alloy gains a larger amount of mass per unit area compared to its NC counterpart within the 240-min time frame. This suggests that the NC alloy readily formed the protective chromium oxide layer which inhibited further oxidation of the metal resulting in minimal mass gain, unlike the MC alloy which gained mass at a higher rate.

Grain growth of the NC materials at high temperatures was previously reported to limit their use in applications requiring high temperature [49]. However, addition of small amount of Zirconium (Zr) to Fe-Cr based alloys at the temperature range of 600–800 °C could prevent grain growth of the NC materials and enhance their oxidation resistance [50].

5.7.2 *High Temperature*

On exposure to high temperature conditions (e.g., in industrial reactors), metals could be oxidized through reactions with gases such as oxygen, carbon, nitrogen, etc. Over time this oxidation is usually detrimental to the structural integrity of the metal and can lead to severe damage [51]. In order to protect metals against this high temperature oxidation, certain elements such as aluminium and chromium are incorporated into the alloy matrix to improve oxidation resistance as they form protective oxide layers which are also stable at high temperatures [51].

Under corrosive conditions such as high temperature, NC coatings are able to achieve greater outward diffusion through their much smaller grain size and higher density of grain boundaries. This means that with NC coatings less amount of coating element, e.g., chromium or aluminium is required to provide protection against oxidation in high temperature conditions [52]. This can be seen with the Ni-20Cr-Al alloy matrix which is commonly adopted for high temperature conditions and has a critical value of 6 wt.% aluminium for formation of the protective Al_2O_3 oxide layer as reported by Wallwork and Hed [53]. This was also observed by Saji and Thomas [54]. In the Ni-20Cr-Al alloy system, for instance, greater than 6 wt.% Al is required to form a complete Al_2O_3 scale. However, according to studies by Liu et al. [55], by reducing grain size to around 60 nm, the Al content required to prevent external oxidation can be lowered to 2 wt.% which is still capable of forming the protective oxide layer in a high temperature environment of 1000 °C. This ~2 wt.% Al according to Saji and Thomas [54] could form an α-Al_2O_3 scale at 1000 °C in air.

5.8 Tribocorrosion Control Using Nanomaterials

Tribocorrosion refers to degradation of a material due to combined effect of mechanical stress and a corrosive environment. Essentially, it is material degradation due to a combination of wear and corrosion [56]. This occurs in systems known as tribosystems; where two or more bodies are in contact. The observed tribocorrosion behavior of materials depends on a variety of factors such as mechanical/operational, which involve normal force, sliding velocity, alignment, shape, and size of contacting bodies, type of motion, and vibrations. Other factors include materials pin and plate (hardness, plasticity, surface roughness, oxide film properties, inclusions, etc.), solution (viscosity, conductivity, pH, temperature, and corrosivity), and electrochemical (applied potential, ohmic resistance, film growth, active dissolution, and valance) [57].

The most common forms are corrosion wear, microabrasion corrosion, fretting corrosion, and erosion corrosion. Tribocorrosion poses some obvious challenges in the design and smooth running of tribosystems. The degradation it causes can compromise the functionality of the system entirely and so it becomes necessary to improve the corrosion performance of the system's components. This can be achieved with the use of nanomaterials usually in lubricants. Specifically, they are employed as extreme pressure and anti-wear additives in liquid lubricants [58].

Nanoparticles are able to provide a form of lubrication during the process of corrosive wear. In certain tribosystems, the use of nanoparticles is associated with an observed reduction in wear. This reduction can be attributed to the deposition of nanoparticles on the surfaces in contact. As mentioned earlier, tribocorrosion combines the effect of both mechanical stress and a corrosive environment. Therefore, reducing the effect of tribocorrosion needs to be approached from both sides. Fortunately, nanoparticles have also been shown to be adept in mitigating the effect of the corrosive environment as they are able to lower the anodic current in certain tribosystems as reported by Deepika [58].

There are various ways in which nanomaterials additives may function in tribocorrosion systems and the mechanisms by which oils having nano-additives reduce wear and friction differ. These include colloidal effect, small size effect, under boundary lubrication, rolling effect, third body effect, and protective film effect [59]. Granular nanomaterials are cohesion less hard particles, while soft nanomaterials act as a cohesive third body. The soft nanomaterials can exist between two surfaces to reduce wear and friction, which completely deforms under load, thereby accommodating differences in surface velocity, which often adhered to metal surfaces and shearing in the bulk medium. The granular nanomaterials, on the other hand, maintain their spherical geometry under load and accommodate surface velocity differences through sliding and rolling at low shear rates [58]. Table 1 shows reported work on the application of nanomaterials in tribocorrosion mitigation.

5.9 Corrosion Protection of Aerospace Alloys Using Nanomaterials

The airline industry is one that has been hard hit by the adverse effects of corrosion. This is to be expected as aircraft are exposed to a wide range of environmental conditions from the wet and humid to the dry and arid [60]. Other parameters like temperature and pressure also vary greatly. These conditions can create a very corrosive environment. This is why bodies of aircraft are composed primarily of aluminium alloys which form protective oxide layers to guard against corrosion. However, as pointed out by Asmatulu et al. [61] in a previous study in 2007, the harsh atmospheric conditions, which can also include industrial air pollutants, can accelerate the corrosion process such that extra and more significant protective measures have to be put in place to minimize the effects of corrosion. Environments with high humidity and temperatures are particularly damaging for aircraft.

Nanomaterials find use in the airline industry as part of the structural design of various components of aircraft to protect against corrosion among other things. These nanomaterials are employed usually in the form of coatings to serve various purposes [62]. An example is the nanoscale thermal barrier coating used on surfaces of aircraft engine parts that operate at high temperatures. These coatings are porous and help to provide insulation to the metallic compounds beneath from the hot stream of gas generated in the turbines of the engine [60, 63]. The porosity of the coating can result in a temperature drop of as much as 100–300 °C at the metal surface [64].

Nanomaterials are also applied in the form of nanocomposite coatings synthesized by mixing multiple materials on a nanoscale level [65]. Based on a previous study, a combination of carbon nanotubes and graphene-based nanocomposite coatings exhibit desirable hydrophobic properties to guard the aircraft surface underneath against water and other corrosive agents [66].

5.10 Challenges of the Use of Nanomaterials as Corrosion Mitigation Substances

Application of nanomaterials has been seen by toxicologists to be toxic to the environment and human health as well as aquatic lives. This is due to the nanosize of the materials. Recent research has shown significant toxicity of nanoparticles both in in-vivo and in-vitro systems [67]. Thus, the newly found application of nanomaterials in corrosion mitigation could also raise concern by nanotoxicologists.

Study on the toxicity of nanomaterials using lysosomotropic and chelating agents has shown that the toxicity mechanism for various metal containing metal oxides, metallic and semiconductor nanoparticles is associated with the release of corresponding toxic ions to the environment [68].

Another challenge that has been discovered to associate with nanomaterials such as nanostructured metals is their low ductility [69]. The negative aspect of the strength and hardness of nanostructured metals is that they tend to be less ductile. This is seen to cause limitations to their use in corrosion mitigation areas.

Also, nanomaterials have been reported to form agglomeration in matrices when used as they do not disperse well in coating solutions [12]. The development of nanolubricant additives has the challenge of instability in base oil as they form sedimentation from the bulk medium [70].

5.11 Conclusions

Corrosion mitigation research is constantly evolving resulting in innovative control methods. In recent times, this research has been centered around nanotechnology whereby nanomaterials such as nanocomposites, nanogels, nanocrystals, nanocontainers, etc. are developed for corrosion control. Nanomaterials are used in a variety of ways such as coatings and additives to provide protection against corrosion across a range of sectors from construction to the airline industry. Fortunately, it has proven to be a very promising solution that can only be improved by further research as there are certain challenges that are associated with their use. These include instability, low ductility, toxicity, etc. Mesoporous silica nanocontainers are associated with insufficient time for the release of corrosion inhibitor substances from their cavities.

The studies into the potential use of nanomaterials in surface modification to control tribocorrosion have proven challenging as the small size of nanomaterials could pose increased health risks when applied as implants in humans. Despite the challenges reported on nanotechnology application in corrosion control, nanomaterials are found to be promising materials and require further studies.

References

1. A. Stankiewicz, Self-healing nanocoatings for protection against steel corrosion, in *Nanotechnology in Eco-efficient Construction* (Elsevier, 2019), pp. 303–335
2. L.T. Popoola, A.S. Grema, G.K. Latinwo, B. Gutti, A.S. Balogun, Corrosion problems during oil and gas production and its mitigation. Int. J. Ind. Chem. **4**, 35–49 (2013). https://doi.org/10.1186/2228-5547-4-35
3. P.R. Roberge, *Handbook of Corrosion Engineering* (2000)
4. D. Miller, Corrosion control on aging aircraft: what is being done. Mater. Perform. **29**, 10–11 (1990)
5. V.S. Saji, The impact of nanotechnology on reducing corrosion cost (Elsevier, 2012)
6. P. Burhagohain, G. Shrma, A descriptive study on the use of nanomaterials as corrosion inhibitors in oil and gas industry. Int. J. Technol. Sci. Sci. Res. **8**, 3999–4002 (2019)
7. N. Asadi, R. Naderi, Nanoparticles incorporated in silane sol–gel coatings, in *Corrosion Protection at the Nanoscale, Micro and Nano Technologies* (2020), pp. 451–471
8. S. Amiri, S. Amiri, Inhibitors-loaded nanocontainers for self-healing coatings, in *Corrosion Protection at the Nanoscale, Micro and Nano Technologies* (2020), pp. 379–402

9. A. Gavrilović-Wohlmuther, A. Laskos, E. Kny, Corrosion inhibitor–loaded smart nanocontainers, in *Corrosion Protection at the Nanoscale, Micro and Nano Technologies* (2020), pp. 203–223
10. P. Raja, M. Assad, M. Ismail, Inhibitor-encapsulated smart nanocontainers for the controlled release of corrosion inhibitors, in *Corrosion Protection at the Nanoscale, Micro and Nano Technologies* (2020), pp. 91–105
11. M. Shokouhfar, S. Allahkaram, M. Omidi, Nanostructure and nanocomposite MAO coatings and their corrosion properties, in *Corrosion Protection at the Nanoscale, Micro and Nano Technologies* (2020), pp. 423–449
12. M. Anjum, H. Ali, W. Khan, J. Zhao, G. Yasin, Metal/metal oxide nanoparticles as corrosion inhibitors, in *Corrosion Protection at the Nanoscale, Micro and Nano Technologies* (2020), pp. 181–201
13. J. Praveena, R. Rathish, S. Rajendran, S.S. Kumaran, G. Singh, A. Al-Hashem, Corrosion inhibition by self-assembling nanofilms, in *Corrosion Protection at the Nanoscale, Micro and Nano Technologies* (2020), pp. 107–125
14. F. Kazemi, S. Naghib, Smart controlled release of corrosion inhibitor from normal and stimuli-responsive micro/nanocarriers, in *Corrosion Protection at the Nanoscale, Micro and Nano Technologies* (2020), pp. 161–179
15. A.S. Fouda, H. Megahed, Z.M. Mohamed, M. Shaker, Protection of corrosion of carbon steel with core-shell inorganic magnetic nanogels in formation water. Int. J. Electrochem. Sci. **11**, 8950–8963 (2016). https://doi.org/10.20964/2016.11.57
16. J. Jeevanandam, A. Barhoum, Y.S. Chan, A. Dufresne, M.K. Danquah, Review on nanoparticles and nanostructured materials: history, sources, toxicity and regulations. Beilstein J. Nanotechnol. **9**, 1050–1074 (2018)
17. H. Nalwa, *Handbook of Nanostructured Materials and Nanotechnology*, Vol. 1, Synthesis and processing, Academic press, San Diego, 2000)
18. P. Staron, A. Schreyer, H. Clemens, S. Mayer, From Fundamentals to Applications, in *Neutrons and Synchrotron Radiation in Engineering Materials Science* (Wiley-VCH Verlag GmbH & Co. KGaA, 2017), pp. 3–20
19. H. Gleiter, Nanostructured materials: basic concepts and microstructure. Acta Mater. **48**, 1–29 (2000)
20. X. Ma, L. Xu, W. Wang, Z. Lin, X. Li, Synthesis and characterisation of composite nanoparticles of mesoporous silica loaded with inhibitor for corrosion protection of Cu-Zn alloy. Corros. Sci. **120**, 139–147 (2017). https://doi.org/10.1016/j.corsci.2017.02.004
21. A. Chenan, R.P. George, K.M. Mudali, 2-Mercaptobenzothiazole-loaded hollow mesoporous silica-based hybrid coatings for corrosion protection of modified 9Cr-1Mo ferritic steel. Corrosion **70**(5), 496–511 (2014). https://doi.org/10.5006/1090
22. L.M. Calado, M.G. Taryba, M.J. Carmezim, M.F. Montemor, Self-healing ceria-modified coating for corrosion protection of AZ31 magnesium alloy. Corros. Sci. **142**, 12–21 (2018). https://doi.org/10.1016/j.corsci.2018.06.013
23. Y. Feng, Y. Cheng, An intelligent coating doped with inhibitor-encapsulated nanocontainers for corrosion protection of pipeline steel. Chem. Eng. J. **315**, 537–551 (2017)
24. P. Nguyen-Tri, T. A. Nguyen, P. Carriere, C.N. Xuan, Nanocomposite coatings: preparation, characterization, properties, and applications. Int. J. Corros. **19** (2018). https://doi.org/10.1155/2018/4749501
25. R.H. Fernando, Nanocomposite and nanostructured coatings: Recent advancements, in *ACS Symposium Series*, **1008**, (American Chemical Society, 2009), pp. 2–21
26. B. Fotovvati, N. Namdari, A. Dehghanghadikolaei, Manufacturing and materials processing on coating techniques for surface protection: A review. https://doi.org/10.3390/jmmp3010028
27. A. Singh, S. Mittal, D. Mudgal, P. Gupta, Design, development and application of nanocoatings, in *Nanomaterials and Their Applications. Advanced Structured Materials*, Vol. **84**, ed. by Z. Khan (2018), pp. 191–207
28. A.E. Hughes, I. Cole, T.H. Muster, R.J. Varley, Designing green, self-healing coatings for metal protection nano-scale sensing and material responses. NPG Asia Mater. **2**, 143–151 (2010). https://doi.org/10.1038/asiamat.2010.136

29. I.B. Obot, S.A. Umoren, A.S. Johnson, Sunlight-mediated synthesis of silver nanoparticles using honey and its promising anticorrosion potentials for mild steel in acidic environments. J. Mater. Environ. Sci. **4**, 1013–1018 (2013)
30. L.M. Rivera-Grau, J.G. Gonzalez-Rodriguez, L. Martinez, Effect of hydroxyethil imidazoline and Ag nanoparticles on the CO_2 Corrosion of Carbon Steel. Int. J. Electrochem. Sci. **11**, 80–94 (2016)
31. M.M. Solomon, H. Gerengi, S.A. Umeron, N.B. Essien, U.B. Essien, E. Kaya, Gum Arabic-silver nanoparticles composite as a green anticorrosive formulation for steel corrosion in strong acid media. Carbohyd. Polym. **181**, 43–55 (2017). https://doi.org/10.1016/j.carbpol.2017.10.051
32. A.M. Atta, G.A. El-Mahdy, H.A. Al-Lohedan, S.A. Al-Hussain, S. Arabia, Corrosion inhibition of mild steel in acidic medium by magnetite myrrh nanocomposite. Int. J. Electrochem. Sci. **9**, 8446–8457 (2014)
33. A.M. Atta, H.A. Al-Lohedan, A.M. Atta, G.A. El-Mahdy, H.A. Al-Lohedan, S.A. Al Hussain, Corrosion inhibition of nanocomposite based on acrylamide copolymers/magnetite for steel. Digest J. Nanomaterials Biostructures **9**, 627–639 (2014).
34. A.M. Atta, G.A. El-Mahdy, H.A. Al-Lohedan, A.O. Ezzat, Synthesis and application of hybrid polymer composites based on silver nanoparticles as corrosion protection for line pipe steel. Molecules **19**, 6246–6262 (2014). https://doi.org/10.3390/molecules19056246
35. G.A. El-Mahdy, A.M. Atta, H.A. Al-Lohedan, A.M. Tawfeek, A.A. Abdel-Khalek, Synthesis of encapsulated titanium oxide sodium 2-Acrylamido-2-methylpropan sulfonate nanocomposite for preventing the corrosion of steel. Int. J. Electrochem. Sci. **10**, 5702–5713 (2015)
36. E.A. Essien, D. Kavaz, S. Umoren, E.A. Essien, E.B. Ituen, S.A. Umoren, Synthesis, characterization and anticorrosion property of olive leaves extract-titanium nanoparticles composite. J. Adhes. Sci. Technol. **32**, 1773–1794 (2018). https://doi.org/10.1080/01694243.2018.1445800
37. M. Srivastava, S. Srivastava, G. Ji, R. Prakash, Chitosan based new nanocomposites for corrosion protection of mild steel in aggressive chloride media. Int. J. Biol. Macromol. **140**, 177–187 (2019)
38. K.S. Aneja, S. Bohm, A.S. Khanna, H.L.M. Bohm, Graphene based anticorrosive coatings for Cr(VI) replacement. Nanoscale **7**, 17879–17888 (2015). https://doi.org/10.1039/c5nr04702a
39. X. Zhao, G. Meng, F. Han, X. Li, B. Chen, Q. Xu, X. Zhu, Z. Chu, M. Kong, Q. Huang, Nanocontainers made of various materials with tunable shape and size. Sci. Rep. **3** (2013). https://doi.org/10.1038/srep02238
40. R. Noiville, O. Jaubert, M. Gressier, J. Bonino, P. Taberna, Ce(III) corrosion inhibitor release from silica and boehmite nanocontainers. Mater. Sci. Eng. B Solid-State Mater. Adv. Technol. **229**, 144–154 (2018). https://doi.org/10.1016/j.mseb.2017.12.026
41. J.M. Falcón, F.F. Batista, I.V. Aoki, Encapsulation of dodecylamine corrosion inhibitor on silica nanoparticles. Electrochim. Acta **124**, 109–118 (2014). https://doi.org/10.1016/j.electacta.2013.06.114
42. M. Yeganeh, M. Omidi, S.H.H. Mortazavi, A. Etemad, M.H. Nazari, S.M. Marashi, Application of mesoporous silica as the nanocontainer of corrosion inhibitor, in *Corrosion Protection at the Nanoscale* (Elsevier, 2020), pp. 275–294
43. I.A. Kartsonakis, L.L. Danilidis, G.S. Pappas, G.C. Kordas, Encapsulation and release of corrosion inhibitors into titania nanocontainers. J. Nanosci. Nanotechnol. **10**, 5912–5920 (2010). https://doi.org/10.1166/jnn.2010.2571
44. B.V. Mahesh, R.K.S. Raman, ole of nanostructure in electrochemical corrosion and high temperature oxidation: a review. Metall. Mater. Trans. A **45**, 5799–5822 (2014). https://doi.org/10.1007/s11661-014-2452-5
45. R. Raman, P. Singh, Moderate temperature oxidation protection using nanocrystalline structures, Saji, Viswanathan and Ronald Cook, in *Corrosion Protection and Control Using Nanomaterials* (Woodhead Publishing, 2012), pp. 129–145
46. D. Kopeliovich, Advances in manufacture of ceramic matrix composites by infiltration techniques, in *Advances in Ceramic Matrix Composites: Second Edition* (Elsevier Inc., 2018), pp. 93–119

47. D. Young, *High Temperature Oxidation and Corrosion of Metals* (Elsevier, 2008)
48. Z.B. Wang, N.R. Tao, W.P. Tong, J. Lu, K. Lu, Diffusion of chromium in nanocrystalline iron produced by means of surface mechanical attrition treatment. Acta Material. **51**, 4319–4329 (2003). https://doi.org/10.1016/S1359-6454(03)00260-X
49. N. Birbilis, J. Zhang, R. Gupt, Oxidation resistance of nanocrystalline alloys, in *Corrosion Resistance* (InTech, Rijeka, 2012)
50. K.A. Darling, R.N. Chan, P.Z. Wong, J.E. Semones, R.O. Scattergood, C.C. Koch, Grain-size stabilization in nanocrystalline FeZr alloys. Scripta Materlialia **59**, 530–533 (2008). https://doi.org/10.1016/j.scriptamat.2008.04.045
51. W. Gao, Z. Li, Y. He, High temperature oxidation protection using nanocrystalline coatings, in *Corrosion Protection and Control Using Nanomaterials* (Elsevier, 2012), pp. 146–166
52. G. Chen, H. Lou, Effect of γ' precipitation on oxide formation on the Ni-3Cr-20Al and Ni-10Cr-11Al-8Ti nanocrystalline coatings. Corros. Sci. **42**, 1185–1195 (2000). https://doi.org/10.1016/S0010-938X(99)00139-0
53. G.R. Wallwork, A.Z. Hed, Some limiting factors in the use of alloys at high temperatures. Oxid. Met. **3**, 171–184 (1971). https://doi.org/10.1007/BF00603485
54. V.S. Saji, J. Thomas, Nanomaterials for corrosion control. Curr. Sci. **92**, 51–55 (2007). https://doi.org/10.2307/24096821
55. L. Liu, Y. Li, F. Wang, The effect of micro-structure on the oxidation behavior of a Ni-based superalloy in water vapor plus oxygen. Mater. Lett. **62**, 4081–4084 (2008). https://doi.org/10.1016/j.matlet.2008.05.068
56. T.N. Sankara Narayanan, Nanocoatings to improve the tribocorrosion performance of materials, in *Corrosion Protection and Control Using Nanomaterials, Series in Metals and Surface Engineering* (Woodhead Publishing , 2012), pp. 167–212
57. D. Landolt, S. Mischler, M. Stemp, Electrochemical methods in tribocorrosion: a critical appraisal. Electrochim. Acta **46**, 3913–3929 (2001). https://doi.org/10.1016/S0013-4686(01)00679-X
58. Deepika, Nanotechnology implications for high performance lubricants, **2**, 1128 (2020). https://doi.org/10.1007/s42452-020-2916-8
59. X. Li, Z. Cao, Z. Zhang, H. Dang, Surface-modification in situ of nano-SiO_2 and its structure and tribological properties. Appl. Surf. Sci. **2**, 1128–1139 (2020)
60. R. Asmatulu, Nanocoatings for corrosion protection of aerospace alloys, in *Corrosion Protection and Control Using Nanomaterials* (Elsevier, 2012), pp. 357–374
61. R. Asmatulu, R.O. Claus, J.B. Mecham, S.G. Corcoran, Nanotechnology-associated coatings for aircrafts. Mater. Sci. **43**, 415–422 (2007). https://doi.org/10.1007/s11003-007-0047-7
62. J. Mathew, J. Joy, S.C. George, Potential applications of nanotechnology in transportation: a review. J. King Saud Univ. Sci. **31**, 586–594 (2019). https://doi.org/10.1016/j.jksus.2018.03.015
63. A.G. Evans, D.R. Mumm, J.W. Hutchinson, G.H. Meier, F.S. Pettit, Mechanisms controlling the durability of thermal barrier coatings. Prog. Mater Sci. **46**, 505–553 (2001)
64. M. Shourgeshty, M. Aliofkhazraei, and M.M. Alipour, Introduction to high-temperature coatings, in *High Temperature Corrosion*, ed. by Zaki Ahmad, (InTech, Rijeka, 2016).https://doi.org/10.5772/64282
65. R. Asmatulu, S. Revuri, Synthesis and characterization of nanocomposite coatings for the prevention of metal surfaces, SAMPE Fall Technical Conference, Memphis, TN, September 8–11, 2008, pp. 13
66. R. Asmatulu, G.A. Mahmud, C. Hille, H.E. Misak, Effects of UV degradation on surface hydrophobicity, crack, and thickness of MWCNT-based nanocomposite coatings. Prog. Org. Coat. **72**, 553–561 (2011). https://doi.org/10.1016/j.porgcoat.2011.06.015
67. V. Kumar, N. Sharma, S.S. Maitra, In vitro and in vivo toxicity assessment of nanoparticles, **7**, 243–256 (2017).https://doi.org/10.1007/s40089-017-0221-3
68. S. Sabella Randy, P. Carney, V. Brunetti, M.A. Malvindi, N. Al-Juffali, G. Vecchio, S.M. Janes, O.M. Bakr, R. Cingolani, F. Stellacci, P.P. Pompa, A general mechanism for intracellular toxicity of metal-containing nanoparticles. Nanoscale **6**, 7052–7061 (2014). https://doi.org/10.1039/C4NR01234H

69. Y.T. Zhu, X.L. Wu, Ductility and plasticity of nanostructured metals: differences and issues. Mater. Today Nano **2**, 15–20 (2018). https://doi.org/10.1016/j.mtnano.2018.09.004
70. M. Sarno, D. Scarpa, A. Senatore, W. Ahmed Abdalglil Mustafa, rGO/GO nanosheets in tribology: from the state of the art to the future prospective. Lubricants **8**, 31 (2020).https://doi.org/10.3390/lubricants8030031

Chapter 6
Application of Surface-Modified Electrode Materials in Wastewater Treatment

Onoyivwe Monday Ama, Khotso Khoele, Penny Poomani Govender, and Suprakas Sinha Ray

Abstract Photo-active metal oxides (PMOs) have outstanding physical and chemical properties which are ideal to disintegrate wastewater pollutants. Titanium oxide (TiO_2) initially found popularity in wastewater treatment. TiO_2 utilization on wastewater degradation was attributed particularly to its wider bandgap. Nonetheless, TiO_2 retains antibacterial activity during the application and that renders it to rapid recombination of photogenerated electron–hole pairs. In a subsequent search of alternative PMO, zinc oxide (ZnO) was obtained, and it was found to have a wider bandgap equivalent to that of TiO_2. However, ZnO suffers from photo-corrosion and poor response to visible light. This rigorously proved that an application of a single PMO leads to both inefficiency and ineffectiveness in wastewater treatment. This phenomenon necessitates the hybridization of photocatalysts and improvement of their surface properties.

The present chapter details organic pollutants which are found in wastewater and the methods which are used to remove them from the wastewater. Further discussions are made intensively on photocatalysis and advance oxidation methods. Furthermore, photocatalysts and their advancements are clearly stated and elaborated. Finally, surface-modified photoanodes and their applications using the photoelectrochemical technique have been thoroughly explained. From overall analyses, several deductions have been documented:

- POMs on their singular existence are packed with pros and cons, and that makes wastewater treatment dynamic.
- AOMs, particularly the photoelectrochemical technique, are worthwhile for the degradation of wastewater pollutants.

K. Khoele
Department of Chemical, Metallurgical and Materials Engineering, The Tshwane University of Technology, Pretoria, South Africa

O. M. Ama (✉) · P. P. Govender · S. S. Ray
Department of Chemical Science, University of Johannesburg, DoornfonteinJohannesburg 2028, South Africa

S. S. Ray
DST-CSIR National Center for Nanostructured Materials Council for Scientific and Industrial Research, Pretoria 0001, South Africa

O. M. Ama et al. (eds.), *Modified Nanomaterials for Environmental Applications*, Engineering Materials, https://doi.org/10.1007/978-3-030-85555-0_6

- Factors that affect degradation processes on wastewater pollutants include light captivation properties, reduction and oxidation rates on the surface by the photogenerated electrons and holes and a recombination rate of such charges.
- Surface modification of photoanodes is carried through nanostructured materials, the addition of metals particularly noble ones such as gold (Au), silver (Ag), platinum (Pt), and through the use of novel titanium alloys and cubic double-perovskite.

6.1 Introduction

The rapid increase of organic pollutants (OPs) in wastewater is linked to various health hazards and environmental pollution. Major OPs found in wastewater are from azo dyes, and 15% of their annual production goes into the wastewater [1–5]. Hazardous minerals from organic pollutants such as sodium, potassium, calcium, chloride, and bromide have been reported to be recalcitrant during wastewater treatment, and their direct contact with living species leads to various chronic illnesses [6–15]. Hence, it is a matter of urgency to remove organic pollutants from the wastewater.

Photo-active materials (PMs), particularly in a form of metal oxides, have various significances [16, 17] which make them able to degrade wastewater pollutants. Some of the predominantly utilized PMs are ZnO, CU_2O, MnO_2, $Ag_2O/Ag_3VO_4/AgVO_3$, TiO_2, Fe_2O_3, NiO, $SrZrO_3$, $BiOI/Ag_3VO_4$, MoS_2, ZnS and WS_2 [18–28]. In recent applications, PMs are synthesized from nanostructured materials through rigorous steps. Typically, PMs are fabricated to form a conductive photoanode that directly oxidizes and degrade wastewater pollutants from the wastewater using electrochemistry techniques.

Presently, the photoelectrochemical technique (PT) is regarded as a superior electrochemistry method for the degradation of wastewater pollutants. The utilization of PT on degradation goes along with various photoanodes. Typically, PT is engaged in wastewater treatment through the aid of bias potential which is supplied from the potentiostat. Herein, photoanode is immobilized and hydroxyl species form from occurring reactions. Ultimately, electrons from the photoanode react with some oxygen, and that consequently develop oxygen reactive radicals ($O_2{}^-$) which react with protons to produce H_2O^-. Both $O_2{}^-$ and H_2O^- are then able to degrade wastewater pollutants [29, 30].

In light of the wastewater pollutants' nature and their dynamic processes, this chapter elucidates factors that affect degradation processes on wastewater pollutants. Three factors that generally affect degradation processes have been discussed. Most importantly, surface modifications of different photoanodes which make in-roads on wastewater treatment are detailed.

Fig. 6.1 Different organic pollutants consisted in wastewater under treatment [32]

6.2 Wastewater and Organic Pollutants

6.2.1 Wastewater

Wastewater comprises organic pollutants from a wide range of applications: domestic, industrial, commercial, and agricultural activities. Most notably, wastewater consists of about 15% of annually produced organic dyes [31, 32] and some other pollutants. Figure 6.1 demonstrates the wastewater profile consists of different organic pollutants.

6.2.2 Organic Pollutants and Their Impact

Dyes are used mainly in the production of food, textiles, leather, fabrics, paper, cosmetic, electroplating, distillation, and pharmaceutical products [33–40]. With global industrialization, the uses of dyes increase at an alarming rate. The presence of pollutants in wastewater causes major problems on wastewater treatment and leads to environmental pollution and hazardous to living species [41–44]. In particular, hazardous minerals from organic pollutants such as sodium, potassium, calcium, chloride, and bromide are recalcitrant to wastewater treatment, and their direct contact with living species could lead to chronic diseases [45].

6.3 Photocatalysists and Their Applications

6.3.1 Photocatalysts

PMs are capable of captivating light under either direct sunlight or a solar application. Due to their nature, PMs are commonly known as photocatalysts. There are quite a several photocatalysts, and some of the heavily engaged photocatalysts include ZnO, CU_2O, MnO_2, $Ag_2O/Ag_3VO_4/AgVO_3$, TiO_2, Fe_2O_3, NiO, $SrZrO_3$, $BiOI/Ag_3VO_4$, MoS_2, ZnS and WS_2 [18–28]. Among other significant characteristics listed photocatalysts and the likes, photocatalysts generally possess bandgaps which enable them to possess a high level of photodegradation to remove pollutants from the wastewater.

6.3.2 Nanostructured Materials and Photoanodes

6.3.2.1 Nanostructured Materials

Nanotechnology has recently found vast application across the world. Among the most prominent applications, nanotechnology is used on medical equipment, car paintings, chronic illness treatment, etc. In wastewater treatment, nanotechnology is applied in a form of nanostructured materials. Nanostructured materials are commercially purchased as nanoparticles (powders less than 100 nm). From their purchases to laboratories, they are molded into conductive semiconductors which are popularly termed photoanodes.

6.3.2.2 Photoanodes

Photoanodes are fabricated from well-mixed nanostructured materials [37] in five major steps which are shown in Fig. 6.2. Firstly, nanostructured composite is taken to a hydraulic press, and it is subjected to higher temperatures. Secondly, the compressed pellet of about 1.3 cm diameters is obtained as a photoanode [38]. Thirdly, a copper wire is inserted and passed out through a glass tube with an opening at both ends and coiled onto the fabricated nanocomposite pellet. Fourthly, conduction between the Cu wire and the pellet is created with the use of conductive silver paste. Finally, epoxy resin is applied to the glass tube to cover up all the ends.

6.3.2.3 Coupling of Photoanode to an Electrochemical Cell for Wastewater Pollutants Degradation

Besides an electrolyte which simulates wastewater pollutant, an electrochemical cell is comprised of three conductive electrodes. The simulated electrolyte is mostly made

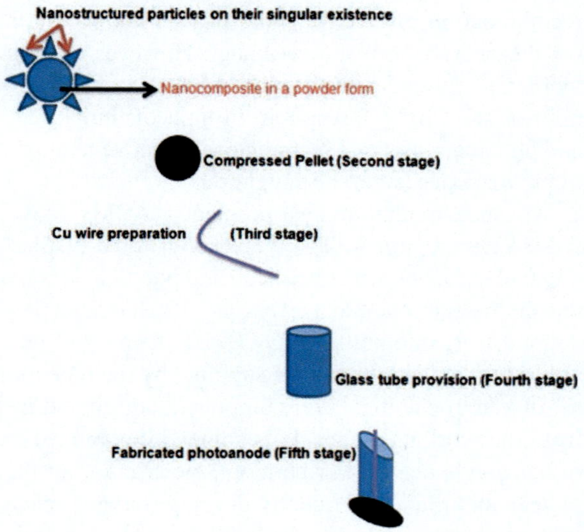

Fig. 6.2 Preparation of photoanode from nanostructured particles

up of potassium ferricyanide, ferrocyanide, and potassium chloride. The incompleteness of wastewater pollutant, sodium sulfate, sodium hydroxide, and hydrochloric acid are added as complementary solutions to make wastewater pollutant conductive [45]. The photoanodes is popularly known as the working electrode (WE) as it is the one degrading organic pollutants from the wastewater. Then, there is a reference electrode (RE), and it is specifically used to measure the potential difference produced from the electrochemical cell. Lastly, the counter electrode (CE) is used to count a current that is produced as the photoanode is applied to wastewater pollutant (electrolyte).

6.4 Photocatalysis and Advance Oxidation Measurements

Photocatalysis is a fundamental technique in both wastewater treatment and renewable energy systems. Photocatalysis is carried out both spontaneously and synthetically. Naturally, direct sunlight is exposed to the photoanode, and consequently, hydrogen and oxygen are produced, and they are used to degrade wastewater pollutants [46]. Synthetically, however, the light from solar light is directed to the PMs, and that produces electrical energy. Relative to sunlight utilization, this method is comprised of insufficient energy, difficult storage and transportation [46].

Although all photo-active materials ideal for wastewater degradation, their singular existence is packed with distinctive limitations and that makes their uses on wastewater treatment to be dynamic. For instance, TiO_2 has a wider bandgap and that makes it provide inefficient degradation processes. On the other hand, ZnO suffers from photo-corrosion, has a poor response to visible light, and has higher recombination of electron–hole pairs. Photocatalysis has been a standard method for the degradation of wastewater. However, photocatalysis has been reported to inefficient due to an occurrence of higher recombination of electron–hole pairs and inferior solar light absorption. In light of limitations, singular photocatalysts have and photocatalysis as a technique, alternative methods which work on hybridization of photoanodes had to be sought out.

Advance oxidation measurements (AOMs) have so far been worthwhile on wastewater treatment. There are about three distinct AOMs for wastewater treatment: electrochemical, chemical, and photoelectrochemical technique. Nonetheless, the photoelectrochemical technique (PT) has been proven to be superior in the degradation of organic pollutants [47]. PT is engaged in wastewater treatment with the aid of bias potential which is supplied by the potentiostat. As the potential on a full set of electrochemical cell components discussed in Sect. 3.2.3, a nanocomposite from the working electrode becomes immobilized and that results in a formation of hydroxyl species. Furthermore, electrons from the photoanode react with some oxygen and that consequently develop (oxygen reactive radicals) which react with protons to produce H_2O^-. Both $O_2{}^-$ and H_2O^- are then able to degrade wastewater pollutants [31].

6.5 Factors that Affect Degradation Processes on Wastewater Pollutants Using a Photoelectrochemical Technique

There are about three factors that affect degradation processes on wastewater pollutants, and they are light captivation properties, reduction and oxidation rates on the surface by the photogenerated species, and a recombination rate of such charges.

6.5.1 Light Captivation Properties

PMs absorb and transmit light differently, and they are normally characterized using UV measurements to dictate light absorption of a certain photo-active material within a specific wavelength range. As can be seen in Fig. 6.3, three photoanodes exhibit different curves on the absorption of light. Cautiously, photo-active materials have different significances and limitations. In the hybridization of photoanodes, improved light captivation is obtained. Significantly, faster degradation processes are also reached under the hybridization of photoanodes [48].

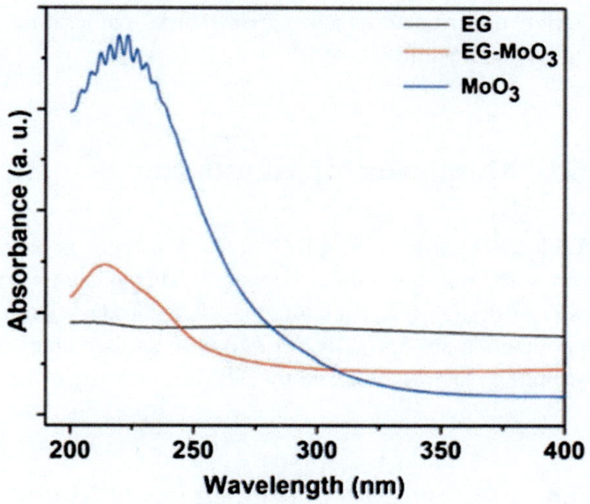

Fig. 6.3 UV–Vis spectra absorbance for three photoanodes: EG, EG-MoO_3, and MoO_3 [48]

6.5.2 The Reduction and Oxidation Rates on the Surface by the Photo-generated Electrons and Holes

In applications of PET, a composite from the photoanode is oxidized (electrons are emitted from it), and then, the wastewater pollutant, in a form of liquid, is degraded. The faster the electrons emission from the photoanode, the higher the wastewater pollutant is being degraded (reduced). This is the most important factor which is particularly examined from the photoanode material [49].

6.5.3 Recombination Rate of Electron Charges

As mentioned earlier that an extreme recombination rate of electrons occurs particularly from TiO_2, most surface modification has been carried on TiO_2 [15, 24, 29, 39, 41, 42]. With such modifications, better degradation efficiencies have been reached.

6.6 Surface-Modified Photoanodes

Surface modification of photoanodes is mainly carried out in three ways. The first step is on morphological adjustments. Herein, nanostructured materials are added particularly in the form of noble metals such as Au, Ag, and Pt. Furthermore, novel titanium alloys and cubic double-perovskite are sometimes used. In the second step, synthetic techniques are varied on different novel materials. Finally, nanostructured materials in the forms of nanotubes, nanorods, nanowire, and nanoclusters are applied in a variety of parameters such as growth time, temperature, initial reactant concentration, acidity, and additives [61].

6.7 Morphological Adjustments

With the hybridization of photoanodes, there is a particular form of structure in terms of morphologies and phases. This phenomenon leads to a new characteristic of the fabricated photoanode. The surface characteristics on newly fabricated photoanode are normally verified with surface characterization techniques such as scanning electron microscopy (SEM) and X-ray diffraction (XRD).

6.8 Synthetic Technique on Novel Materials

PT being reckoned as the most efficient technique on wastewater treatment, its outstanding application has been corroborated by a considerable number of studies on improvement of limitations possessed by major photoanodes (TiO_2 and ZnO). Firstly, Chakrabarti et al. [62] combined both TiO_2 and ZnO to improve degradation efficiency on methylene blue (MB). An enhanced degradation efficiency was recorded to be 99.41% within 3 h of measurements. With the addition of novel material, Mirzaei et al. [63] established TiO_2 nanotube arrays enfolded with g-C_3N_4 to degrade Phenol. Enhanced degradation efficiency of about 90% was obtained from the TiO_2/g-C_3N_4 photoanode. Similarly, Ahmed et al. [64] invented A/RTiO2251 photoanode to degrade orange II dye, and 96% degradation efficiency was obtained. This, therefore, proves that the application of PT with an appropriately hybridized photoanode, on specific wastewater pollutant, provides improved degradation processes.

6.9 Variation of Synthetic Process Parameters

Synthetic parameters could positively or negatively affect degradation processes. Some of the most crucial synthetic parameters affecting degradation processes are listed below.

6.9.1 Temperature

Generally, nanostructured materials are affected by temperature treatment [65]. Synthetic processes which form photoanode composites have been reported to go with changes in phases and crystalline structures. This has been attributed to temperature, particularly higher temperatures.

6.9.2 pH Range

An impact of pH on degradation processes was initially profiled on TiO_2 as the standard photoanode. From some analyses, a similar temperature range has been realized to negatively affect degradation processes. For instance, the pH range between 3 and 10 corresponds was found to produce higher degradation efficiency in one study. In attest from the other study, TiO_2 has been reported to agglomerates and possesses reduced surface area under acidic conditions. Therefore, a particular effect on the surface area of the photoanode is due to the pH range. Pinpointing the effect of pH range on degradation processes, Baran et al. [66] lowered the pH from 8.0 to 4.5 during their degradation processes. Significantly, magnificent degradation efficiency was attained at the lower pH values. Moreover, from the other similar work [67], higher degradation efficiency was also attained from lower pH values. This, therefore, proves that acidic pH is more favorable on most photoanodes for effective degradation processes.

6.10 Summary

6.10.1 Wastewater Pollutants and Their Impacts

- Wastewater pollutants come mainly from domestic, industrial, commercial, and agricultural activities.
- Major wastewater pollutant is organic dyes.
- Wastewater pollutants cause environmental pollution and chronic illnesses on people and animals.

6.10.2 Photocatalysis and Advanced Oxidation Methods

- The photoelectrochemical technique is a state-of-the-art electrochemistry technique for the degradation of wastewater pollutants and the enhancement of renewable energy systems.
- Photoelectrochemical technique engages an immobilization of photoanode electrode through an application of bias potential.

6.10.3 Factors that Affect Degradation Processes on Wastewater Pollutants Using a Photoelectrochemical Technique

There are mainly three factors that affect degradation processes on wastewater pollutants, and these are light captivation properties, reduction and oxidation rates on the surface by the photogenerated species, and a recombination rate of electron charges during electrochemical measurements.

6.10.4 Methods Modifying Surface Properties of Photoanodes and Their Applications

Surface modification of photoanodes is mainly carried out through morphological adjustments, synthetic technique on novel material, and variation of synthetic process's parameters.

6.11 Conclusions

The present chapter has outlined organic pollutants which are found in wastewater and methods which are used to remove them from the wastewater. Further discussions have been made on photocatalysis and advance oxidation methods. Photocatalysts and their advancements have also been clearly stated and elaborated. Most importantly, surface-modified photoanodes and their applications using the photoelectrochemical technique have carefully been explained. Overall, the following conclusions are drawn up from the main findings:

- Photo-active materials (POMs) in their singular existence are packed with advantages and limitations and that makes wastewater treatment to be dynamic.
- Advance oxidation methods (AOMs), particularly photoelectrochemical technique, are worthwhile for the degradation of wastewater pollutants.
- Factors that affect degradation processes on wastewater pollutants include light captivation properties, reduction and oxidation rates on the surface by the photogenerated electrons, and holes and a recombination rate of such charges.
- Variation of synthetic process parameters such as temperature and ph can either be positive or be negative on degradation processes. Higher temperatures and higher pH ranges are toxic to degradation processes.
- Surface modification of photoanodes is carried through nanostructured materials, the addition of metals particularly noble ones such as gold (Au), silver (Ag), platinum (Pt), and using novel titanium alloys and cubic double-perovskite.

References

1. R. Saravanan, V.K. Gupta, E. Mosquera, F. Gracia, V. Narayanan, A. Stephen, Visible light-induced degradation of methyl orange using β-Ag0. $333V_2O_5$ nanorod catalysts by facile thermal decomposition method. J. Saudi Chem. Soc. **19**(5), 521–527 (2015)
2. P. Li, G. Zhao, K. Zhao, J. Gao, T. Wu, An efficient and energy-saving approach to photocatalytic degradation of opaque high-chroma methylene blue wastewater by electrocatalytic pre-oxidation. Dyes Pigm. **92**(3), 923–928 (2012)
3. M.A. Mahadik, G.W. An, S. David, S.H. Choi, M. Cho, J.S. Jang, Fabrication of A/R-TiO_2 composite for enhanced photoelectrochemical performance: solar hydrogen generation and dye degradation. Appl. Surf. Sci. **426**, 833–843 (2017)
4. X. Yuan, J. Yi, H. Wang, H. Yu, S. Zhang, F. Peng, New route of fabricating BiOI and Bi_2O_3 supported TiO_2 nanotube arrays via the electrodeposition of bismuth nanoparticles for photocatalytic degradation of acid orange II. Mater. Chem. Phys. **196**, 237–244 (2017)
5. Y.M. Hunge, M.A. Mahadik, S.S. Kumbhar, V.S. Mohite, K.Y. Rajpure, N.G. Deshpande et al., Visible light catalysis of methyl orange using nanostructured WO_3 thin films. Ceram.442 Int. **42**(1), 789–798 (2016)
6. O.M. Ama, K. Khoele, D.J. Delport, S.S. Ray, P.O. Osifo, Synthesis and fabrication of photoactive nanocomposites electrodes for the degradation of wastewater pollutants. Nanostruct. Metal-oxide Electrode Mater. Water Purif. Eng. Mater.
7. L.M. Reid, T. Li, Y. Cao & C.P. Berlinguette, Organic chemistry at anodes and photoanodes, NOPAGES & V (2018)
8. D. Rawat, V. Mishra, R.S. Sharma, Detoxification of Azo dyes in the context of environmental processes. Chemosphere **155**, 591–605 (2016)
9. X.L. He, C. Song, Y.Y. Li, N. Wang, L. Xu, X. Han, D.S. Wei, Efficient degradation of Azo dyes by a newly isolated fungus Trichoderma tomentosum under non-sterile conditions. Ecotoxicol. Environ. Saf. **150**, 232–239 (2018)
10. V.M. Daskalaki, M. Antoniadou, G. Li Puma, D.I. Kondarides, P. Lianos, Solar light-responsive Pt/CdS/TiO_2 photocatalysts for hydrogen production and simultaneous degradation of inorganic or organic sacrificial agents in wastewater. Environ. Sci. Technol. **44**(19), 7200–7205 (2010)
11. C. Yu, Y. Shu, X. Zhou, Y. Ren, Z. Liu, Multi-branched Cu_2O nanowires for photocatalytic degradation of methyl orange. Mater. Res. Exp. **5**(3), 035046 (2018)
12. Q. Zheng, C. Lee, Visible light photoelectrocatalytic degradation of methyl orange using anodized nanoporous WO_3. Electrochim. Acta **115**, 140–145 (2014)
13. N. Chaukura, W. Gwenzi, N. Tavengwa, M.M. Manyuchi, Biosorbents for the removal of synthetic organics and emerging pollutants: opportunities and challenges for developing countries. Environ. Dev. **19**, 84–89 (2016)
14. S. Garcia-Segura, S. Dosta, J.M. Guilemany, E. Brillas, Solar photoelectrocatalytic degradation of acid orange 7 Azo dye using a highly stable TiO_2 photoanode synthesized by atmospheric plasma spray. Appl. Catal. B **132**, 142–150 (2013)
15. N. Lezana, F. Fernández-Vidal, C. Berríos, E. Garrido-Ramírez, Electrochemical and photo-504 electrochemical processes of methylene blue oxidation by Ti/TiO_2 electrodes modified with 505 Fe-allophane. J. Chil. Chem. Soc. **62**(2), 3529–3534 (2017)
16. H. Wu, Z. Zang, Photoelectrochemical water splitting and simultaneous photoelectrocatalytic degradation of organic pollutant on highly smooth and ordered TiO_2 nanotubearrays. J. Solid State Chem. **184**(12), 3202–3207 (2011)
17. R.M. Fernández-Domene, R. Sánchez-Tovar, B. Lucas-granados, M.J. Munoz-Portero, J. García-Antón, Elimination of pesticide atrazine by photoelectrocatalysis using a photoanode based on WO_3 nanosheets Chem. Eng. J. **350**, 1114–1124 (2018)
18. X. Meng, Z. Zang, X. Li, Synergetic photoelectrocatalytic reactors for environmental remediation: a review. J. Photochem. Photobiol. C **24**, 83–101 (2015)

19. S. Garcia-Segura, S. Dosta, J.M. Guilemany, E. Brillas, Solar photoelectrocatalytic degra-397 dation of acid orange 7 Azo dye using a highly stable TiO_2 photoanode synthesized by atmospheric plasma spray. Appl. Catal. B **132**, 142–150 (2013)
20. Y. Huang, H. Cai, D. Feng, D. Gu, Y. Deng, B. Tu et al., One-step hydrothermal synthesis of ordered mesostructured carbonaceous monoliths with hierarchical porosities. Chem.478 Commun. **23**, 2641–2643 (2008)
21. A. Ray. Electrodeposition of thin films for low-cost solar cells, in *Electroplating of Nanostructures* (IntechOpen, 2015)
22. O.J. Ilegbusi, S.N. Khatami, L.I. Trakhtenberg, Spray pyrolysis deposition of single and mixed oxide thin films. Mater. Sci. Appl. **8**(02), 153 (2017)
23. N. Liu, S.P. Albu, K. Lee, S. So, P. Schmuki, Water annealing and other low-temperature treatments of anodic TiO_2 nanotubes: a comparison of properties and efficiencies in dye-sensitized solar cells and for water splitting. Electrochim. Acta **82**, 98–102 (2012)
24. K. Nakata, A. Fujishima, TiO_2 photocatalysis: design and applications. J. Photochem. Photobiol. C **13**(3), 169–189 (2012)
25. S. Li, J. Qiu, M. Ling, F. Peng, B. Wood, S. Zhang, Photoelectrochemical characterization of hydrogenated TiO_2 nanotubes as photoanodes for sensing applications. ACS Appl. Mater.490 Interfaces. **5**(21), 11129–11135 (2013)
26. J.H. Pan, H. Dou, Z. Xiong, C. Xu, J. Ma, X.S. Zhao, Porous photocatalysts for advanced water purifications. J. Mater. Chem. **20**(22), 4512–4528 (2010)
27. C. Fu, M. Li, H. Li, C. Li, X. Guo Wu, B. Yang, Fabrication of Au nanoparticle/TiO_2 hybrid494 filmsfor photo electrocatalytic degradation of methyl orange. J. Alloy. Compd. **692**, 727–733 (2017)
28. C. Fu, M. Li, H. Li, C. Li, X. Guo Wu, B. Yang, Fabrication of Au nanoparticle/TiO_2 hybrid filmsfor photo electrocatalytic degradation of methyl orange. J. Alloy. Compd. **692**, 727–733468 (2017)
29. G.P. Awasthi, S.P. Adhikari, S. Ko, H.J. Kim, C.H. Park, C.S. Kim, Facile synthesis of ZnO flowers modified graphene-like MoS_2 sheets for enhanced visible-light-driven photocatalytic activity and antibacterial properties. J. Alloy. Compd. **682**, 208–215 (2016)
30. Y. Li, C. Ji, Y.C. Chi, Z.H. Dan, H.F. Zhang, F.X. Qin, Fabrication and photocatalytic activity of Cu_2O nanobelts on nanoporous Cu substrate. Acta Metallurgica Sinica (English Letters) 341 **32**(1), 63–73 (2019)
31. H.H. Cheng, S.S. Chen, S.Y. Yang, H.M. Liu, K.S. Lin, Sol-gel hydrothermal synthesis and visible light photocatalytic degradation performance of Fe/N co-doped TiO_2 catalysts. Materials **11**(6), 939 (2018)
32. T. Wright, State reviewing controversial wastewater treatment technique. Environment (2019)
33. M. Vaara, New approaches in peptide antibiotics. Curr. Opin. Pharmacol. **9**(5), 571–576 (2009)
34. L.T. Nguyen, E.F. Haney, H.J. Vogel, The expanding scope of antimicrobial peptide structures and their modes of action. Trends Biotechnol. **29**(9), 464–472 (2011)
35. V. Vega-Sánchez, F. Latif-Eugenín, E. Soriano-Vargas, R. Beaz-Hidalgo, M.J. Figueras, M.G. Aguilera-Arreola, G. Castro-Escarpulli, Re-identification of Aeromonas isolates from rainbow trout and incidence of class1 integronand β-lactamase genes. Vet. Microbiol. **172**(3–4), 528–533 (2014)
36. M. Mahboubi, G. Haghi, Antimicrobial activity and chemical composition of Mentha pulegium L. essential oil. J. Ethnopharmacol. **119**(2), 325–327 (2008)
37. R. Gothwal, T. Shashidhar, Antibiotic pollution in the environment: a review. Clean–Soil, Air, 386 Water **43**(4), 479–489 (2015)
38. H, Wu, Z. Zang, Photoelectrochemical water splitting and simultaneous photoelectrocatalytic degradation of organic pollutant on highly smooth and ordered TiO_2 nanotube arrays. J. Solid State Chem. **184**(12), 3202–3207 (2011)
39. G.P. Awasthi, S.P. Adhikari, S. Ko, H.J. Kim, C.H. Park, C.S. Kim, Facile synthesis of ZnO flowers modified graphene like MoS_2 sheets for enhanced visible-light-driven photocatalytic activity and antibacterial properties. J. Alloy. Compd. **682**, 208–215 (2016)

40. D. Liu, J. Zhou, J. Wang, R. Tian, X. Li, E. Nie et al., Enhanced visible light photoelectrocatalytic degradation of organic contaminants by F and Sn co-doped TiO_2 photoelectrode. 498 Chem. Eng. J. **344**, 332–341 (2018)
41. D. Liu, R. Tian, J. Wang, E. Nie, X. Piao, X. Li, Z. Sun, Photoelectrocatalytic degradation of methylene blue using F doped TiO_2 photoelectrode under visible light irradiation. Chemosphere **185**, 574–581 (2017)
42. D. Cao, Y. Wang, X. Zhao, Combination of photocatalytic and electrochemical degradation of organic pollutants from water. Current Opin. Green Sustain. Chem. **6**, 78–84 (2017)
43. J. Tao, Z. Gong, G. Yao, Y. Cheng, M. Zhang, J. Lv et al., Enhanced photocatalytic and photoelectrochemical properties of TiO_2 nanorod arrays sensitized with CdS nanoplates. Ceram. Int. **42**(10), 11716–11723 (2016)
44. X.D. Li, T.P. Chen, P. Liu, Y. Liu, K.C. Leong, Effects of free electrons and quantum confinement in ultrathin ZnO films: a comparison between undoped and Al-doped ZnO. Opt. Express330 **21**(12), 14131–14138 (2013)
45. G. Li, N.M. Dimitrijevic, L. Chen, T. Rajh, K.A. Gray, J. Phys. Chem. C **112**, 19040–19044 (2003)
46. D. Cao, Y. Wang, X. Zhao, Combination of photocatalytic and electrochemical degradation536 of organic pollutants from water. Curr. Opin. Green Sustain. Chem. **6**, 78–84 (2017)
47. M.E. Osugi, G.A. Umbuzeiro, M.A. Anderson, M.V.B. Zanoni, Degradation of metallophtalocyanine dye by combined processes of electrochemistry and photoelectrochemistry. Electrochim. Acta **50**(25–26), 5261–5269 (2005)
48. O.M. Ama, N. Kumar, F.V. Adams, S.S. Ray, Efficient and cost-effective photoelectrochemical degradation of dyes in wastewater over an exfoliated graphite-MoO_3 nanocomposite electrode. Electrocatalysis (2018)
49. O.M Ama, O.A Arotiba , Exfoliated graphite/titanium dioxide for enhanced photoelectrochemical degradation of methylene blue dye under simulated visible light irradiationJ. Electroanalytical Chemistry, 157-164 (2017)
50. D. Pletcher, R.A. Green, R.C. Brown, Flow electrolysis cells for the synthetic organic chemistry laboratory. Chem. Rev. **118**(9), 4573–4591 (2017)
51. A. Goshadrou, A. Moheb, Continuous fixed bed adsorption of C.I. Acid blue by exfoliated graphite: an experimental and modeling study. Desalination **269**, 170–176 (2011)
52. G.W. An, M.A. Mahadik, W.S. Chae, H.G. Kim, M. Cho, J.S. Jang, Enhanced solar photoelectrochemical conversion efficiency of the hydrothermally-deposited TiO_2 nanorod arrays: Effects of the light trapping and optimum charge transfer. Appl. Surf. Sci. **440**, 688–699 (2018)
53. S. Chakrabarti, B.K. Dutta, Photocatalytic degradation of model textile dyes in wastewater using ZnO as semiconductor catalyst. J. Hazard. Mater. **112**(3), 269–278 (2004)
54. A. Mirzaei, Z. Chen, F. Haghighat, L. Yerushalmi, Removal of pharmaceuticals from water by homo/heterogonous Fenton-type processes–a review. Chemosphere **174**, 665–688 (2017)
55. M.J. Ahmed, B.H. Hameed, Removal of emerging pharmaceutical contaminants by adsorption in a fixed-bed column: a review. Ecotoxicol. Environ. Saf. **149**, 257–266 (2018)
56. T. Krishnakumar, N. Pinna, K.P. Kumara, K. Perumal, R. Jayaprakash, Microwave-assisted synthesis and characterization of tin oxide nanoparticles. Mater. Lett. **62**, 3437–3440 (2008)
57. W. Baran, A. Makowski, W. Wardas, The effect of UV radiation absorption of cationic and anionic dye solutions on their photocatalytic degradation in the presence of TiO_2. Dyes Pigm. **76**, 226–230 (2008)

Chapter 7
Photoelectrochemical Application of Nanomaterials

Seyi Philemon Akanji, Onoyivwe Monday Ama, Omotayo A. Arotiba, Duduzile Nkosi, Idris Azeez Olayiwola, Uyiosa Osagie Aigbe, Robert Birundu Onyancha, and Kingsley Eghonghon Ukhurebor

Abstract Tremendous progress has been made in the aspect of material design, especially in nanomaterials fabrication over the past decades. And as such, the rapid development of nanomaterials through nanotechnology has opened up new avenues for their use in photoelectrochemical applications. Their unique characteristics at nano dimensions compared to their bulk counterparts make them very much attractive. Largely, optical properties and functionalities in emerging technologies, flexibility and semiconducting properties to mention a few make the use of nanomaterials indispensable for photoelectrochemical applications. Owing to this effect, the present chapter deals with the photoelectrochemical (PEC) application of nanomaterials in specific areas such as photoelectrochemical biosensing, hydrogen production via solar water splitting and photovoltaic devices. The chapter also addresses the progress in the utilization of the nanomaterials and illustrates the designs for their integration in biosensors, highly efficient and stable photoelectrode in PEC water splitting for hydrogen production and outstanding smart solar energy harvesting devices for third-generation photovoltaics.

S. P. Akanji (✉) · O. M. Ama · O. A. Arotiba · D. Nkosi
Department of Chemical Science, University of Johannesburg, Doornfontein 2028, South Africa

I. A. Olayiwola
Nanotechnology for Water Sustainability (NaNOWS) Research Unit, College of Science, Engineering and Technology, University of South Africa, Private Bag, Johannesburg 1709, South Africa

U. O. Aigbe
Department of Mathematics and Physics, Faculty of Applied Sciences, Cape Peninsula University of Technology, Cape Town, South Africa

R. B. Onyancha
School of Physical Sciences and Technology, Technical University of Kenya, Nairobi, Kenya

K. E. Ukhurebor
Climatic/Environmental/Telecommunication Unit, Department of Physics, Edo University Iyamho, Okpella, Edo, Nigeria

O. M. Ama et al. (eds.), *Modified Nanomaterials for Environmental Applications*, Engineering Materials, https://doi.org/10.1007/978-3-030-85555-0_7

Graphical Abstract

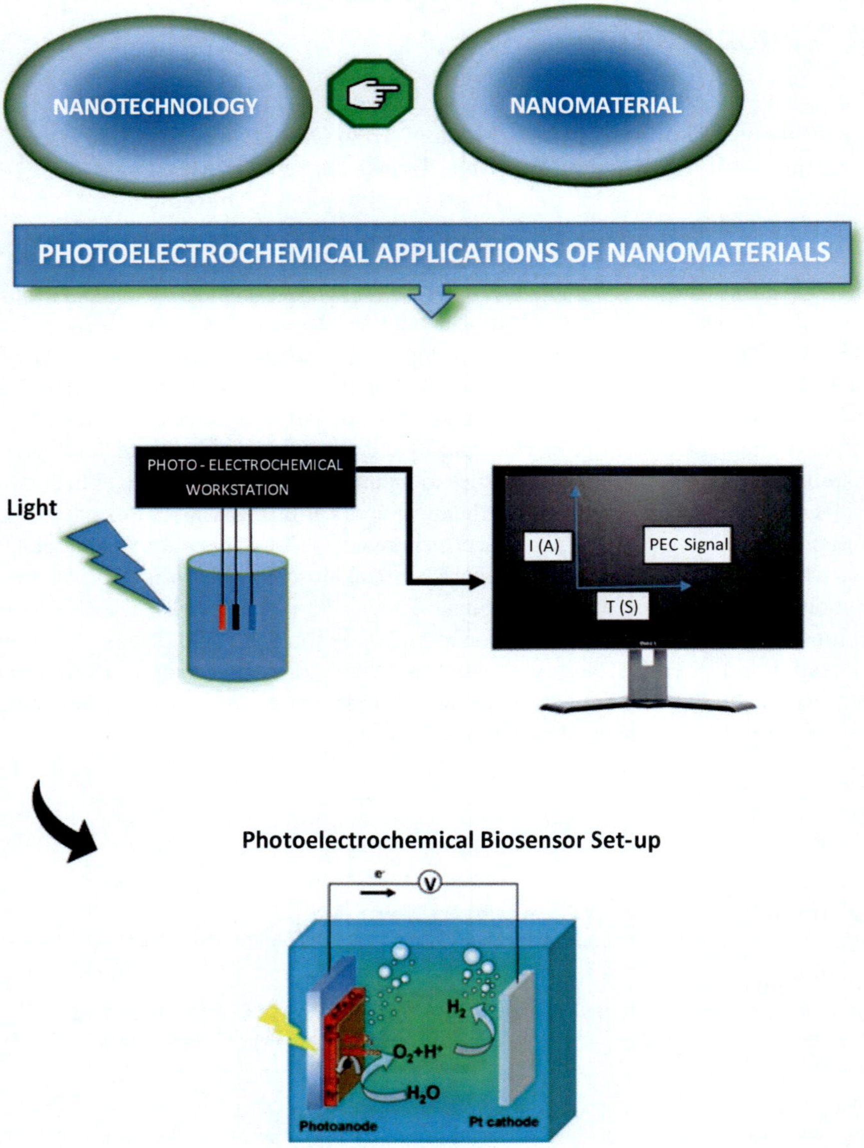

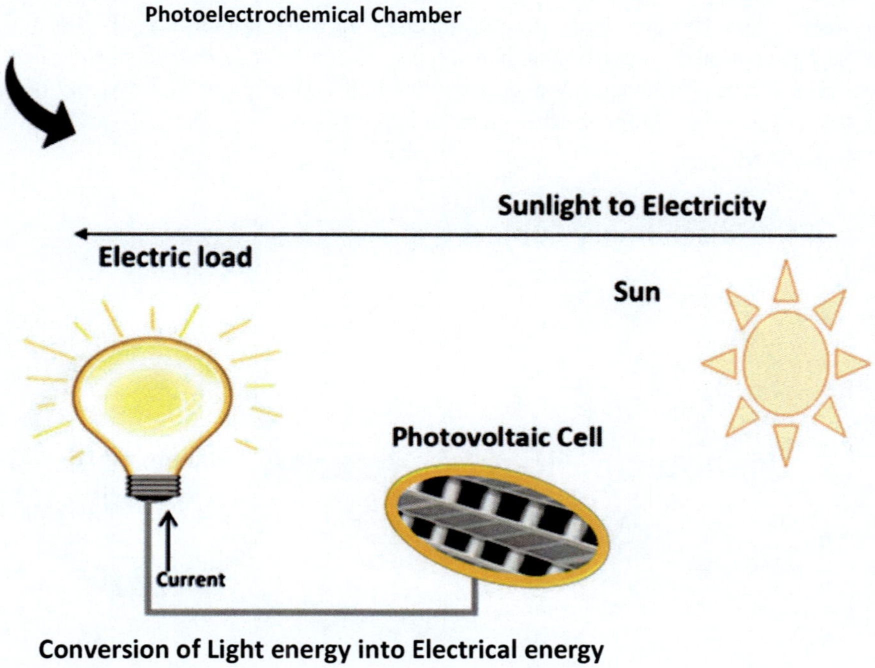

Highlights

The following are discussed in this chapter:

- Brief introduction on the concept of nanotechnology with respect to nanomaterials as well as the useful properties that facilitate their broad range of applications.
- Brief introduction on photoelectrochemical process and its unique advantage.
- Three broad ranges of photoelectrochemical applications of nanomaterials—biosensors (photoelectrochemical biosensors), photoelectrochemical water splitting and photovoltaics where specific application is directed toward the third-generation photovoltaics such as organic photovoltaic cells (OPVs), perovskite solar cells (PSCs) and dye-sensitized solar cells (DSSCs).

Nanotechnology

Synopses I

Nanomaterials came into existence through a multidisciplinary subject termed nanotechnology. Materials in this category are classified as nanomaterials based on convention because they consist of nanoparticles with at least one dimension between 1 and 100 nm. Interestingly, materials in these dimensions (1–100 nm) possess exceptional properties (such as large surface area and electronic properties) which give them preference in usage above their bulk counterparts. These exceptional abilities impart their usage in wide range of applications.

Graphical Abstract 1

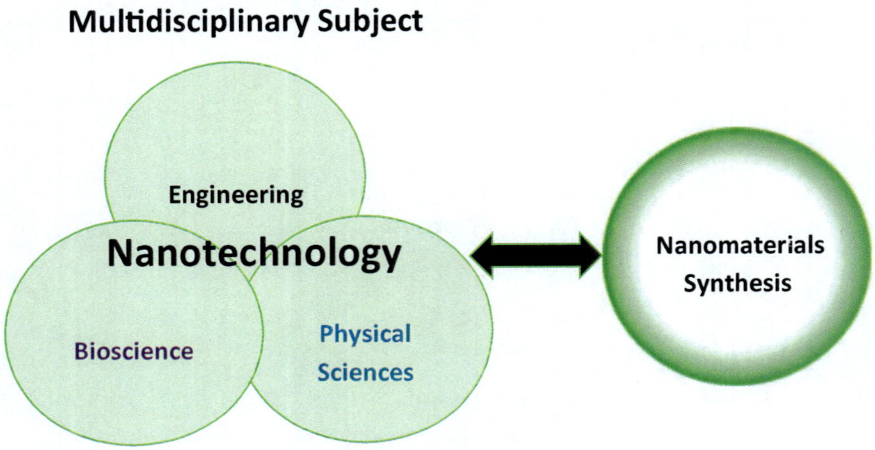

Photoelectrochemical process

Synopses 2

The separation and transfer of charges upon the absorption of photons during illumination normally lead to the conversion of the absorbed photons to electricity. This kind of conversion process is termed photoelectrochemical process. To a large extent, photoelectrochemical process combines the unique advantages of being both optical and electrochemical. The unique advantages of the photoelectrochemical process, coupled with the exceptional features of nanomaterials, led to the substantial progress in technological advancement. This chapter therefore delineates the substantial progress in the photoelectrochemical application of nanomaterials by channeling it toward three major applications, namely biosensors, hydrogen production through photoelectrochemical water splitting and the design of renewable energy devices, generally known as photovoltaics.

Schematic Illustration of the Photoelectrochemical Process & Its Application Using Nanomaterials

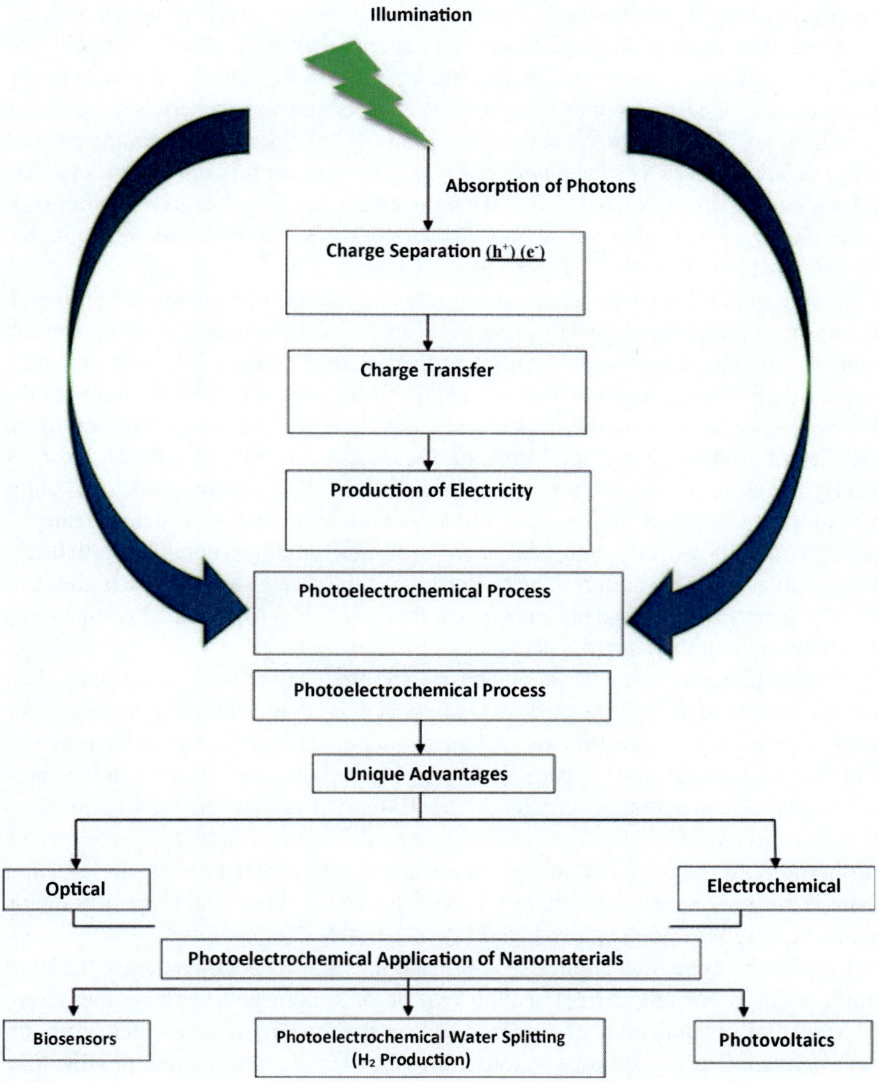

7.1 Introduction

Nanotechnology is an interdisciplinary area of research which provides answers to systematic, engineering and business challenges by way of advancement and implementation [1, 2] of new materials with new, interesting, attractive [3] and useful properties [2]. These new materials, however, are called nanomaterials, made from nanoparticles [2]. The European Commission (2017) EU defined *nanomaterials as natural, incidental or synthetic material with particles not less than one dimension in nanosized scale between 1 and 100 nm, in a boundless state or as a combined or an agglomerate as well as where 50% or more of the particles are in the nanosized grouping* [4].

Nanomaterials are useful in a broad variety of operations owing to their high ratio to surface volume, the prospect for surface functionalization and favorable thermal features [5]. The attractive and useful properties of nanomaterials also include a variety of structural and non-structural applications, however, but not limited to the following [3]: better insulation materials, kinetic energy interlopers with improved lethality, phosphors for high-performance TV, inexpensive flat panel shows, stronger and rigid cutting equipment, removal of impurities, highly energetic magnets, autos with more fuel strength, aerospace machineries with improved performance features, superior and prospect ammunitions channels, durable satellites, durable health facilities, flexible-workable potteries, huge electrochromic display appliances, high sensitivity sensors [3] including biosensors and their photoelectrochemical counterparts [6] and energy storage devices [7].

The broad applications of nanomaterials are made possible by controlling their physico-chemical properties through analytical selectivity and method reliability, integration of useful constituents or transformation of their surfaces by practical fragments, structural design, particulates or polymers that can strengthen the separation and preconcentration efficiency [5]. Extensive reviews as well as research covering numerous applications of nanomaterials are well documented [2, 5, 8]. Furthermore, it is imperative to note that while single-channel nanomaterials offer a single purpose, hybrid nanomaterials can enhance the functional properties owing to their synergistic effects [5, 9–12].

Despite the numerous applications of nanomaterials as documented in the literature, emphasis of this current section will be on established work on the photoelectrochemical application of nanomaterials in specific applications such as photoelectrochemical water splitting, photoelectrochemical biosensors and photovoltaic devices.

The innovatory finding of the photoelectric impact by Becquerel in 1839 led to the substantial use of photoelectrochemistry in various fields comprising photovoltaics, sensing and photocatalysis [13]. The photoelectrochemical (PEC) process is a photo-to-electric conversion emanating from excitation of electron and successive charge transfer of a photoexcited material produced by an applied light. The PEC process is a favorable inexpensive systematic device that comprises the movement of charge responses between a photoactive solid, test object and electrode in the course of exposure to light rays. The PEC process has the exclusive merit of optical and electrochemical through combination of photoexcitation with electrochemical process. Besides, the dissociation between the excitation supplies which is light and photocurrent in the PEC procedure provides high sensitivity with low background signal [13, 14]. Owing to the exclusive merit of both optical and electrochemical in PEC procedure, the remaining part of this section will dwell on the photoelectrochemical application of nanomaterials in various identified fields as highlighted below.

7.2 Photoelectrochemical Applications of Nanomaterials

Nanomaterials have varieties of photoelectrochemical applications such as biosensors, photoelectrochemical water splitting and photovoltaic devices among others as will be highlighted below.

7.2.1 Applications in Biosensors

The study of photoelectron transfer processes at biomodified electrode/solution boundary layers is a novel detection technique in PEC biosensing. The electron donor/acceptor produced during a target recognition process in the course of exposure to light rays at a suitable wavelength is a reflection of the resulting photocurrent. The express operation of PEC biosensing is well documented in numerous perspectives including early clinical analysis, environmental monitoring and gene testing [15]. By way of illustration, photoactive materials with greater photoelectron transfer competence, a core function in the design of PEC biosensing facilities, were used to prepare a treble CdS@Au-g-C_3N_4 heterojunction-supported PEC immunosensor for detection of prostate specific antigen (PSA) [16], the best recognized cancer biomarker commonly used by physicians for early detection of prostate cancer [17]. AuNPs was inserted between CdS and g-C_3N_4 to play a dual role of both photosensitivity and electron transmission. More so, the AuNPs significantly intensify the absorption of light and accelerates transfer of charge from g-C_3N_4 to CdS, which mutually added to the reinforcement of photoelectric conversion efficiency. Graphene oxide-CuS (GO-CuS) conjugates on the other hand augment the PSA recognition sign by competently reducing the PEC reaction of the treble heterojunction. Overall, the synergistic effect of the outstanding PEC attributes of CdS@Au-g-C_3N_4 and the notable reducing impacts of GO-CuS led to plethora of signal amplification mechanisms with a detection limit of 0.6 pg mL^{-1} over an extensive uninterrupted magnitude from 1.0 pg mL^{-1} to 10 ng mL^{-1} [16].

Tian and her colleagues developed a PEC biosensing platform for glutathione (γ-glutamyl-cysteinyl-glycine, (GSH) detection by means of Pt-IrO_2/TiO_2NTs/Ti electrode [18]. Concisely, GSH is a paramount tripeptide which normally takes part in numerous biological activities including cell protection and immune regulation to mention a few [19, 20]. A range of diseases which affect humans is traceable to cellular concentration [19, 21], and hence, detection of GSH through a PEC biosensor remains a top-priority owing to its prominence in physiological settings. The design of the proposed PEC biosensor involves the uniform distribution of Pt (via photoreductive deposition) and IrO_2 (chemical deposition) on TiO_2NTs/Ti (Scheme 7.1). At this point, it is also noteworthy to mention that despite the vast research on the use of TiO_2 nanomaterials owing to superior PEC performance, excellent bioaffinity, clear surface effect and steady performance, TiO_2 implementation in PEC detection has limitations, viz broad band gap and swift recoupling of light-induced charges. The implications of the limitations on PEC biosensors include reduction of photoenergy transformation efficiency and impact on the sensitiveness of PEC biosensors. A conventional way of overcoming these limitations is by modifying the TiO_2 photoelectrode with noble metals [22–24] or metal oxides [25, 26]. The combined effect of modifying TiO_2 photoelectrode with noble metals and metal oxides is the establishment of a Schottky barrier at the boundary layer of the metal–semiconductor. Schottky barrier establishment is capable of minimizing the recoupling of light-induced charges in addition to fostering the splitting of light-induced charge transporters [27–29]. There are numerous reports in the literature containing more practical illustration of this procedure [30–33]. However, in this report, Tian and her colleagues employed the same technique by uniformly distributing Pt and IrO_2 on TiO_2NTs/Ti, and they recorded an amazing result. Overall, the synergy of Pt and IrO_2 in the Pt-IrO_2/TiO_2NTs/Ti PEC biosensor displayed a recognizable (Fig. 7.1) and consistent photocurrent signal (Fig. 7.2) in contrast to the TiO_2NTs/Ti or the single-reformed ones. The consistency of the Pt-IrO_2/TiO_2NTs biosensor was tested by measuring the photocurrent reaction over 30 days. No relevant fall was observed showing the electrode's excellent consistency for GSH detection. Additionally, a linear photocurrent signal to GSH concentration was observed for the Pt-IrO_2/TiO_2NT/Ti biosensor in the magnitude 1–10 μM and a detection limit of 0.8 μM [18].

Zhang and co-workers newly presented a photoelectrochemical biosensing strategy for standard scrutiny of a cancer biomarker, carcinoembryonic antigen (CEA) [6]. The PEC biosensing system (Scheme 7.2) comprises copper indium disulfide ($CuInS_2$) sensitized with graphitic carbon nitride (g-C_3N_4), the photosensitizerand cobalt oxyhydroxide (CoOOH), acting as the light-blocking material. The

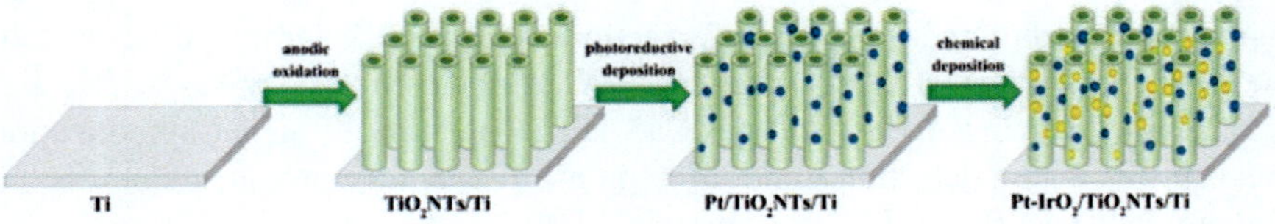

Scheme 7.1 Pictorial model of development process of Pt-IrO_2/TiO_2NTs/Ti PEC biosensor. Adapted from [18] (Open Access)

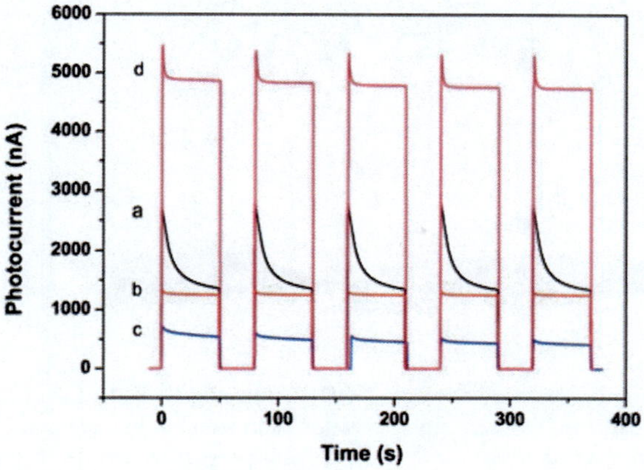

Fig. 7.1 Photocurrent reactions of differing electrodes **a** TiO_2NTs/Ti, **b** Pt/TiO_2NTs/Ti, **c** IrO_2/TiO_2NTs/Ti, **d** Pt-IrO_2/TiO_2NTs/Ti electrodes in 0.1 M PBS solution under 365 nm LED light rays. Adapted from [18] (Open Access)

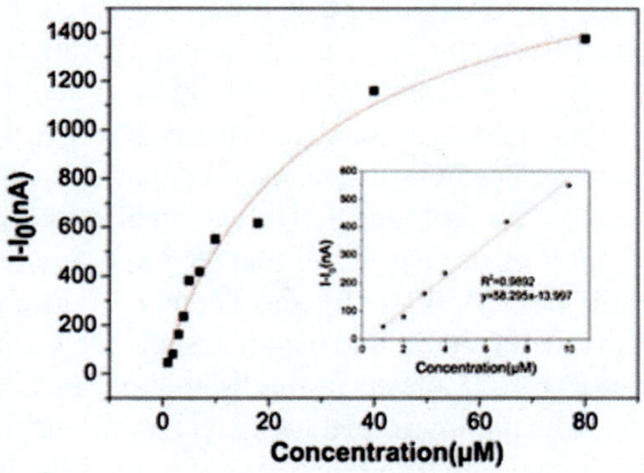

Fig. 7.2 Arch of Pt-IrO_2/TiO_2NTs/Ti (over 30 days) for the determination of various concentration for GSH. Adapted from [18] (Open Access)

system was allied with hybridization chain reaction (HCR) and etching reaction. For emphasis, CEA, a tumor-related glycoprotein, was first discovered by Dr Joseph Gold in 1965 [6, 17]. It is found in healthy grown-up serum at small concentrations between 2.5–5.0 ng mL^{-1} [6]. Beyond the normal CEA concentration in healthy grown-ups, its elevated concentration is an indication that normal cells in certain tissues are cancerous. Meanwhile, CEA is a broad-spectrum biomarker for cancerous tumors—bowel cancer, lung cancer, ovarian carcinoma, cystadenocarcinoma [6] and breast cancer [17]. As a result, quantification and super-sensitive determination of CEA have an amazing medical relevance in early point-of-care, treatment appraisal and transition checks of cancers despite the inability of CEA to serve as a diagnose for one form

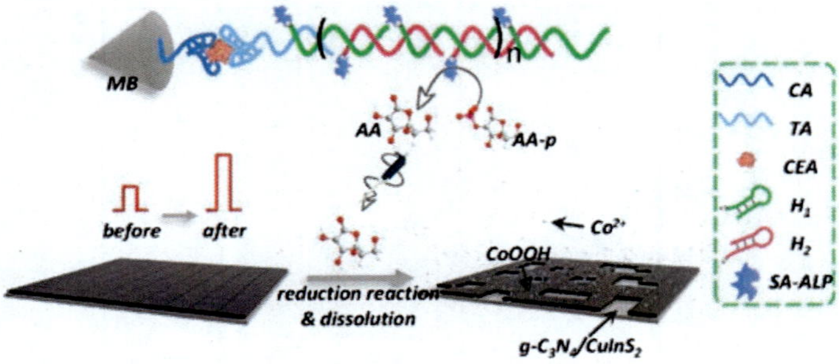

Scheme 7.2 Descriptive representation of CoOOH-covered g-C_3N_4/$CuInS_2$ nanohybrids-derived photoelectrochemical (PEC) biosensing system for carcinoembryonic antigen (CEA) by bonding hybridization chain reaction (HCR) with ascorbic acid (AA)-activated disintegration of CoOOH. Adapted from [6] with permission

of pandemic. By and large, test results of the proposed PEC biosensor recently discovered an enhancement in light uptake when g-C_3N_4 and $CuInS_2$ were combined as compared to pure g-C_3N_4 nanosheets. Further, the proposed PEC biosensor displayed high sensitivity toward CEA with a detection limit of 5.2 pg mL^{-1} within the magnitude of 0.02–40 ng mL^{-1}. In addition, the PEC biosensor is favorably selective (Fig. 7.3) and remarkably stable with satisfactory precision [6].

A photoelectrochemical (PEC) immunosensor termed as "signal–on" was reported by Chen et al. [34] for profound determination of human epididymis protein 4 (HE4)—a revolutionary serum biological marker for ovarian cancer with molar mass of nearly 25 kDa [34, 35, 46–48]. Considerably, ovarian cancer is a prevalent cancerous tumor of female reproductive organs and also a paramount cause of cancer death in women [34, 36]. Conversely, the "signal-on" PEC immunosensor was invented in accordance with porphyrin-metal–organic framework (MOF) nanosphere (nPCN-224) and nanobody (Nb). The functionality of the PEC immunosensor was strengthened by employing nanohybrids consisting of synthesized AuNPs embellished with polydopamine-functionalized multiwalled carbon nanotubes (MWCNTs-PDA-AuNPs) as the matrix material for modifying the ITO semiconductor [34, 37] (Scheme 7.3). By taking the advantages of MOF attributes—extremely ordered porosity, large surface area and photoactive property—consolidated with carboxyl-functionalized porphyrin as ligands to form a porphyrin-based MOF nanosphere (nPCN-224) [34, 38], the PEC immunosensor presented a LOD of 0.560 pg mL^{-1} over a linear magnitude of 1.00 pg mL^{-1} – 10.0 ng mL^{-1} owing to the small molar mass of the biological marker HE4, superiority of nPCN-224 and the grand specific feature and connectivity of Nbs [34].

Acquired immunodeficiency syndrome (AIDS) is another serious problem in this millennium and is therefore instantaneously increasing [39]. AIDS was first reported by the US Center for Diseases Control in 1981 succeeded by the recognition of human immunodeficiency virus (HIV) as the source of the disease in 1983 [40]. Statistical information showed that nearly 36.9 million people lived with HIV at year end of

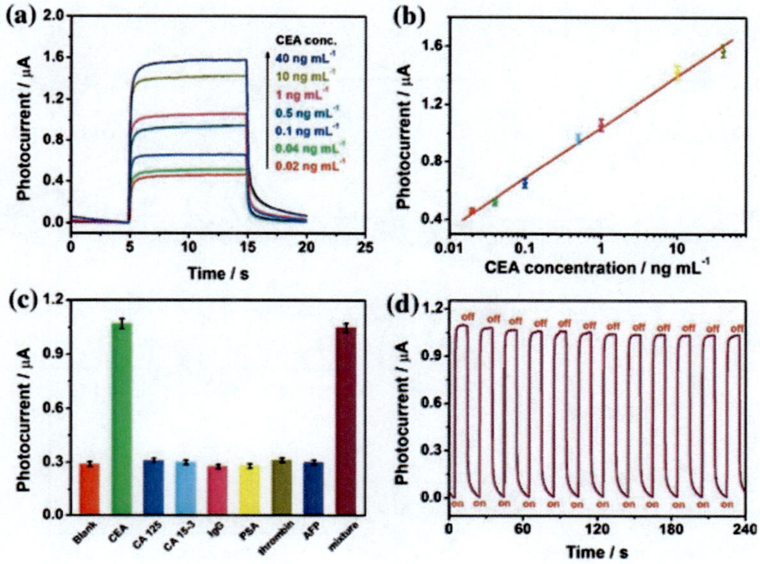

Fig. 7.3 **a** Photocurrents of the designed PEC biosensing scheme over ranged concentration of CEA standards in 0.1 M Na_2SO_4; **b** the resultant standardization trend; **c** the specificity of established system against CEA and miscellaneous consisting of CA 125, CA 15–3, human IgG, PSA, thrombin and AFP; **d** the stability of the biosensing layout under irradiation. Adapted from [6] with authorization

2017 with an overall HIV predominance of 0.8% among grown-ups [41]. An estimated deaths of 940,000 were recorded from HIV-related sources universally in 2017, respectively. The dangers presented by AIDS syndrome have been identified as serious threats to humanity. As a result, the early discovery of HIV infection is significant to all scientists on a global scale [39]. The significance of the early discovery of HIV infection led Fan et al. to develop a novel PEC biosensor for HIV DNA (T_{HIV}) detection [42]. The PEC biosensor was built by integrating a catalyzed hairpin assembly (CHA) and biocatalytic precipitation (BCP) (Scheme 7.4). Beta-cyclodextrin (β-CD) was synthesized by solvothermal means and then functionalized with CdS nanorods to serve as potent building blocks (β-CD@CdS NRs). The β-CD@CdS NRs were designed as alert unit for photoelectric current production and to hold adamantine-labeled hairpin DNA2 (ADA-H_2) probe through host–guest interaction. Overall, β-CD@CdS NRs considerably increase the film-forming attribute of CdS NRs and speed up the electron transfer and dissociation performance of light-induced charge transporters. The β-CD@CdS NRs were also found to support the linking of DNA check via host–guest communication. More so, the favorable transition of ADA-H2 onto the electrode covering activated CHA reusability to yield G-quadruplex in the occurrence of Hi (hairpin DNA1) and HIV DNA leading to the formation of hemin/G-quadruplex DNAzyme after the insertion of hemin. The photoelectric current signal was considerably attenuated when insulative materials

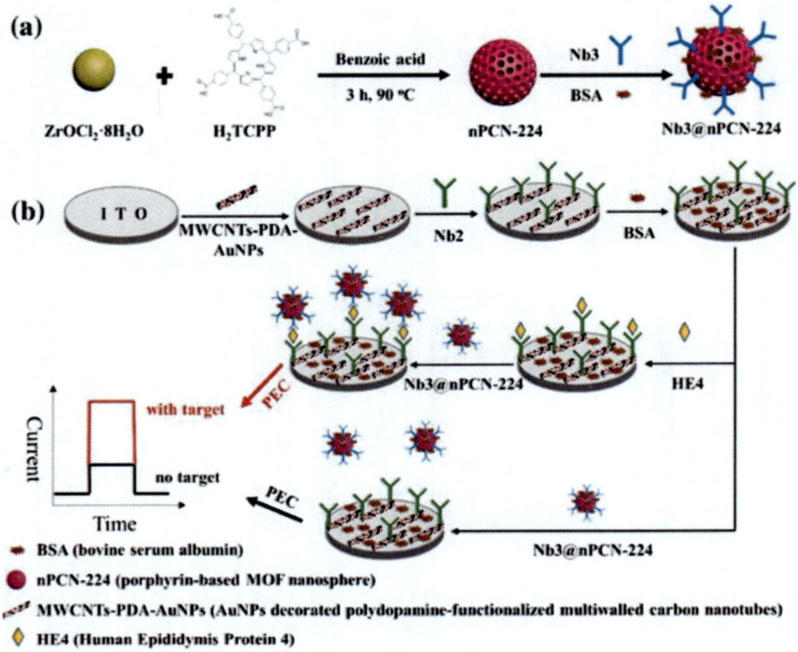

Scheme 7.3 Blueprint for **a** the formulation of Nb3@nPCN-224 and assemblage of photoelectrochemical HE4 immunosensor **b**. Adapted from [34] with authorization

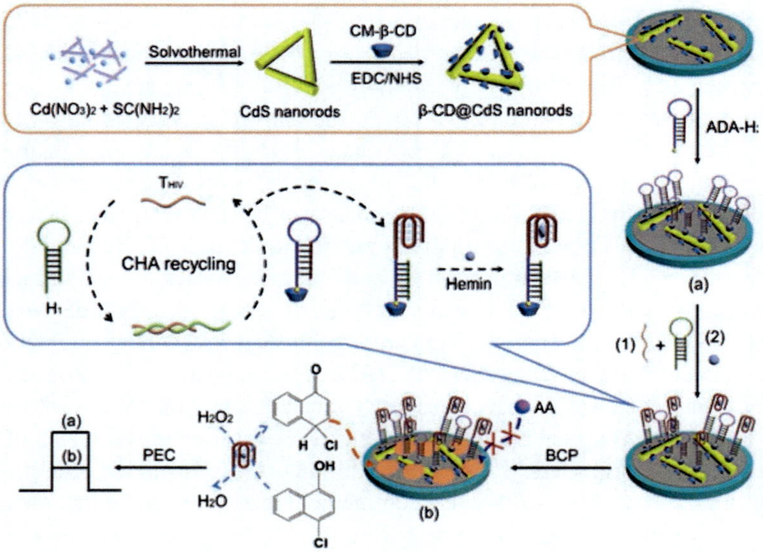

Scheme 7.4 Diagrammatic print of the development of β-CD@CdS NRs and step-by-step arrangement approach for photoelectrochemical bioassay of T_{HIV} built upon CHA-facilitated enzymatic reaction. Adapted from [42] with authorization

created by enzyme catalysis reaction excellently obstructed the boundary diffusion and electron transport of ascorbic acid (AA). The designed PEC biosensor was able to detect trace HIV DNA over a linear magnitude of 10 fM–1 nM and low detection limit of 1.16 fM. The proposed PEC biosensor was satisfactorily utilized in actual human serum research exhibiting substantial assurances in medical test.

Diabetes has been a long standing serious common health challenge [43, 44]. It has caused severe harm to biological systems including nervous, retinal, circulatory and renal owing to rise and fall in blood glucose level. It is therefore imperative to develop credible and exact assessment systems for glucose monitoring. This monitoring is extremely important in clinical and non-clinical application fields [44, 45]. To corroborate this fact, a single-step electrochemical deposition system utilizing permeable anodized aluminum modules for instantaneous synthesis of AuNi nanodendrite arrays (NDAs) was reported [44]. The as-synthesized AuNi NDAs were applied for developing a non-semiconductor photoelectrochemical biosensor for glucose and hydrogen peroxide determination. Owing to the presence of sufficient sites of action and interactive catalysis of Au and Ni, the plasmonic AuNi nanodendrite arrays exhibited superior photo electrocatalytic actions for the oxidation of glucose as well as the reduction of hydrogen peroxide response under illumination. Further, the identification sensibility for glucose was reported to be 3.7277 mA mM^{-1} cm^{-2} under illumination. This value was found to be around 3.3 folds improvement in contrast to that which was obtained in the dark with a magnitude of 1.1287 mA mM^{-1} cm^{-2} plus high accuracy and low detection limit of 3 μM. The appreciably advance operation of AuNi nanodendrite network was ascribed to its sequence order through sufficient sites of action and plasmonic impact of Au with consistent absorption band in the visual range. This established mechanism has been reported to have an amazing importance in the design of plasmonic-assisted photoelectrochemical biosensors.

Sun and co-workers reported a photoelectrochemical glucose biosensor comprising nanoparticle-based CdS solid-state crystals [46]. They exploited a generation 4 amino-terminated polyamidoamine (PAMAM) dendrimer as internal design in the synthesis of CdS nanoparticles. This was followed by immobilizing CdS nanoparticles and glucose oxidase (GOD) using sandwiching approach on a platinum (Pt) electrode to devise a biological-inorganic mixed pattern. The charge transfer of the mixed system was strengthened by mixing Pt nanoparticles with nafion coating to hold the CdS/enzyme composites and nafion (Pt-PAMAM)/$(CdS\text{-}PAMAM/GOD)_n$. The PEC biosensing system expressed a sensitivity of 1.83 $\mu A/mM\ cm^2$ and a detection limit of 1 μM. It also proved a satisfactory repeatability of a RSD rate of 1% current responses over five continuous measurements and was found to be stable over a 2-month period at 4 °C.

One-dimensional (1D) core/shell structure has been proven to increase charge transfer across boundary areas. To support this fact, this attribute was employed in the durable architecture of a high sensitivity PEC glucose biosensor [47]. The architecture of the PEC glucose biosensor was based on titanium dioxide/polydopamine core/shell nanorod arrangement (TiO_2/PDA core/shell). The PDA shell-encapsulated TiO_2 nanorods were made using water-based chemical growth methodology, while fluorine transparent oxide (FTO) glass was employed as the electrical conduction

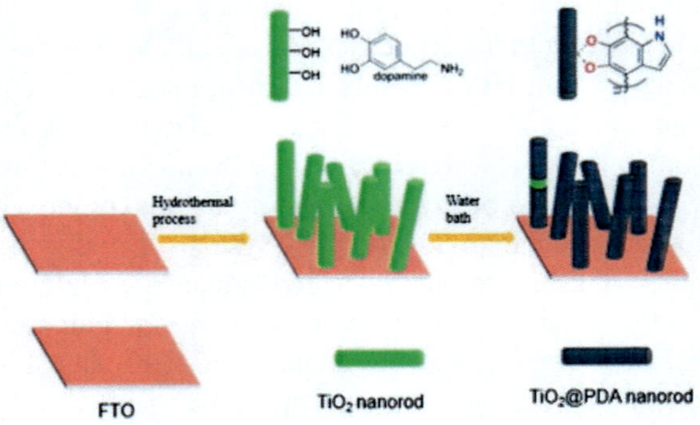

Scheme 7.5 Vivid model of the integration of main TiO_2/PDA shell nanorod arrangements on FTO glass support materials. Adapted from [47]. Published by RSC advances

substrate on which the nanorods arrangement was prepared during the PEC biosensor invention process (Scheme 7.5).

Furthermore, PDA is a p-type semiconducting material with suitable band sides for type II heterostructure architectures employing TiO_2. In addition, the hydroxyl groups present on the area of TiO_2 can function with catechol units of PDA to create a coupled link of bidentate as lead electron transfer passages in the midst of PDA and TiO_2. More so, PDA aids the splitting of light-induced charge transporters and prevents the disintegration of enzymes triggered by elevated oxidative holes produced by the main TiO_2 under illumination. By and large, the proposed PEC glucose biosensor displayed a super-high sensitivity of 57.72 $\mu A\ mM^{-1}\ cm^{-2}$ over a linear magnitude of 0.2–1.0 mM. The biosensor detection limit for glucose concentration was found to be 0.0285 mM ($S/N = 3$) with a significant sensitivity of 8.75 $\mu A\ mM^{-1}\ cm^{-2}$ over a dynamic magnitude of 1.0–6.0 mM [47].

7.2.2 Applications in Photoelectrochemical Water Splitting

Photoelectrochemical water splitting (PEC WS) is a technique that entails producing hydrogen from water, employing light and distinguished semiconductor materials known as photoelectrochemical materials. The PEC materials make use of the energy from light (light energy) to instantaneously split up molecules of water into hydrogen together with oxygen. Basically, in the PEC WS technique, the semiconductor materials are used to transform solar power swiftly to chemical power as hydrogen. The semiconductive materials employed in the PEC technique closely resemble the materials utilized in photovoltaic systems; however, in favor of PEC demands, the semiconductive material is dipped amidst water-related electrolyte wherein light assists the water splitting technique [48].

Illustratively, a characteristic water splitting chamber device comprises three main components, namely semiconductors known as photoelectrodes (photoanode or photocathode) with appropriate band gap to absorb light, conductive substrates

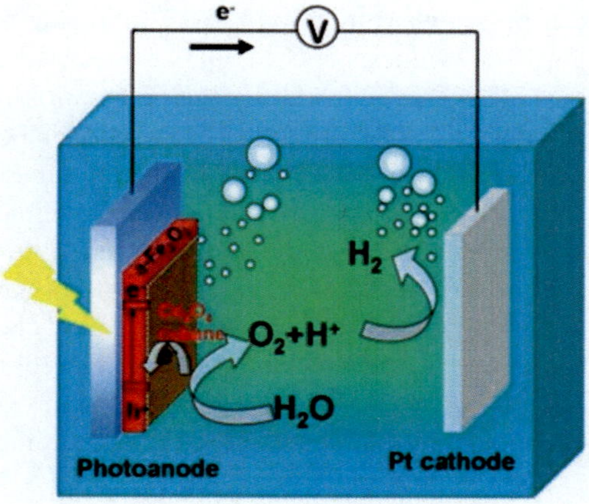

Fig. 7.4 PEC chamber containing semiconductor photoanode with a Pt cathode. The surface of the photoanode was modified with a cobalt-based catalyst. Adapted from [50] with authorisation. Copyright @ American chemical society

to transport charges and electrolyte to support the reactions (Fig. 7.4). Fundamentally, upon irradiation of photoelectrodes, electrons (e^-) and holes (h^+) are generated, which are then transferred to the conduction and valence bands, individually. The photogenerated electrons (e^-) flow to the surface of the cathode where they reduce H^+ to H_2. Simultaneously, photogenerated holes (h^+) move to the surface of the anode and oxidize water to O_2. General reaction for water splitting is written as [49]

$$2H^+ + 2e \rightarrow H_2(g)$$
$$2H_2O \rightarrow O_2 + 4H^+4e^-$$
$$\text{Overall}: 2H_2O \rightarrow O_2 + 2H_2 \text{ (light condition)}$$

Importantly, catalysts are often needed to boost the reaction movement on both of the photoanode and photocathode surface so as to match the rates of charge generation and separation with the photoelectrodes [49]. Overall, essential considerations required in the selection of a material for solar water splitting include potential requirement, appropriate band structure, high crystallization and surface area, stability and low cost. Further information can be found in the literature [49]. Furthermore, it is noteworthy to mention that each step in the PEC WS technique (photon absorption, charge separation and removal of e^- and h^+ to the photoelectrodes for H_2 and O_2 evolution reactions) has main effects on the overall performance of the system [49]. Further information on the principle of photoelectrochemical water splitting can be found in the literature [49, 51, 52].

Miscellaneous semiconducting nanomaterials have now been explored for the purpose of PEC WS and widely reported in the literature. Some, among many, are highlighted below.

Researchers have reported the use of zinc oxide two-dimensional (2D) nanosheets decorated with graphite-like carbon nitride (g-C_3N_4) quantum dots (QDs) as photoanodes in photoelectrochemical water splitting [53]. Being inspired by previous reports, here we have developed a methodology for the synthesis of ZnO 2D nanosheets that can be grown vertically on FTO following a simple electrode position technique [54]. The synthesis of C_3N_4 QDS was carried out using a procedure described earlier [55]. This was followed by the preparation of ZnO/C_3N_4 QDs by immersing the as-prepared zinc oxide nanosheets on the FTO in a solution containing 3 mL C_3N_4 QDS periodically for 10, 15, 30 and 60 min (Scheme 7.6).

In regard to experimental work as reported by the researchers, the sensitization of ZnO nanosheets with C_3N_4 QDs led to an improvement in PEC performance in contrast to bare ZnO (Fig. 7.5). This performance was ascribed to increased absorption of light, higher charge-transport dissociation efficiency and reduced recombination rate owing to establishment of heterojunction in the midst of C_3N_4 QDs and ZnO. More so, ZnO/C_3N_4-QDs at 15 min (ZnO/C_3N_4–15) performed best among the photoanodes prepared at the reported time intervals during optimization owing

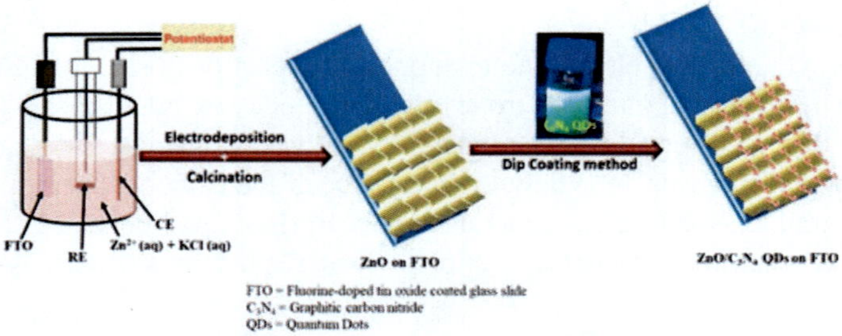

Scheme 7.6 Vivid model of the synthesis of ZnO/C_3N_4 QDs on FTO by means of electrodeposition accompanied by a dip-coating technique. Adapted from [53] with approval. Copyright @ American chemical society

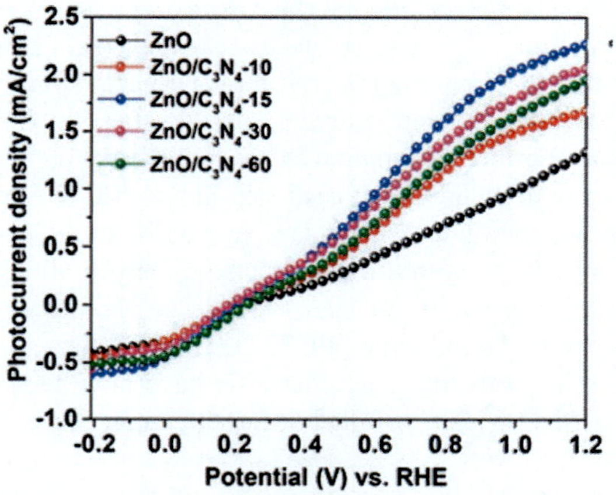

Fig. 7.5 Linear sweep voltammetry (LSV) curved lines displaying alteration in the photocurrent density with the employed potentials of ZnO and ZnO/C_3N_4–10, –15, –30 and –60. Adapted from [53] with authorisation. Copyright @ American chemical society

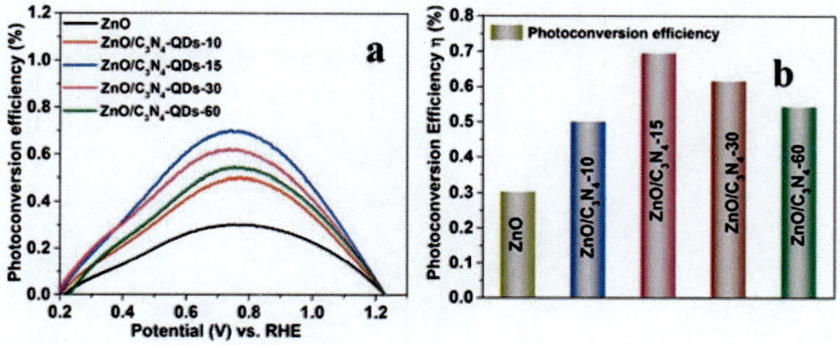

Fig. 7.6 **a, b** Photoconversion efficiency for ZnO and ZnO/C_3N_4–n (n = 10, 15, 30 & 60). Adapted from [53] with authorisation. Right of ownership @ American chemical society

to higher charge-transport dissociation and reduced recombination rate (Figure not showed) [53].

Further, 2.3 times photocurrent density was recorded for the photoconversion efficiency (η) of ZnO/C_3N_4–15 photoanode in contrast to bare ZnO (Fig. 7.6a). Additionally, a photocurrent density of 0.952 mA cm^{-2} was recorded for ZnO/C_3N_4–15, while for bare ZnO, 2D sheet 0.414 mA cm^{-2} was recorded, both around 0.5994 V with reference to reversible hydrogen electrode (RHE) during constant medium of light irradiation. The magnitude of the enhancement in PEC activity of the heterostructure (ZnO/C_3N_4–15) was calculated to be 1.20×10^{21} cm^{-3} owing to an improved carrier density with magnitude of about 2.2 times in the heterostructure in contrast to bare ZnO sheet with a magnitude of 5.45×10^{20} cm^{-3}. Notably, ZnO/C_3N_4 displayed an optimal photoconversion efficiency (η) of 0.70% (Fig. 7.6b). Interestingly, each one of ZnO 2D sheets and ZnO/C_3N_4 heterostructure displayed active consistency during disconnected light radiation for 1000 s. Conversely, ZnO/C_3N_4 was found to be consistent for 1 h during steady luminosity (Figures not showed) [53].

An active PEC water splitting operation underlying layered double hydroxide nanosheet arrangements anchoring permeable $BiVO_4$ to fabricate $BiVO_4$/LDH photoanode was described [56]. Dual NiFe-LDH nanosheet template was chosen as the standard oxygen evolution cocatalysts (OECs) to devise $BiVO_4$/NiFe-LDH photoanode owing to their simple formulation requirements and remarkably active oxygen evolution reaction (OER) performance [56]. $BiVO_4$ photoanode with permeable layer was synthesized to reduce the mobile gap of light-induced holes as well as the recoupling chance of charge transporters. This was followed by deposition of NiFe-LDH nanosheet templates on the permeable $BiVO_4$ photoanode via a solution chemical evolvement (Fig. 7.7). Notably, the $BiVO_4$/NiFe-LDH photoanode displayed a photocurrent density magnitude of 4.02 mA cm^{-2} at 1.23 V with reference to the RHE, and this was found to be 2.8 times greater than the pure $BiVO_4$.

In addition, based on the LSV curve supposing 100% Faradaic efficiency, an improved applied bias photo-to-current efficiency (ABPE) of 1.07% at 0.75 V with reference to RHE was recorded. Notably, H_2 and O_2 gas formation rate for

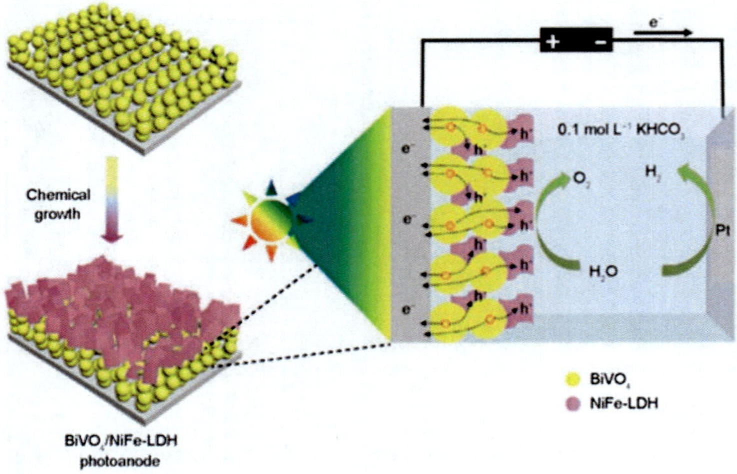

Fig. 7.7 Pictorial sketch presenting the building of $BiVO_4$/NiFe-LDH photoanode, inherent charge separation step and PEC water splitting system. Adapted from [56] with approval

$BiVO_4$/NiFe-LDH photoanode—Pt PEC system—was about 2:1 with oxygen evolution estimate at the $BiVO_4$/NiFe-LDH photoanode surface reach of 9.6 μmol h^{-1} cm^{-2} initially for 3 h. Eventually, more than 30 h activity was recorded for water oxidation in $KHCO_3$ weak alkaline electrolyte. Ultimately, Co^{2+} cation was incorporated into the NiFe-LDH as OECs so as to upgrade the operation of the fabricated photoanode for PEC water splitting, and then, this yielded a photocurrent density of 4.45 mA cm^{-2} at 1.23 V with reference to RHE [56].

The performance of PEC water splitting system has also been made more effective by employing a graphene-based ternary blend [57, 58]. The blend comprises a molybdenum-doped Bismuth vanadate @reduced graphene oxide (Mo-$BiVO_4$/rGO) nanocomposites, prepared in a single wet impregnation procedure intended to be used as per photoelectrode material in PEC water splitting. The pasty blend was affixed onto the FTO layer as described [58]. Notably, the PEC operation for water oxidation as well as the photocurrent was amplified via reduction of charge recoupling after introducing Mo-dopant and rGO in $BiVO_4$. Consequently, a photocurrent density of 8.51 mA cm^{-2} at 1.23 V with reference to the RHE was recorded for the Mo-$BiVO_4$@rGO blend photoanode. This result was reported to double Mo-$BiVO_4$ (5.3 mA cm^{-2} at 1.23 V with reference to the RHE) and four-fold increase of the pure $BiVO_4$ (2.01 mA cm^{-2} at 1.23 V with reference to the RHE) light-sensitive electrodes. More so, 2.45% at 0.6 V with reference to the RHE was obtained as the value for optimum solar-to-hydrogen transformation efficiency of the investigated photoanode blend. The researchers of this work claimed that this value (2.45% at 0.6 V) is greater than $BiVO_4$-supported photoanodes stated in the literature. The photocatalytic water splitting responses also affirmed that greatest hydrogen evolution was reached for Mo-BiVO4@rGO. Additionally, Mott–Schottky (M-S) analysis

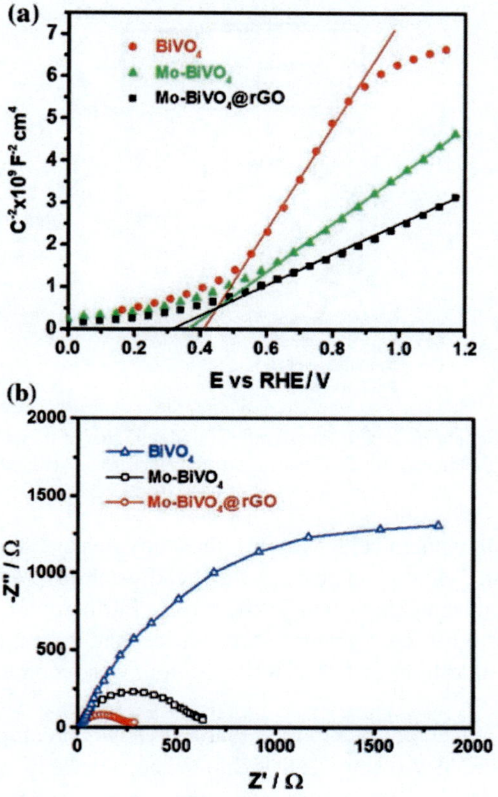

Fig. 7.8 **a** Mott–Schottky (M-S) charts of the pure $BiVO_4$, Mo-$BiVO_4$ and Mo-$BiVO_4$@rGO blend photoanodes in a Na_2SO_4 electrolyte solution (around 1000 Hz). Adapted from [58] with authorisation. **b** Nyquist charts of the pure $BiVO_4$, Mo-$BiVO_4$ and Mo-$BiVO_4$@rGO blend photoanodes in a Na_2SO_4 electrolyte solution during light irradiation. Adapted from [58] with authorisation

revealed this blend to display significant charge carrier density owing to a noticed lower dimension of the slope (Fig. 7.8a).

Similarly, the electrochemical impedance spectroscopy (EIS) results showed that this blend exhibited good carrier transport and dissociation, an indication that the overall water splitting performance was enhanced (Fig. 7.8b).

Based on the aforementioned, it can be said that the synergy of Mo-doping and rGO sheets in the Mo-BiVO4@rGO blend assisted in the improvement of the charge carrier density as well as the proficient charge movement of photogenerated electrons, causing a substantial rise in the PEC operation for water splitting. Scheme. 7.7 demonstrated the photographic model of the Mo-$BiVO_4$/rGO blend photoelectrode charging approach for photocatalytic water splitting responses.

Researchers have proven hematite (α-Fe_2O_3) to be supplementary optimistic photoanode material. However, for this study, references are made to self-biased PEC water splitting owing to its limited energy bandgap (around 2.0 eV), chemically stable in strong base dissolving agent, earth-ample elements [59]. As a case study, Urbain et al. reported a study based on multifaceted hematite nanowires with thin-film silicon photovoltaics in an all-earth-ample hybrid tandem apparatus for solar water splitting. Fe_2O_3 nanowire photoanode as well as Si solar cells was used to set

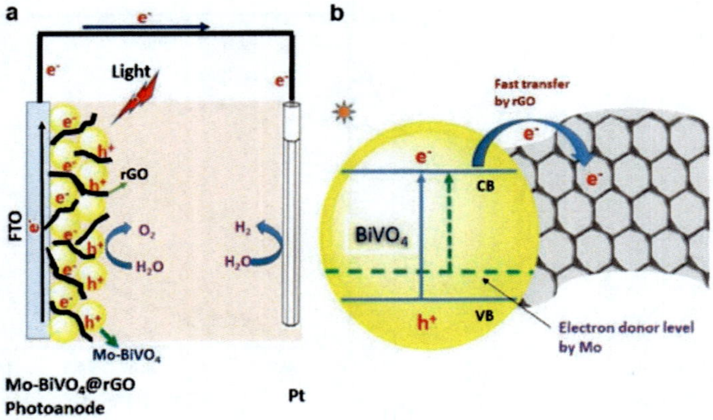

Scheme 7.7 **a**, **b** Photographic model of charge-transfer approach of the Mo-$BiVO_4$@rGO blend photoanode for PEC water splitting responses. Adapted from [58] with authorisation

up tandem cells. Notably, the study revealed that multiple connection framework of the solar panel possesses great adjustability with adjustable photovoltages and spectral matching for the tandem cells. Furthermore, a self-biased photocurrent density of 4.6 mA cm^{-2}, which stayed more than the entire day and night using manufactured solar light in 1 M NaOH solution, was recorded for the ideal Fe_2O_3/Si tandem cell [59, 60].

Chen et al. reported that $BiVO_4$ when compared to WO_3- and Fe_2O_3-supported PEC/PV tandem cells exhibited considerably less solar-to-hydrogen (STH) efficiencies owing to their small photovoltages and unfavorable charge-transfer characteristics. Therefore, in order to regulate the band architecture of the metal oxides, one effective strategy is to introduce oxygen vacancies [59]. For instance, Kim and fellow researchers proposed a simple chemical reduction procedure [61] to methodologically regulate oxygen vacancies in $BaSnO_3$ crystal and its energy band gap. Remarkably, a superior light uptake as well as improved charge dissociation abilities was demonstrated by $BaSnO_3$ photoanode with an ideal oxygen vacancy concentration of 8.7%. Furthermore, the photoanode expressed an impressive photocurrent density of 7.32 mA cm^{-2} at a bias potential of 1.23 V after introducing FeOOH/NiOOH oxygen evolution cocatalysts on its surface. The overall system comprising the devised photoanode along with a perovskite solar cell led to the development of a proficient tandem instrument of an STH efficiency 7.92% which was found to be stable for 100 h of operation [59, 61].

7.2.3 Applications in Photovoltaics

The swift expansion in overall population and current technologies has increased our day-to-day energy consumption up to an extraordinary level. The overall energy consumption is estimated to augment by just about 50% by 2040. Currently, 80% of the world's energy delivery is obtained from fossil fuels (coal, oil and gas). The continuous increase in fossil fuel price is accelerated through the release of harmful gases during fossil fuels combustion which results in greenhouse effect; this is the reason for climate change. As a result, developing renewable energy sources that can satisfy both existing and forthcoming energy requirements is a rising importance [62, 63].

Renewable energy resources including solar, wind, biomass, hydro and geothermal are promising choices because they are unlimited in supply, exist in infinite amounts and can be replaced within a short period of time. In recent times, wind energy and hydroelectric power have been the fastest rising means; however, they are limited by distribution tasks and cost problems. Building devices such as photovoltaic (PV) devices (i.e., solar cells), which convert energy from sunlight (solar energy) into electricity (electrical energy), are reasonably smart, because sunlight is an endless source of energy [63–65]. Improvements in solar technologies have appreciated in terms of their efficiency and power density, and this has conceded fast rise in their distribution [63, 66]. There are three classifications of PV technologies for background claims: (i) The "first-generation PV", founded on silicon (Si) semiconductors [67], is practically advanced providing the leading power conversion efficiency (PCE) of just about 26% [68] within experimental stage. This method has taken over the PV market-based with 91% of the commercialized unit (ii) The "second-generation PV", commonly founded on thin-film inorganic semiconductors [69], has merely a 9% commercialized unit owing to high-rise of production cost. The "second-generation PV" has a maximum efficiency of just about 23% [63, 68]. Despite the commercial availability of these first two generation PVs, their engineering procedure is difficult and expensive [63, 70]. Since there is a need to provide great quantity of electricity, then techniques requiring cost-effective PVs are desperately needed. (iii) The "third-generation PVs" comprising perovskite solar cells (PSCs), dye-sensitized solar cells (DSSCs), organic photovoltaic (OPV) as well as quantum-dot sensitized solar cells (QDSSCs), were implemented to cover expenses and operation targets [63, 71]. The foremost characteristics of the "third-generation PVs" include affordability, high level of performance plus simple built-up. More so, other characteristics such as lightweight and flexibility can enable further applications for instance interior light collectors, solar-powered wings in favor of drones and portable power source [63].

The exploitation of nanostructured materials in photovoltaics has gained so much focus owing to the various benefits that cannot be accomplished by their bulk associates, for example, increased photon-to-exciton conversion efficiency plus economical processability founded on solution treatment approaches [63, 72–74]. Thus far, an extensive variety of nanomaterials comprising unidimensional (1D), bidimensional (2D), tridimensional (3D), solid-state oxides and nanocarbons has been utilized in interfacial modification coatings, buffer coatings and photoactive coatings. Figure 7.9 displays the outline of the three PV generations in addition to several nanomaterials and nanostructures that have been well used in the third-generation PVs [63]. Here, we highlight the photoelectrochemical applications of the nanomaterials in emerging photovoltaics comprising OPVs, PSCs and DSSCs.

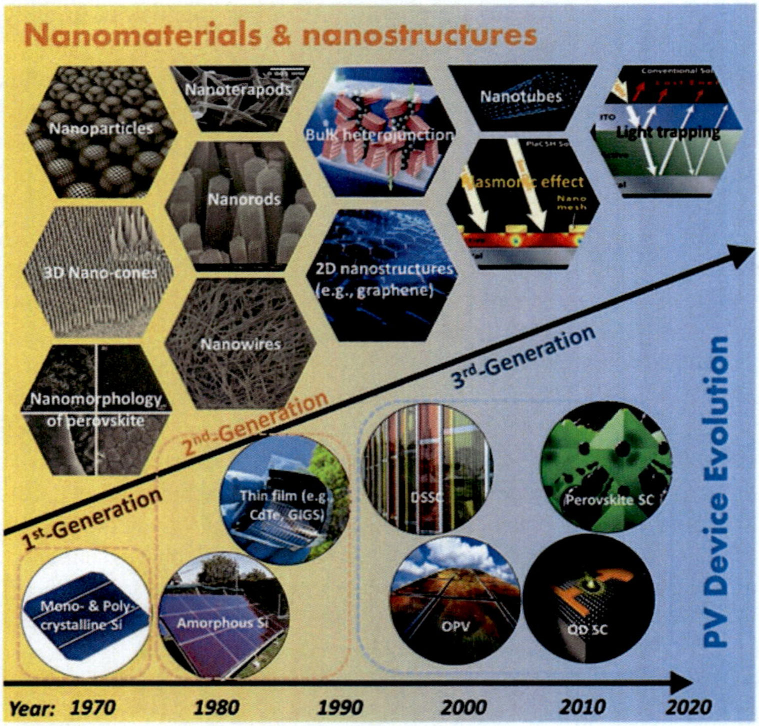

Fig. 7.9 Outline of the three PV generations accompanied by several nanomaterials and nanostructures that have been successfully employed in the third-generation PVs comprising DSSCs, OPVs, perovskite PVs and QD PVs. Adapted from [63] with authorisation

Unidimensional (1D) nanostructured materials, for example, nanotubes, nanowires and nanorods, have found use as replacement for nanoparticles. This is because unidimensional nanomaterials can offer leading electron transport routes, as well as lower recombination of charges owing to surface defects and grain boundaries. As a matter of fact, ID materials have been thoroughly examined and practically employed in dye-sensitized solar cells (DSSCs) [63]. For practical illustration, the preparation of anatase titania nanotubes (TNTs) and its use in flexible dye-sensitized solar cells were reported [75]. Synthesis of the titania nanomaterial was carried out using two-steps hydrothermal growth method, first at 160 °C for 4 h and then 110 °C for 20 h as reported. The prepared anatase nanomaterial was found to be 12 nm in diameter and was used to prepare a uniform and steady combined colloid coated on an ITO/PEN flexible substrate to fabricate a flexible dye-sensitized solar cell (DSSC) at reduced temperature (Fig. 7.10, schematic diagram of the flexible DSSC). Notably, under optimized conditions, the flexible DSSC yielded a light-to-electric energy conversion efficiency of 4.0% during an artificial solar light radiation of 100 mWcm^{-2} [75].

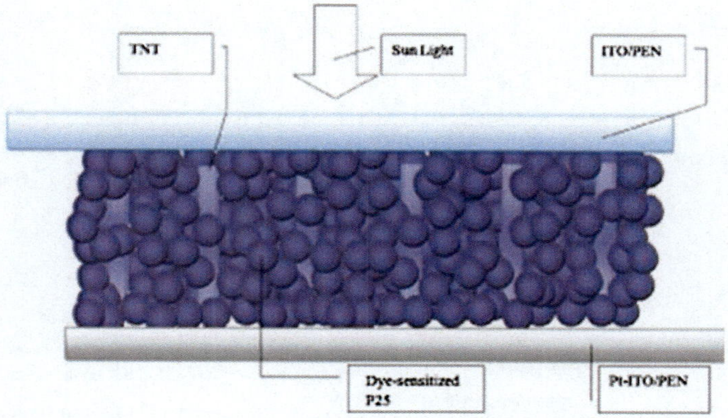

Fig. 7.10 Pictorial model of the flexible DSSC. Adapted from [75] with authorisation

According to literature, the most commonly used method to devise unidimensional nanomaterials is the "hydrothermal method" [63]. However, Kasuga and co-workers first introduce "chemical method" to prepare TiO_2 nanotubes. In brief, powdered TiO_2–SiO_2 was treated with an aqueous solution of 10 M NaOH at 110 °C for 20 h followed by HCl aqueous solution for the interchange of Na^+. This yielded nanotubes with an outer diameter of 8 nm as well as a length of 100 nm. Treatment of the caustic solution led to the production of Ti–O–Na and Ti–OH sheets throughout this method. Afterward, the lack of moisture in the Ti–OH bonds through HCl caused a reduction in the bond distance from a Ti to the neighboring Ti across the surface ensuing the turnover of the sheets into nanotubes [76, 77]. With this method, TiO_2 nanotubes with high specific surface area have been devised with the PCE of the resultant DSSC being just about 7% [63, 78, 79].

The advantage of nanocomposites was put into use via the synergistic effect of unidimensional ZnO nanorods and TiO_2 nanoparticles. While unidimensional ZnO nanorods offer swift electron transport routes, the enveloped TiO_2 nanosized particles provided adequate working surface for dye loading [63]. As a result, researchers reported a procedure for integrating multiple layered networks of perpendicularly arrayed ZnO nanowires plainly on clear conductive oxides [80]. Afterward, cycles of TiO_2 monolayer were grown and self-assembled onto the ZnO nanowires. The nanocomposite structure assisted in the prevention of the wire- fusion concern with the purpose of increasing the working surface of ZnO nanowires. Notably, the attainable PCE was 5.65% under solid-state electrolyte of Spiro-OMeTAD [2, 2′, 7, 7′-tetrakis-(N, N-di-*p*-methoxyphenylamine) 9, 9′-spirobifluorene] [80].

Perovskite solar cells (PSCs), among the third-generation solar cells, have gained considerable responsiveness simultaneously from science-based plus manufacturing group during the last couple of years. PSCs belong to a novel group of PV practical application. They have also appeared as favorable inexpensive solar power collecting means as against traditional silicon solar cells [63]. Illustratively, a complete printable processed mesoscopic methylammonium lead iodide

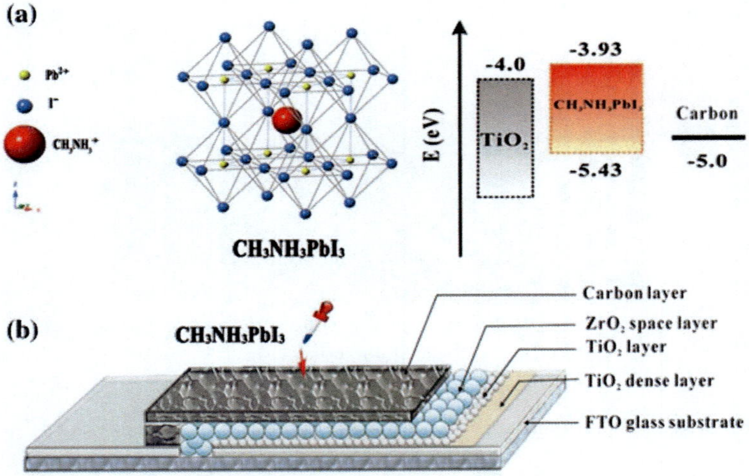

Fig. 7.11 **a** The crystalline framework of $CH_3NH_3PbI_3$ perovskite and the related levels of energy of TiO_2, $CH_3NH_3PbI_3$ and carbon. **b** A pictorial design of a carbonated monolithic apparatus. Adapted from [81]

($CH_3NH_3PbI_3$) perovskite/TiO_2 heterojunction solar cells (PSC), composed of economical carbon black counter electrode (CE), was fabricated [81]. Notably, $CH_3NH_3PbI_3$ has a perovskite structure (Fig. 7.11a) consisting of single Pb^{2+}, single $CH_3NH_3^+$ and ternary iodine anions in the device chamber leading to the preparation of iodide perovskite original solution via the intermixture of $CH_3NH_3PbI_3$ and PbI_2 in 1:1 molar proportion to satisfy the atom proportion in $CH_3NH_3PbI_3$. Additionally, the carbon black/graphite supported monolithic apparatus (Fig. 7.11b) comprises FTO glass substrate, a coated dense layer of TiO_2, a 1 μm nanoporous TiO_2 film applied via screen-printing (PASOL HPW-18 NR), a 1 μm ZrO_2 layer acting as the insulant against power surge, then eventually a carbon black/spheroidal graphite applied above the device. Owing to the presence of carbon black/spheroidal graphite counter electrode, the mesoscopic heterojunction layer solar cell displayed an extremely stable and PCE of 6.64% [81].

Importantly, two kinds of charge transporting materials (CTMs) perform incredibly important tasks in PSCs. These include electron transporting materials (ETMs) and hole transporting materials (HTMs). The role of CTMs is to extract light-induced charges out of perovskite and transfer them to resultant conductors. For the record, Li et al. described the application of SnS quantum dots (QDs) as novel inorganic hole transporting material (HTM) for PSCs [82]. A conventional one-pot hot injection procedure was employed in the preparation of SnS QDs nonaqueous octadecene (ODE) at an injection temperature of 180 °C. The source materials for the preparation of SnS QDs ($SnCl_2$ and thioacetamide) were captured by the massive bonding agents of oleylamine (OAm), oleic acid (OA) and trioctylphosphine (TOP) leading to a suitable dispersion in chlorobenzene. Remarkably, the as-prepared SnS

QDs after being processed on the surface structure of the ternary cation perovskite exhibited a suitable surface coverage as well as a higher photoluminescence (PL) quenching efficiency of the SnS hole carrier suggesting additional competent hole removal capability. Furthermore, a safe device technical practice as regards layer thickness, thermal process and ligand interchange on the SnS layer yielded a PCE of 13.7%. The SnS-incorporated perovskite photovoltaic cell also exhibited a greater stability in air, displaying perfect device operation after being conserved for 1000 h under environmental conditions against Spiro-OMeTAD [2,2′,7,7′-tetrakis(N,N-di-p-methoxyphenylamine)-9,9′-spirobifluorene]-incorporated command system [82].

Well-known bidimensional materials such as black phosphorous, carbon nanomaterials like CNTs and graphene oxide have been successfully used as HTMs in PSCs. For instance, researchers employed black phosphorous quantum dots (BPQDs) as the hole transporting layer (HTL) of reversed bidimensional PSCs (Fig. 7.12). Notably, the BPQD-integrated PSC devices displayed a remarkable PV efficiency of 16.69% owing to their high carrier movement as well as compatible band alignment. The efficiency of the BPQD-integrated PSC devices was found to be significantly higher

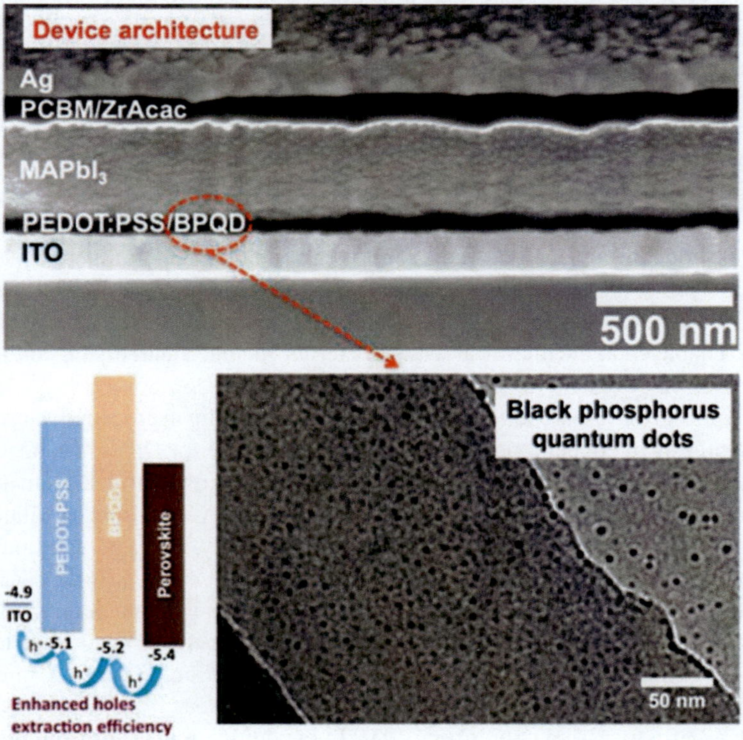

Fig. 7.12 Transverse section and energy chart of BPQDs integrated PEDOT: PSS HTM established reversed PSC device. Adapted from [63, 83] with consent of © 2017 American chemical society

than that of the poly (3, 4-ethylenedioxythiophene) poly (styrene-sulfonate) (PEDOT: PSS)-based device with a PCE of 14.10% [63, 83].

Organic photovoltaics (OPVs) possess striking features like cost effective, easy processing capability, mechanical flexibility and environmental pleasantness [84, 85]. The progress of OPVs can be traditionally found from the early innovation of super-high speed electron movement between polymers and fullerene in 1992 [86, 87]. The foundation of the bulk heterojunction (BHJ) organic solar cells is traceable to this super-high speed charge transfer between a supposed "electron donor" and "electron acceptor". OPVs have now become the most typically established blueprint for the advanced technology OPVs. Heeger and co-researchers founded the BHJ OPV in 1995 [63, 88, 89] using unidimensional nanomaterial made of fullerene derived products, PCBM ([6, 6]-phenyl-C_{61}-butyric acid methyl ester) as the electron acceptor, which now has application both in OPVs and halide perovskite PVs as competent electron removal film. A nanoscopic BHJ structure of donor–acceptor system with uninterrupted permeable stage characteristics is of appreciable significance for efficient OPVs. Wu and co-researchers have done extensive reviews on this area and may be consulted for more details [63]. For the record, Li and co-workers described a material made of Cs_2CO_3 with a layer of 1 nm thick for OPV application. Notably, a PCE of 2.25% in addition to a short circuit current density (J_{sc}) of 8.42 mA/cm^2 and open circuit voltage (V_{oc}) of 0.56 V was established using a device construction of ITO/Cs_2CO_3 (1 nm)/P3HT: PCBM/V_2O_5/Al [90].

Similarly in 2008, Liao and fellow researchers recounted the use of a low temperature annealed Cs_2CO_3 layer for the design of a reversed OPVs made of PCBM. In their report, they claimed that the approach significantly improves the PCE from 2.3% to 4.2% in addition to a short circuit current (J_{sc}) of 11.17 mA/cm^2 and a fill factor (FF) of 63% under light radiation [91]. Notably, alkali compounds such as ultrathin layer of metals—LiF, NaF, KF, CsF, Li_2O, Na_2O, Cs_2CO_3—as illustrated in the OPV applications above [90, 91] aided efficient interfacial modification layer and also enable charge withdrawal. Other materials include small organic molecules and conjugated polymers [63, 91].

Siddiqui and co-researchers also recounted a moist chemical procedure on account of the synthesis of copper oxide nanoparticles (CuO-NPs) for OPVs. The CuO-NPs were utilized as the working surface in the hybrid BHJ photovoltaic cell. Remarkably, a PCE of just about 2.85% was realized for pure P3HT: $PC_{70}BM$ photovoltaic cells with a photocurrent density of 9.43 mA/cm^2. Conversely, with the addition of an ideal quantity of CuO-NPs to the blend (P3HT: CuO5wt%: $PC_{70}BM$, i.e., 50% at 550 nm), a higher PCE of 3.82% was achieved in addition to a photocurrent density of about 11.32 mA/cm^2, and no harm was done to the open circuit voltage and fill factor (FF). This was achievable owing to adequate excitation production, improved light uptake and a photoexcited charge splitting and recovery [92].

A blend of poly (3-hexylthiophene)(P3HT) in addition to an inorganic nanomaterial comprising narrow band gap GaAs nanowires (NWs) for organic/inorganic photovoltaic cell architecture was reported [93]. The steadily sized GaAs NWs was blended with a closely bound polymer P3HT in a single liquid to create a standardized film involving scattered nanowires in a polymeric template. Contrary to introducing the polymer steadily into NW arrangements, the structure and alignment were designed by using spin coating followed by thermal treatment techniques. Noticeably, an enhancement in the charge carrier extraction was observed through automatic adjustment of nanowires above the working area and the subsequent creation of a P3HT-rich under layer. This enhancement (i.e., P3HT molecular ordering) was attributed to the unidimensional structure of the GaAs nanowires. The device performance was further enhanced through modification of the organic/inorganic intermediate layer for effective charge splitting; hence, GaAs nanowires were coated with 3 nm of TiO_x (50 wt. %). The PCE of the devised P3HT/NW photovoltaic cells was improved by 20%, yielding an efficiency of 2.36% [93].

The electronic and optoelectronic attributes of graphene quantum dots (GQDs) have also been explored by regulating its size and functional groups. For the record, Kim et al. recounted the synthesis of reduction regulated GQDs as well as their usage in BHJ solar cells [94]. Three varieties of GQDs with diverse oxidation degrees were synthesized, namely the oxidized GQDs (GOQD), 5 h reduced GQDs (GQD 5) and 10 h reduced GQDs (GQD 10) (Fig. 7.13). The researchers claimed that no further reduction was considered owing to the fact that saturation of as-synthesized materials was observed after a duration of just about 10 h. The as-synthesized materials were then added to the BHJ OPVs with PTB7: PC71BM (PTB7: PC71BM in 1:1.5 in weight). Notably, the positive effect of GQDs was found to vary along with reduction time of GQDs, where short circuit current density (J_{sc}) rises with oxidation, whereas the fill factor rises with reduction. Thus, the GQD 5 led to a rise in PCE from 6.7% to 7.6% as against the standard device excluding GQDs (6.70%) by balancing optical absorptivity and electrical conductivity. This implies that the enhanced PCE of

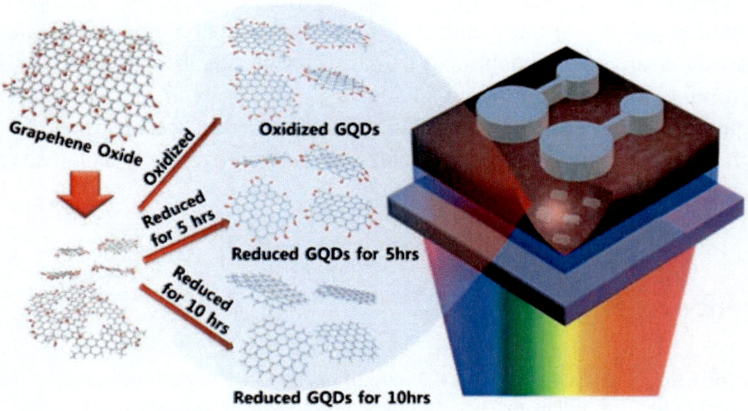

Fig. 7.13 Graphic image of a BHJ photovoltaic cell including three distinct kinds of GQDs, such that the functional groups at the side are regulated by thermal reduction time. Adapted from [94] with consent of © 2013 American chemical society

GQD-incorporated OPV systems is significant to charge carrier transmission and the light-absorption capacity of the various GQDs, thus further describing QD materials as potential materials that can be utilized for enhancing the efficiency of OPVs.

7.3 Conclusions and Future Scope

This topic has clearly identified and demonstrated the different properties of nanostructured materials as well as their photoelectrochemical applications in emerging technologies—photoelectrochemical biosensors, PEC water splitting devices for hydrogen generation and third-generation photovoltaics. It has also demonstrated the advantages of both unidimensional and nanocomposites when put into use. While unidimensional nanomaterials can offer leading electron transport routes as well as lower recombination of charges owing to surface defects and grain boundaries, nanocomposites/hybrid nanomaterials can also enhance the functionalities of the emerging technologies owing to their synergistic effects compared to monofunctional nanomaterials. The topic also apparently found that modification of organic/inorganic interfaces will help to improve effective charge separation. More so, the synergistic effect of nanostructured materials has been proven to possess many advantages such as fast electron transport pathways, large surface area, efficient excitation generation, improved light absorption and photoexcited charge separation and collection. Owing to these reasons, better sensitivity and specificity as well as improved detection limits will be achievable in photoelectrochemical biosensing; more so, high PCE as well as photocurrent density will be achievable in solar-to-hydrogen generation and third-generation photovoltaic technologies. Further work is needed to be done through the exploration of synergies of nanomaterials to fully realize their potentials in biosensors especially in the area of cancer research. On that note, future concentration will be on the wonders of phthalocyanines as emerging materials in biomedical research with much focus on early cancer diagnosis. In PEC water splitting and photovoltaics, continuous increases in efficiency, durability and cost are still required for market sustainability. In order to make such giant strides, combined research efforts in multidisciplinary fields such as photovoltaics, nanotechnologies and computational materials are needed.

References

1. L.J. Johnston, N. Gonzalez-Rojano, K.J. Wilkinson, B. Xing, Key challenges for evaluation of the safety of engineered nanomaterials. NanoImpact **18**, 100219 (2020)
2. A. Bratovcic, Different applications of nanomaterials and their impact on the environment. Int. J. Mater. Sci. Eng. **5**, 1–7 (2019)
3. AZO Materials, The Applications of Nanomaterials 1–9 (2001). https://www.azom.com/article.aspx?ArticleID=1066

4. C. Rodríguez-Ibarra, A. Déciga-Alcaraz, O. Ispanixtlahuatl-Meráz, E.I. Medina-Reyes, N.L. Delgado-Buenrostro, Y.I. Chirino, International landscape of limits and recommendations for occupational exposure to engineered nanomaterials. Toxicol. Lett. **322**, 111–119 (2020)
5. K. Pyrzynska, Nanomaterials in speciation analysis of metals and metalloids. Talanta **212**, 120784 (2020)
6. K. Zhang, S. Lv, Q. Zhou, D. Tang, CoOOH nanosheets-coated g-C_3N_4/$CuInS_2$ nanohybrids for photoelectrochemical biosensor of carcinoembryonic antigen coupling hybridization chain reaction with etching reaction. Sensors Actuators. B Chem. **307**, 127631 (2020)
7. S. Ge, F. Lan, F. Yu, J. Yu, Applications of graphene and related nanomaterials in analytical chemistry New. J. Chem. **39**, 2380–2395 (2015)
8. S.K. Bhullar, H.P. Singh, G. Kaur, H.S. Buttar, An overview of the applications of nanomaterials and development of stents in treating cardiovascular disorders. Rev. Adv. Mater. Sci. **44**, 286–296 (2016)
9. T. Taguchi, S. Yamamoto, H. Ohba, Synthesis of novel hybrid carbon nanomaterials inside silicon carbide nanotubes by ion irradiation. Acta Mater. **173**, 153–162 (2019)
10. M. Martinelli, M.C. Strumia, Multifunctional nanomaterials: design, synthesis and application properties. Molecules **22**, 1–11 (2017)
11. A.M. El-Toni, M.A. Habila, J.P. Labis, Z.A. ALOthman, M. Alhoshan, A.A. Elzatahry, F. Zhang, Design, synthesis and applications of core-shell, hollow core, and nanorattle multifunctional nanostructures. Nanoscale **8**, 2510–2531 (2016)
12. W.-C. Lu, C.M. Quezada, K.-S. Chang, Two-step solvothermal synthesis of $BaZnO_2$ films on indium tin oxide substrates and their piezo-related and photoelectrochemical performance. Mater. Chem. Phys. **247**, 122880 (2020)
13. A. Devadoss, P. Sudhagar, C. Terashima, K. Nakata, A. Fujishima, Photoelectrochemical biosensors: New insights into promising photoelectrodes and signal amplification strategies. J. Photochem. Photobiol. C Photochem. Rev. **24**, 43–63 (2015)
14. W.W. Zhao, J.J. Xu, H.Y. Chen, Photoelectrochemical DNA biosensors. Chem. Rev. **114**, 7421–7441 (2014)
15. X. Ma, C. Wang, F. Wu, Y. Guan, G. Xu, TiO_2 nanomaterials in photoelectrochemical and electrochemiluminescent biosensing. Top. Curr. Chem. **378**, 1–17 (2020)
16. J.T. Cao, Y.X. Dong, Y. Ma, B. Wang, S.H. Ma, Y.M. Liu, A ternary CdS@Au-g-C_3N_4 heterojunction-based photoelectrochemical immunosensor for prostate specific antigen detection using graphene oxide-CuS as tags for signal amplification. Anal. Chim. Acta **1106**, 183–190 (2020)
17. S.K. Chatterjee, B.R. Zetter, Cancer biomarkers: knowing the present and predicting the future. Futur. Oncol. **1**, 37–50 (2005)
18. J. Tian, P. Zhao, S. Zhang, G. Huo, Z. Suo, Z. Yue, S. Zhang, W. Huang, B. Zhu, Platinum and iridium oxide co-modified TiO_2 nanotubes array based photoelectrochemical sensors for glutathione. Nanomaterials **10**, 1–10 (2020)
19. X. Li, Y. Xu, Y. Chen, C. Wang, J. Jiang, J. Dong, H. Yan, X. Du, Dual enhanced electrochemiluminescence of aminated Au@SiO_2/CdS quantum dot superstructures: electromagnetic field enhancement and chemical enhancement. ACS Appl. Mater. Interfaces **11**, 4488–4499 (2019)
20. M. Sui, Y. Zhao, Z. Ni, X. Gu, Photoelectrochemical performance and biosensor application for glutathione (GSH) of W-doped $BiVO_4$ thin films. J. Mater. Sci. Mater. Electron. **29**, 10109–10116 (2018)
21. Z. Li, J. Zhang, Y. Li, S. Zhao, P. Zhang, Y. Zhang, J. Bi, G. Liu, Z. Yue, Carbon dots based photoelectrochemical sensors for ultrasensitive detection of glutathione and its applications in probing of myocardial infarction. Biosens. Bioelectron. **99**, 251–258 (2018)
22. V. Subramanian, E. Wolf, P.V. Kamat, Semiconductor-metal composite nanostructures. To what extent do metal nanoparticles improve the photocatalytic activity of TiO_2 films? J. Phys. Chem. B **105**, 11439–11446 (2001)
23. Z. Zhang, Z. Wang, S.W. Cao, C. Xue, Au/Pt nanoparticle-decorated TiO_2 nanofibers with plasmon-enhanced photocatalytic activities for solar-to-fuel conversion. J. Phys. Chem. C **117**, 25939–25947 (2013)

24. N. Chandrasekharan, P.Y. Kamat, Improving the photoelectrochemical performance of nanostructured TiO_2 films by adsorption of gold nanoparticles. J. Phys. Chem. B **104**, 10851–10857 (2000)
25. A. Akita, H. Kobayashi, H. Tada, Ultrathin silicon oxide film-induced enhancement of charge separation and transport of nanostructured titanium(IV) oxide photoelectrode. ChemPhysChem **20**, 2054–2059 (2019)
26. M. Wang, L. Sun, Z. Lin, J. Cai, K. Xie, C. Lin, p-n heterojunction photoelectrodes composed of Cu_2O-loaded TiO_2 nanotube arrays with enhanced photoelectrochemical and photoelectrocatalytic activities. Energy Environ. Sci. **6**, 1211–1220 (2013)
27. L. Ji, D. Spanu, N. Denisov, S. Recchia, P. Schmuki, M. Altomare, A dewetted-dealloyed nanoporous Pt Co-catalyst formed on TiO_2 nanotube arrays leads to strongly enhanced photocatalytic H_2 production. Chem—An Asian J. **15**, 301–309 (2020)
28. A.L. Linsebigler, G. Lu, J.T. Yates Jr., Photocatalysis on TiO_2 surfaces: principles, mechanisms, and selected results. Chem. Rev. **95**, 735–758 (1995)
29. X.Z. Li, F.B. Li, Study of Au/Au^{3+}-TiO_2 photocatalysts toward visible photooxidation for water and wastewater treatment. Environ. Sci. Technol. **35**, 2381–2387 (2001)
30. Y. Guo, J. He, S. Wu, T. Wang, G. Li, Y. Hu, H. Xue, X. Sun, J. Tang, M. Liu, Effects of platinum on photo-assisted electrocatalytic activity of fringe-shaped highly ordered mesoporous titanium dioxide film. J. Power Sources **208**, 58–66 (2012)
31. B.H. Meekins, P.V. Kamat, Role of water oxidation catalyst IrO_2 in shuttling photogenerated holes across TiO_2 interface. J. Phys. Chem. Lett. **2**, 2304–2310 (2011)
32. P.V. Kamat, Manipulation of charge transfer across semiconductor interface. A criterion that cannot be ignored in photocatalyst design. J. Phys. Chem. Lett. **3**, 663–672 (2012)
33. D. Wang, W. Wang, Q. Wang, Z. Guo, W. Yuan, Spatial separation of Pt and IrO_2 cocatalysts on SiC surface for enhanced photocatalysis. Mater. Lett. **201**, 114–117 (2017)
34. K. Chen, J. Xue, Q. Zhou, Y. Zhang, M. Zhang, Y. Zhang, H. Zhang, Y. Shen, Coupling metal-organic framework nanosphere and nanobody for boosted photoelectrochemical immunoassay of human epididymis protein 4. Anal. Chim. Acta **1107**, 145–154 (2020)
35. C. Wang, X. Ye, Z. Wang, T. Wu, Y. Wang, C. Li, Molecularly imprinted photo-electrochemical sensor for human epididymis protein 4 based on polymerized ionic liquid hydrogel and gold nanoparticle/ZnCdHgSe quantum dots composite film. Anal. Chem. **89**, 12391–12398 (2017)
36. D. Holmes, Ovarian cancer: beyond resistance. Nature **527**, S217–S219 (2015)
37. H. Liu, S. Xu, Z. He, A. Deng, J.J. Zhu, Supersandwich cytosensor for selective and ultrasensitive detection of cancer cells using aptamer-DNA concatamer-quantum dots probes. Anal. Chem. **85**, 3385–3392 (2013)
38. X. Ren, J. Yan, D. Wu, Q. Wei, Y. Wan, Nanobody-based apolipoprotein e immunosensor for point-of-care testing. ACS Sensors **2**, 1267–1271 (2017)
39. L. Farzin, M. Shamsipur, L. Samandari, S. Sheibani, HIV biosensors for early diagnosis of infection: the intertwine of nanotechnology with sensing strategies. Talanta **206**, 120201 (2020)
40. M.A.M. Rodrigo, Z. Heger, N. Cernei, A.M.J. Jinemez, O. Zitka, V. Adam, R. Kizek, HIV biosensors—the potential of the electrochemical way. Int. J. Electrochem. Sci. **9**, 3449–3457 (2014)
41. http://aidsinfo.unaids.org/, Number of people living with HIV. **43**, 167–173 (2016).
42. J. Fan, Y. Zang, J. Jiang, J. Lei, H. Xue, Beta-cyclodextrin-functionalized CdS nanorods as building modules for ultrasensitive photoelectrochemical bioassay of HIV DNA Biosens. Bioelectron. **142**, 111557 (2019)
43. G. Danaei, M.M. Finucane, Y. Lu, G.M. Singh, M.J. Cowan, C.J. Paciorek, J.K. Lin, F. Farzadfar, Y-H. Khang, G.A. Stevens, M.K. Ali, L.M. Riley, C.A. Robinson, M. Ezzati, National, regional, and global trends in fasting plasma glucose and diabetes prevalence since 1980: Systematic analysis of health examination surveys and epidemiological studies with 370 country-years and 2·7 million participants Lancet **378,** 31–40 (2011)
44. L. Wang, W. Zhu, W. Lu, L. Shi, R. Wang, R. Pang, Y. Cue, F. Wang, X. Xu, One-step electrodeposition of AuNi nanodendrite arrays as photoelectrochemical biosensors for glucose and hydrogen peroxide detection Biosens. Bioelectron. **142**, 111577 (2019)

45. W.V. Tamborlane, R.W. Beck, B.W. Bode, B. Buckingham, H.P. Chase, R. Clemons, R. Fiallo-Scharer, L.A. Fox, L.K. Gilliam, I.B. Hirsch, E.S. Huang, C. Kollman, A.J. Kowalski, L. Laffel, J.M. Lawrence, J. Lee, N. Mauras, M. O'Grady, K.J. Ruedy, M. Tansey, E. Tsa-likian, S. Weinzimer, D.M. Wilson, H. Wolpert, T. Wysocki, D. Xing, Continuous glucose monitoring and type 1 diabetes N. Engl. J. Med. **360**, 1464–1476 (2009)
46. J. Sun, Y. Zhu, X. Yang, C. Li, Photoelectrochemical glucose biosensor incorporating CdS nanoparticles. Particuology **7**, 347–352 (2009)
47. W. Xu, W. Yang, H. Guo, L. Ge, J. Tu, C. Zhen, Constructing a TiO_2/PDA core/shell nanorod array electrode as a highly sensitive and stable photoelectrochemical glucose biosensor. RSC Adv. **10**, 10017–10022 (2020)
48. https://www.energy.gov/eere/fuelcells/hydrogen-production-photoelectrochemical-water-splitting, Hydrogen Production: Photoelectrochemical Water Splitting Research Focuses on Overcoming Challenges.
49. P. Ma, D. Wang, The principle of photoelectrochemical water splitting. Nanomater. Energy Convers. Storage. 1–61 (2017)
50. Q.-X. Zhang, H. Wen, D. Peng, Q. Fu, X.-J. Huang, Interesting interference evidences of electrochemical detection of Zn(II), Cd(II) and Pb(II) on three different morphologies of MnO_2 nanocrystals. J. Electroanal. Chem. **739**, 89–96 (2014)
51. J.W. Ager, Photoelectrocehmical approach for water splitting, in *Solar to Chemical Energy Conversion: Lect. Notes in Energy*, vol. 32, ed. by M. Sugiyama, K. Fujii, S. Nakamura, (2016) pp. 249–260
52. D.J. Martin, Investigation into high efficiency visible light photocatalysts for water reduction and oxidation. 1–53 (2015)
53. C. Mahala, M.D. Sharma, M. Basu, ZnO nanosheets decorated with graphite-like carbon nitride quantum dots as photoanodes in photoelectrochemical water splitting. ACS Appl. Nano Mater **3**, 1999–2007 (2020)
54. C. Mahala, M.D. Sharma, M. Basu, ZnO@CdS heterostructures: an efficient photoanode for photoelectrochemical water splitting. New J. Chem. **43**, 7001–7010 (2019)
55. P. Fageria, S. Uppala, R. Nazir, S. Gangopadhyay, C.H. Chang, M. Basu, S. Pande, Synthesis of monometallic (Au and Pd) and bimetallic (AuPd) nanoparticles using carbon nitride (C_3N_4) quantum dots via the photochemical route for nitrophenol reduction. Langmuir **32**, 10054–10064 (2016)
56. Y. Huang, Y. Yu, Y. Xin, N. Meng, Y. Yu, B. Zhang, Promoting charge carrier utilization by integrating layered double hydroxide nanosheet arrays with porous $BiVO_4$ photoanode for efficient photoelectrochemical water splitting. Sci. China Mater. **60**, 193–207 (2017)
57. A.R. Marlinda, N. Yusoff, S. Sagadevan, M.R. Johan, Recent developments in reduced graphene oxide nanocomposites for photoelectrochemical water-splitting applications. Int. J. Hydrogen Energy **45**, 11976–11994 (2020)
58. P. Subramanyam, T. Vinodkumar, D. Nepak, M. Deepa, C. Subrahmanyam, Mo-doped $BiVO_4$ @reduced graphene oxide composite as an efficient photoanode for photoelectrochemical water splitting. Catal. Today **325**, 73–80 (2019)
59. Y. Chen, X. Feng, Y. Liu, X. Guan, C. Burda, L. Guo, Metal oxide-based tandem cells for self-biased photoelectrochemical water splitting. ACS Energy Lett. **5**, 844–866 (2020)
60. F. Urbain, P. Tang, V. Smirnov, K. Welter, T. Andreu, F. Finger, J. Arbiol, J.R. Morante, Multilayered hematite nanowires with thin-film silicon photovoltaics in an all-earth-abundant hybrid tandem device for solar water splitting. Chemsuschem **12**, 1428–1436 (2019)
61. M. Kim, B. Lee, H. Ju, J.Y. Kim, J. Kim, S.W. Lee, Oxygen-vacancy-introduced $BaSnO3-\delta$ photoanodes with tunable band structures for efficient solar-driven water splitting. Adv. Mater. **31**, 1–9 (2019)
62. M.I. Hoffert, K. Caldeira, A.K. Jain, E.F. Haites, L.D.D. Harvey, S.D. Potter, M.E. Schlesinger, S.H. Schneider, R.G. Watts, T.M.L. Wigley, D.J. Wuebbies, Energy implications of future stabilization of atmospheric CO_2 content. Nature **395**, 881–884 (1998)
63. C. Wu, K. Wang, M. Batmunkh, A.S.R. Bati, D. Yang, Y. Jiang, Y. Hou, J.G. Shapter, S. Priya, Multifunctional nanostructured materials for next generation photovoltaics. Nano Energy **70**, 104480 (2020)

64. L. Wang, H. Liu, R.M. Konik, J.A. Misewich, S.S. Wong, Carbon nanotube-based heterostructures for solar energy applications. Chem. Soc. Rev. **42**, 8134–8156 (2013)
65. N.S. Lewis, Toward cost-effective solar energy use. Science **315**, 798–801 (2007)
66. N. AL-Rousan, N.A.M. Isa, M.K.M. Desa, Advances in solar photovoltaic tracking systems: a review. Renew. Sustain. Energy Rev. **82**, 2548–2569 (2018).
67. D.M. Chapin, C.S. Fuller, G.L. Pearson, A new silicon p-n junction photocell for converting solar radiation into electrical power. J. Appl. Phys. **25**, 676–677 (1954)
68. P.K. Nayak, S. Mahesh, H.J. Snaith, D. Cahen, Photovoltaic solar cell technologies: analysing the state of the art. Nat. Rev. Mater. **4**, 269–285 (2019)
69. M. Bosi, C. Pelos, The potential of III-V semiconductors as terrestrial photovoltaic devices. Prog. Photovoltaics Res. Appl. **15**, 51–68 (2007)
70. T.W. Hamann, R.A. Jensen, A.B.F. Martinson, H. Van Ryswyk, J.T. Hupp, Advancing beyond current generation dye-sensitized solar cells. Energy Environ. Sci. **1**, 66–78 (2008)
71. G. Conibeer, Third-generation photovoltaics. Mater. Today **10**, 42–50 (2007)
72. I.J. Kramer, J.C. Minor, G. Moreno-Bautista, L. Rollny, P. Kanjanaboos, D. Kopilovic, S.M. Thon, G.H. Carey, K.W. Chou, D. Zhitomirsky, A. Amassian, E.H. Sargent, Efficient spray-coated colloidal quantum dot solar cells. Adv. Mater. **1404**, 116–121 (2014)
73. N.E. Hjerrild, D.C.J. Neo, A. Kasdi, H.E. Assender, J.H. Warner, A.A.R. Watt, Transfer printed silver nanowire transparent conductors for PbS-ZnO heterojunction quantum dot solar cells. ACS Appl. Mater. Interfaces. **7**, 6417–6421 (2015)
74. C.-H.M. Chuang, P.R. Brown, V. Bulovic, M. Bawendi, Improved performance and stability in quantum dot solar cells through band alignment engineering. Nat. Mater. **13**, 796–801 (2014)
75. Y. Xiao, J. Wu, G. Yue, G. Xie, J. Lin, M. Huang, The preparation of titania nanotubes and its application in flexible dye-sensitized solar cells. Electrochim. Acta **55**, 4573–4578 (2010)
76. T. Kasuga, M. Hiramatsu, A. Hoson, T. Sekino, K. Niihara, Titania nanotubes prepared by chemical processing. Adv. Mater. **11**(15), 1307–1311 (1999). https://doi.org/10.1002/(SICI)1521-4095(199910)11:15%3c1307::AID-ADMA1307%3e3.0.CO;2-H
77. T. Kasuga, M. Hiramatsu, A. Hoson, T. Sekino, K. Niihara, Formation of titanium oxide nanotube. Langmuir **14**, 3160–3163 (1998)
78. M. Wei, Y. Konishi, H. Zhou, H. Sugihara, H. Arakawa, Utilization of titanate nanotubes as an electrode material in dye-sensitized solar cells. J. Electrochem. Soc. **153**, A1232–A1236 (2006)
79. Y. Ohsaki, N. Masaki, T. Kitamura, Y. Wada, T. Okamoto, T. Sekino, K. Niihara, S. Yanagida, Dye-sensitized TiO_2 nanotube solar cells: fabrication and electronic characterization. Phys. Chem. Chem. Phys. **7**, 4157–4163 (2005)
80. C. Xu, J. Wu, U.V. Desai, D. Gao, High-efficiency solid-state dye-sensitized solar cells based on TiO_2-coated ZnO nanowire arrays. Nano Lett **12**, 2420–2424 (2012)
81. Z. Ku, Y. Rong, M. Xu, T. Liu, H. Han, Full printable processed mesoscopic $CH_3NH_3PbI_3$ /TiO_2 heterojunction solar cells with carbon counter electrode. Sci. Rep. **3**, 3132 (2013)
82. Y. Li, Z. Wang, D. Ren, Y. Liu, A. Zheng, S.M. Zakeeruddin, X. Dong, A. Hagfield, M. Grätzel, P. Wang, SnS quantum dots as hole transporter of perovskite solar cells. ACS Appl. Energy Mater. **2**, 3822–3829 (2019)
83. W. Chen, K. Li, Y. Wang, X. Feng, Z. Liao, Q. Su, X. Lin, Z. He, Black phosphorus quantum dots for hole extraction of typical planar hybrid perovskite solar cells. J. Phys. Chem. Lett. **8**, 591–598 (2017)
84. C. Liu, K. Wang, X. Gong, A.J. Heeger, Low band gap semiconducting polymers for polymeric photovoltaics. Chem. Soc. Rev. **45**, 4825–4846 (2016)
85. K. Wang, C. Liu, T. Meng, C. Yi, X. Gong, Inverted organic photovoltaic cells. Chem. Soc. Rev. **45**, 2937–2975 (2016)
86. N.S. Sariciftci, L. Smilowitz, A.J. Heeger, F. Wudl, Photoinduced electron transfer from a conducting polymer to buckminsterfullerene. Science **258**, 1474–1476 (1992)
87. B. Kraabel, C.H. Lee, D. McBranch, D. Moses, N.S. Sariciftci, A.J. Heeger, Ultrafast photoinduced electron transfer in conducting polymer-buckminsterfullerene composites. Chem. Phys. Lett. **213**, 389–394 (1993)

88. G. Yu, A.J. Heeger, Charge separation and photovoltaic conversion in polymer composites with internal donor/acceptor heterojunctions. J. Appl. Phys. **78**, 4510–4515 (1995)
89. G. Yu, J. Gao, J.C. Hummelen, F. Wudl, A.J. Heeger, Polymer photovoltaic cells: enhanced efficiencies via a network of internal donor-acceptor heterojunctions. Science **270**, 1789–1791 (1995)
90. G. Li, C.W. Chu, V. Shrotriya, J. Huang, Y. Yang, Efficient inverted polymer solar cells. Appl. Phys. Lett. **88**, 1–4 (2006)
91. H.H. Liao, L.M. Chen, Z. Xu, G. Li, Y. Yang, Highly efficient inverted polymer solar cell by low temperature annealing of Cs_2CO_3 interlayer. Appl. Phys. Lett. **92**, 1–4 (2008)
92. H. Siddiqui, M.R. Parra, P. Pandey, M.S. Qureshi, F.Z. Haque, Utility of copper oxide nanoparticles (CuO-NPs) as efficient electron donor material in bulk-heterojunction solar cells with enhanced power conversion efficiency. J. Sci. Adv. Mater. Devices **5**, 104–110 (2020)
93. S. Ren, N. Zhao, S.C. Crawford, M. Tambe, V. Bulović, S. Gradečak, Heterojunction photovoltaics using GaAs nanowires and conjugated polymers. Nano Lett. **11**, 408–413 (2011)
94. J.K. Kim, M.J. Park, S.J. Kim, D.H. Wang, S.P. Cho, S. Bae, J.H. Park, B.H. Hong, Balancing light absorptivity and carrier conductivity of graphene quantum dots for high-efficiency bulk heterojunction solar cells. ACS Nano **7**, 7207–7212 (2013)

Chapter 8
Electrode Materials for Pharmaceuticals Determination

Azeez Olayiwola Idris, Onoyivwe Monday Ama, Kabir Opeyemi Otun, Seyi Philemon Akanji, Usisipho Feleni, Bhekie Mamba, Robert Birundu Onyancha, Uyiosa Osagie Aigbe, and Kingsley Eghonghon Ukhurebor

Abstract The presence of pharmaceuticals in surface waters called for urgent concern in recent years due to their prospective environmental effects. Various analytical methods including chemiluminescence, high-performance liquid chromatography, capillary electrophoresis-mass spectrometry, spectrophotometry and liquid chromatography have been employed for the determination of various pharmaceuticals. However, all these techniques are time-consuming, complicated and require expensive equipment. On the contrary, the electrochemical technique resolved these

K. O. Otun
Institute for the Development of Energy for African Sustainability, University of South Africa, Private Bag, Johannesburg 1709, South Africa

Department of Chemical, Geological and Physical Sciences, College of Pure and Applied Sciences, Kwara State University, Malete, PMB. 1530, Ilorin, Nigeria

A. O. Idris (✉) · U. Feleni · B. Mamba
Institute for Nanotechnology and Water Sustainability (iNanoWS), Florida Campus, College of Science Engineering and Technology, University of South Africa, Johannesburg 1709, South Africa

O. M. Ama · S. P. Akanji
Department of Chemical Science, University of Johannesburg, Doornfontein, South Africa

R. B. Onyancha
Department of Physics and Space Science, School of Physical Sciences and Technology, Technical University of Kenya, Nairobi, Kenya

U. O. Aigbe
Department of Mathematics and Physics, Faculty of Applied Sciences, Cape Peninsula University of Technology, Cape Town, South Africa

K. E. Ukhurebor
Climatic/Environmental/Telecommunication Unit, Department of Physics, Edo University Iyamho, Edo State, Nigeria

O. M. Ama et al. (eds.), *Modified Nanomaterials for Environmental Applications*, Engineering Materials, https://doi.org/10.1007/978-3-030-85555-0_8

problems owing to its low cost, fast response, simplicity and ease of on-site application. Considering this, various electrodes have played significant roles in the determination of different drugs in biological, urine and pharmaceuticals formulations. Various electrodes are modified with various nanomaterials to improve the sluggish electron migration and electrode fouling, which reduces their selectivity and sensitivity. Considering this, the present chapter discusses the applications of various electrodes for different electrochemical analyses.

Graphical Abstract

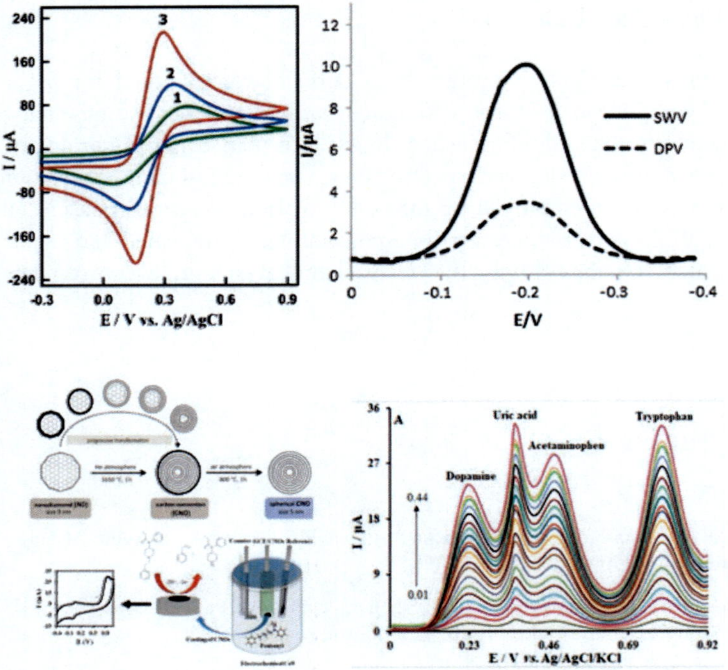

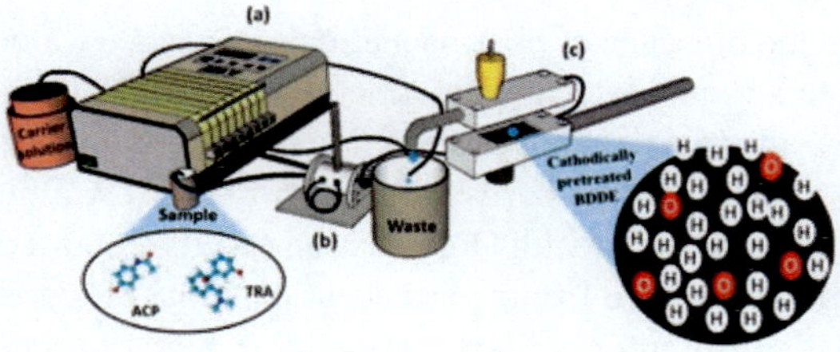

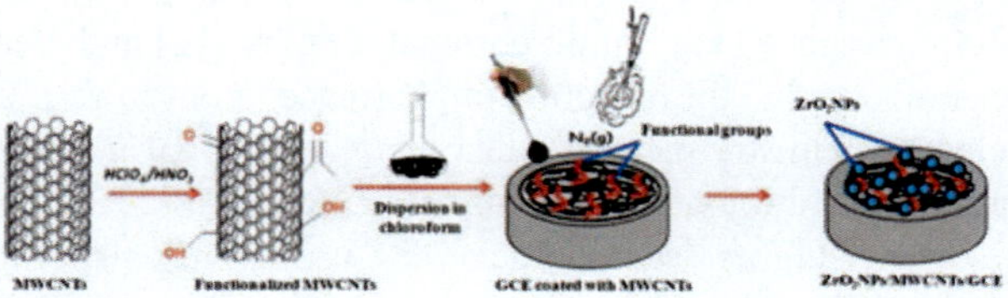

Highlights

1. The application of various electrodes for electrochemical analysis.
2. The surface of boron-doped diamond electrodes can be cathodically and anodically pretreated.
3. Various nanomaterials are networked on the electrode surface to promote electron communication and migration.
4. Co-detection of various pharmaceuticals in various matrices.
5. Different binders are used to prevent the leaching of nanomaterials on the electrode surface.

Synopsis
Several analytical techniques are used in the detection of pharmaceuticals owing to their presence in various water bodies. Electrochemical methods have been the technique of choice owing to their analytical merits including simplicity, portability and low cost. In electrochemistry, different working electrodes play pivotal roles in performing an electrochemical experiment. Although they have several limitations such as sluggish electron transfer, high background current and electron fouling. These limitations can be overcome using smart and intelligent nanomaterials.

8.1 Introduction

In recent years, pharmaceuticals have attracted scientists, industrial communities and public attention owing to their occurrence in various water bodies including wastewater, groundwater and drinking water [1, 2]. The presence of pharmaceuticals has been reported in water bodies in Europe and worldwide [3, 4]. Although, various sources contributed significantly to the presence of pharmaceuticals in surface water including agricultural runoff, effluents from wastewater treatments plants (WWTPs), pharmaceuticals industrial wastes, pharmaceuticals used for animal and human medicine that are not completely metabolised or partly removed in (WWTPs) are discharged into water bodies [5, 6]. The World Health Organisation (WHO) reported that 25% of the world water pollution emanated from pharmaceutical industries and various studies have confirmed that consumption of water containing high concentrations of pharmaceuticals can lead to various adverse effects such as acute renal failure, vomiting, diarrhoea, hepatic necrosis and hepatic failure [7, 8].

However, it is important to thoroughly monitor the concentrations of pharmaceuticals in the waterbodies. Various analytical techniques such as spectrophotometry [9], titrimetry [10], chemiluminescence [11], liquid chromatography [12] and electrochemistry [13] have been employed in the detection of pharmaceuticals in various matrices. Aside from the electrochemistry method, other techniques involve longer analysis time, tedious sample preparation and costly instrumentation [14, 15]. The electrochemistry technique offers a higher degree of accuracy, precision, sensitivity, selectivity and is suitable for real-time detection of environmental pollutants due to the portability of the instrument [16].

Electrochemistry is predominantly based on electrodes made up of solid materials. These electrodes must possess the following analytical merits including electrochemical stability, fast electron transfer, electrical conductivity, surface reproducibility, chemical stability, lower toxicity, high conductivity and low background current [17, 18]. For this reason, this chapter focuses on the application of various electrodes such as carbon paste electrode, boron-doped diamond electrode, glassy carbon electrode, pencil graphite electrode and screen-printed carbon electrode for monitoring pharmaceuticals in various matrices.

8.1.1 Carbon Paste Electrode (CPE)

It belongs to a unique group of heterogeneous carbon electrodes and its surface can easily be renewed for electron exchange [19]. It is prepared from a mixture of pasting liquid and conducting graphite powder. CPEs have received great attention due to its low ohmic resistance, wide potential window, easy modification process, non-toxicity, good electrical conductivity, stable response, facile renewability of electrode surface after usage, ease of application and regeneration [19]. The uniqueness of CPE is due to its availability at a low cost and ease of modification with nanomaterials. Although bare CPE is not so sensitive for trace level sensing because its surface interacts with matter and air, due to this, different nanomaterials have been used to modify carbon paste electrode to enhance its electron transfer kinetics, catalytic and conductivity properties [19]. In addition, liquid modifiers are used to enhance the signal intensity, improve the limit of detection and increase the analytical performance of CPEs.

For instance, Pardakhty et al. constructed a novel carbon ionic liquid paste electrode for the electrochemical determination of ascorbic acid in pharmaceuticals and food samples using nickel oxide nanoparticles (NiONPs) and 1-butyl-3-methylimidazolium tetrafluoroborate (BMITFB) as a binder [20]. The results obtained revealed that the fabricated sensor (CPE@BMITFB/NiONPs) effectively promote the electrocatalytic oxidation of ascorbic acid and displayed impressive low background current, convenient modification, renewability, wide potential window and satisfactory reproducibility results.

In another report, aluminium silicate (Al_2SiO_5) was immobilised on CPE for the quantification of anti-depressant dothiepin HCl (DTP) in biological fluids and pharmaceutical formulation [21]. It was highlighted that Al_2SiO_5 was employed for the sensor fabrication owing to its analytical merits including low cost, thermodynamically stable, large surface area, high adsorption and ion exchange capacity. The results obtained from the cyclic voltammetry (CV) and square wave voltammetry (SWV) revealed that the sensor Al_2SiO_5/CPE was reported to enhance the electrical conductivity of the bare CPE by three folds.

A nano-molecularly imprinted polymer (NMIP) modified CPE was employed in the electrochemical detection of quetiapine (anti-psychotic drug) [22]. The bare CPE was assembled by adding 0.075 g of graphite powder and 0.025 g of n-eicosane in a watch glass to form a homogenous paste (HP). The HP was carefully placed in a 2.5 mm glass syringe and compressed using a piston. Thereafter, copper wire was dipped inside the HP to activate the electrical conductivity of the electrode. The modified CPE/NMIP was prepared by adding a mixture of carbon powder and polymer powder to n-eicosane powder, the resulting mixture was carefully placed in a glass syringe and compressed using a piston. It was reported that before each measurement, the electrode was polished on a photo-printing paper to form a smooth surface and gently rinsed with distilled water. The sensor was employed in the detection of quetiapine drugs in pharmaceuticals and urine samples.

Furthermore, multiwalled carbon nanotubes (MWCNTs) were networked on CPE for the quantification of gentamicin sulphate (antibiotic drug) in pure form, biological fluids and pharmaceutical samples [23]. The MWCNT was used for the sensor development owing to its analytical merits including high mechanical strength, metallic or semi-metallic behaviour, large surface area, thermal conductivity and rapid electron migration between the electroactive species and the electrode surface. The sensor was not affected by a change in pH from 2.5 to 8.5 and displayed a fast-dynamic response time (8 s) with a detection limit of 2.2×10^{-7} [23].

Similarly, the detection of isoproterenol (asthmatic drug) in pharmaceutical and biological samples was carried out using pyrogallol red multiwalled carbon nanotubes paste electrode (PGRMMCNTPE) [24]. The electrocatalytic oxidation of isoproterenol (ISPR) at the developed electrode was investigated using CV, SWV and chronoamperometry. Different thermodynamic parameters including electron transfer coefficient, catalytic rate constant and catalytic rate constants were calculated using different electrochemical techniques.

The synergy of gold nanoparticles (AuNPs) and functionalised multiwalled carbon nanotubes (*f*-MWCNTs) was to modify glassy carbon paste electrode (GCPE) for

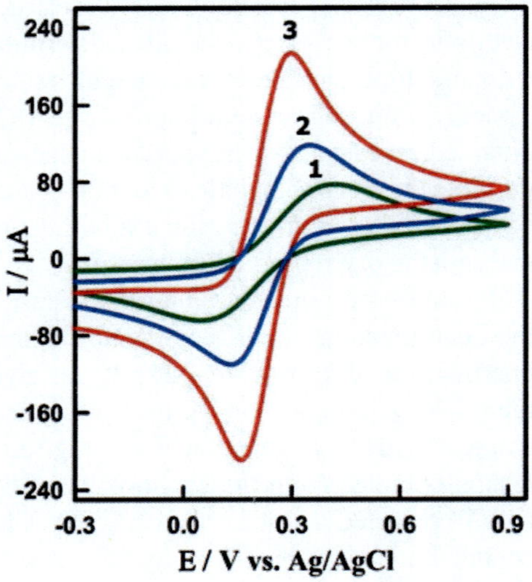

Fig. 8.1 CVs of FCN in 0.1 M KCl obtained at a scan rate of 0.1 Vs^{-1}. Curve 1: bare GCPE, curve 2: f-MWCNTs/GCPE, curve 3: AuNPs/f-MWCNT/GCPE [25]. (Copyright (2018), with permission from Elsevier)

the quantification of cyproterone acetate (steroidal antiandrogens) in pharmaceutical and human body fluids [25]. The cyclic voltammetry was employed to interrogate the electrochemical impact of the modifier on the bare GCPE. The AuNPs/*f*-MWCNTs enhanced the electroactive surface area of the CPE by ˜16 times and facilitated electron transfer for the detection of CPA. There was a significant enhancement in the reduction peak current of CPA because of the reduction of ketone to hydroxyl group. The pitfall of this work was that AuNPs was not used to modify the electrode alone as shown in Fig. 8.1, this oversight makes it very difficult to compare the electrochemical activity and the impact of each modifier in the sensor fabrication.

In another report, ionic liquids (*N*-butyl pyridinium hexafluorophosphate and 1-octylpyridinium hexafluorophosphate were used to modify carbon paste electrodes for the detection of dopamine [26]. The sensor displayed outstanding electrocatalytic properties towards the electrochemical oxidation/reduction of dopamine at micro and sub-micromolar concentration levels. The brilliant electrocatalytic property of the sensor was accredited to the distinct hydrophilicity and the conductivity features of the ionic liquid used to modify the CPE.

Similarly, carbon ionic liquid electrode was assembled through the collaboration of Fe_3O_4@PA-Ni@Pd yolk-shell structure nanoparticles and chitosan for the electrochemical detection of Fluconazole (anti-viral drug) [27]. The nanocomposite of Fe_3O_4@PA-Ni@Pd nanoparticles were used in this study to facilitate electron transfer due to their synergistic electrocatalytic effects while the yolk-shell was used to support the catalyst owing to its low density and large vacant space. The chitosan was used in the construction of the sensor owing to its hydroxyl and amino functional groups, which play critical roles in covalent immobilisation of bioactive molecules. Various electrochemical techniques such as electrochemical impedance spectroscopy (EIS), cyclic voltammetry and chronocoulometry were used to investigate the effect of each modifier on the sensor development, the results obtained for each technique

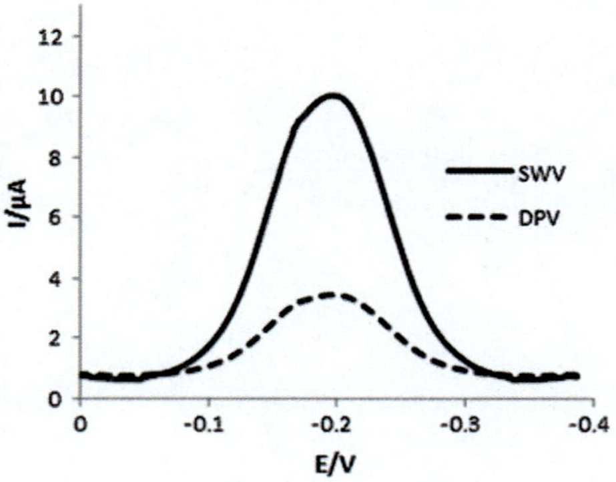

Fig. 8.2 Comparing the SWV and DPV for the detection of flutamide using CPE/ CuO/GO/PANI [28]. (Copyright (2020), with permission from Elsevier)

agree. The sensor was responsive to different concentrations of fluconazole from 0.01 to 400 μmol/L, with a detection limit (DL) of 3.5 nmol/L using differential pulse voltammetry (DPV).

Afzali et al. reported the electrochemical detection of flutamide (an anticancer drug) by modifying carbon paste with CuO/GO/PANI nanocomposite [28]. A comparative study between square wave voltammetry (SWV) and DPV for the detection of flutamide was shown in Fig. 8.2. The result obtained revealed that the peak current signal of SWV is far better than DPV. Hence, it was employed for the detection of flutamide with a DL of 14 pM. The modified electrode (CPE/ CuO/GO/PANI) was used for the quantification of flutamide in pharmaceuticals and urine samples with satisfactory results.

Shetti and co-workers reported the electrochemical analysis of nimesulide (an anti-inflammatory drug) using amberlite XAD-4 resin-modified CPE [29]. The results obtained from the voltammetric techniques (CV and SWV) revealed that the modifier promotes electron migration and improved the sensitivity of the bare CPE. The electrooxidation of nomesulide was an irreversible reaction involving an equal number of protons and electrons in a diffusion-controlled system. The proposed mechanism for the electrooxidation of NIM on the fabricated sensor (XAD/CPE) was illustrated in Scheme 8.1.

In another report, dopamine was polymerised on CPE in the presence of anionic surfactant for the electrochemical oxidation of ranitidine (gastric ulceration drug) [30]. The polymerisation of dopamine monomer was carried out by cycling a potential from −0.4 to +1.5 V for eight cycles at a scan rate of 100 mV/s in acetate buffer solution (pH 5.0) containing 0.001 M dopamine hydrochloride. The polymer film on the electrode surface significantly increases the electrocatalytic properties of the analytes, reduce electrode fouling, increase the reaction rate, improve the reproducibility and stability of the electrode. While the anionic surfactant improved the antifouling properties and the electrocatalytic activities of the sensor. The developed sensor was employed to quantify ranitidine in the concentration range from 1 ×

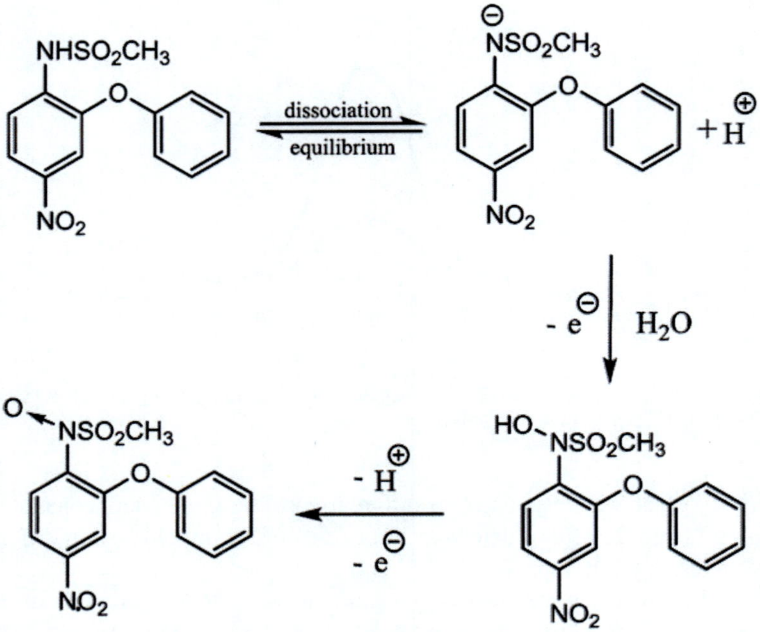

Scheme 8.1 The proposed oxidation mechanism of NIM on XAD/CPE [29]

10^{-7} to 7.5×10^{-6} M, with a detection limit of 1.9×10^{-8} M using square wave voltammetry technique.

Antifouling bi-polymeric membrane was used in modifying carbon nanoparticle paste electrode for the detection of riboflavin (vitamin B2) in biological and pharmaceutical samples [31]. The bi-polymeric membrane was synthesised from fluorosurfactant Zonyl-FSN (FSZ) and perfluorate resin Nafion (PRN). The PRN was used as an antifouling agent owing to its ability to block the anionic neutral species from reaching the electrode matrix and the FSZ enhanced the electrocatalytic activities of the sensor towards the electrooxidation of riboflavin (RF). It is important to note that the electrooxidation of RF involves two protons and two electrons oxidation processes. The developed sensor was used for quantifying riboflavin in urine samples, cough syrup and pharmaceutical tablets with RSD values less than 2.75% revealing the reliability property of the sensor.

Among the most recent developments in the utilisation of CPE, the application that stands out was the fabrication of an electrochemical sensor for the simultaneous detection of dopamine (DA), uric acid (UA), acetaminophen (AC) and tryptophan (Trp) using zeolitic imidazolate (ZI) alongside with cobalt-tannic (Co-Ta) nanocomposite modified CPE [32]. The results obtained from the CV and DPV revealed that the nanocomposite improved the selectivity, increased the surface area, accelerated electron transfer, improved the conductivity and enhanced the electrocatalytic ability of the bare CPE towards the simultaneous detection of the analytes. Interestingly, the peak current signal of the analytes was clearly distinguished by their peak potentials

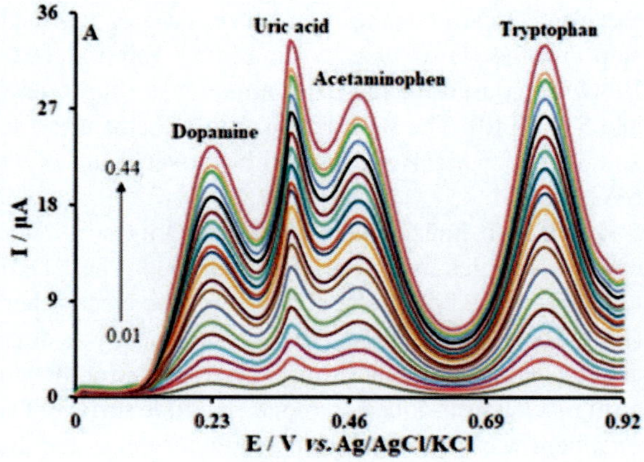

Fig. 8.3 DPV of ZI@Co-Ta/CPE for detection of different concentrations of DA, Ur, Ac and Trp [32]. (Copyright (2020), with permission from Elsevier)

and each analyte was successfully quantified in the quaternary mixture as shown in Fig. 8.3.

Although, the drawback of CPE includes porosity, reproducibility and it takes time before the modifier(s) are packed into the electrode. Yet, we urged scientific and industrial communities to explore and employ CPE for the electrochemical quantification of different pharmaceuticals in water and biological samples due to the exciting results reported from its applications. In the next section, the fabrication of boron-doped diamond electrodes for the quantification of pharmaceuticals is quantitatively discussed.

8.1.2 Boron-Doped Diamond Electrode (BDDE)

It is a new class of electrode material employed for various neurochemistry and electrochemical applications owing to its analytical importance including low background current, high resistance to passivation, chemical inertness, low adsorption capacity, wide potential window, high sensitivity and electrochemical stability in acidic and alkaline media [33, 34]. These analytical properties of BDDEs depend on several factors including crystallographic orientation, surface termination (H, O and F), grain boundaries morphological influences and presence of non-diamond sp^2 carbon, these factors make BDDE surface to be significantly different from other electrodes including CPE and GCE [35, 36].

A BDDE is a powerful electrochemical sensor used for the detection of various biomolecules in food, organic compounds, pharmaceutical samples and in the environment [37, 38]. For instance, Kuzmanovic et al. used unmodified BDDE for the electrochemical detection of pyridoxine (vitamin B_6) in urine and pharmaceuticals samples [39]. The unmodified BDDE sensor was responsive to the detection of vitamin B_6 at a linear range from 7 to 47 μM and a DL of 3.76 μM using differential pulse voltammetry technique. The merits of this developed sensor are attributed to the wide linear concentration range, ease of fabrication without any pretreatment step and the sensor was not modified, this reduced the sensor cost. Similarly, a BDDE was used for the determination of ibuprofen (antipyretic drug) using DPV and SWV [40]. The BDDE sensor was used for the analysis of spiked human urine and pharmaceuticals samples with recovery rates of 97–103% and 99.8–105.0 using SWV.

In addition, benzocaine (skin cancer drug) was detected in biological and pharmaceutical samples using BDDE sensor [41]. The BDDE was anodically and cathodically pretreated before applying it for the electrochemical detection of benzocaine (BC). The cathodic and anodic pre-treatment was done in the presence of 1 M nitric acid by using a potential from −2.0 to 2.0 V for 40 s to obtain hydrogen and oxygen terminated boron-doped diamond electrode surface. The CV signal for the anodic pre-treatment was stable and gave a high current signal than the cathodic pre-treatment. Hence, it was used for the detection of benzocaine. The best optimal experimental parameters were obtained at pH 4 and different supporting electrolytes (acetate, phosphate buffer solution, HCl and NaOH, Britton-Robinson (BR) buffer) were interrogated for the detection of BC. The BR buffer solution was chosen because it gave the lowest background current, well-shaped voltammetry peak current as well as the utmost peak current signal. Under the optimised experimental parameters, the limit of detection values for BC were 80 and 100 nM for DPV and SWV. Similarly, BR buffer was used as the supporting electrolyte for the quantification of codeine in pharmaceutical tablets and human urine using DPV [41].

In another report, BDDE was used for the preparation of electrochemical sensor for the quantification of amlodipine (calcium channel blocker drug) [42]. The CV result for the electrochemical oxidation of amlodipine revealed that it is an irreversible process with a well-defined peak at a potential of +0.75 V. The electrode reaction on BDDE was a two-electron diffusion-controlled system and responsive to different concentration ranges of 0.2–38 μM and 0.2–6 μM for amlodipine.

Ardila et al. used SWV for the detection of bezafibrate (BZF) in pharmaceutical samples using pretreated BDDE [43]. The electrochemical response for the determination of BZF was enhanced by pre-treating BDDE anodically and cathodically. The finding showed that the former displayed a lower oxidation peak potential while the latter presented a higher current intensity and better peak resolution. The cathodically pretreated boron-doped diamond electrode was used for the determination of BZF in pharmaceutical samples and the result obtained was in concord with the spectrophotometric method.

Furthermore, BDDE was used for the electrochemical oxidation of meloxicam (an anti-inflammatory drug) [44]. The electrochemical detection of meloxicam displayed two oxidation peaks, the first peak is the oxidation of the amide functional group and the second represents the oxidation of enol functional group as displayed in Fig. 8.4. Different optimisation parameters such as scan rate, pulse width, pH, supporting electrolyte and pulse height were investigated. A detection limit of 5.9×10^{-8} mol/L was obtained using DPV [44].

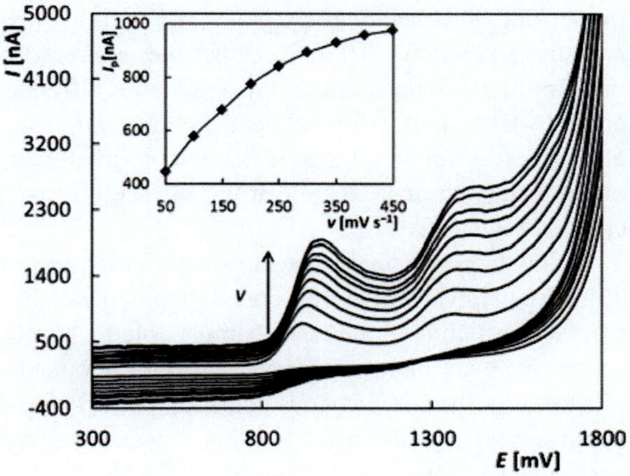

Fig. 8.4 CV of meloxicam on BDDE showing two oxidation peaks at different potentials [44]. (Copyright (2020), with permission from Elsevier)

Selesovska and co-workers investigated the influence of boron to carbon (B/C) ratio for the fabrication of lab-made boron-doped diamond electrodes for the quantification of leucovorin (LCV) [36]. Different B/C ratios (1000, 2000, 4000, 8000 and 20,000 pm) were used for the preparation of BDDEs in the gas phase. The B/C 10,000 pm displayed the best electrochemical properties on the two redox probes $Fe(CN)_6{}^{3-/4-}$ and $Ru(NH_3)_6{}^{2+/3+}$ used to interrogate the constructed BDDEs. Thereafter, it was used to detect different concentrations of LCV with a detection limit of 6.7×10^{-8}.

A comparative study between BDDE and GCE for the electrochemical detection of pefloxacin (PEF) was reported [45]. The cyclic voltammograms for pefloxacin on both electrodes displayed an irreversible oxidation peak and a small irreversible ill-pronounced peak. In addition, the scan rate studies showed that the oxidation step of PEF is irreversible for both electrodes, but it is adsorption and diffusion controlled for GCE and BDDE. Due to the diffusion-controlled process for the BDDE, DPV and SWV were used for the determination of PEF but adsorptive DPV (AdsDPV) was used for the detection of PEF on GCE. Different parameters including accumulation time, stirring rate and potential effect were investigated for the AdsDPV, SWV and DPV using GCE, yet no linear response for the determination of PEF was obtained. This result discouraged the comparative study between GCE and BDDE for the electrochemical detection of PEF. Hence, BDDE was used for the detection of PEF in pharmaceutical and spiked serum and samples.

In another report, the redox properties of lidocaine (LID) and epinephrine (EP) were examined on anodically pretreated unmodified BDDE [46]. The bare BDDE was used for individual and simultaneous detection of LID and EP. The cyclic voltammograms of the binary mixture of EP and LID displayed two irreversible and diffusion-controlled peaks at +0.91 and +1.67 for EP and LID in BR buffer at pH 2.0. The BDDE electrode was anodically pretreated to reduce its hydrophobicity and incorporate negative charge into the electrode surface to promote electrostatic attraction with the positive charge of the analytes. The experimental parameters investigated for the sensing of the analytes are pulse amplitude (10–60 mV), frequency (15–150 Hz) and step potential (4–16 mV). Under the optimised conditions, SWV was used for the simultaneous detection of the analytes at different concentration ranges (0.1–20 and 1.0–100 μmol^{-1} for LID and EP) and DLs of (0.03 and 0.27 μmol^{-1}) were obtained. It is important to note that the pre-treatment of BDDE is done cathodically or anodically to make the electrode surface positive or negative depending on the charge of the analyte.

A flow injection analysis was used for co-detection of acetaminophen (ACP) and tramadol (TRA) using boron-doped diamond electrode [47]. Different optimisation parameters including flow rate, sample volume, pre-treatment, applied potential and supporting electrolyte were investigated. Under the optimised parameters, the sensor was used for the simultaneous detection of ACP and TR in commercial pharmaceutical samples and the results obtained were compared with the result from high-performance liquid chromatography (HPLC). A t-test at a confidence level of 95% was used for the comparison. The $t_{experimental}$ values obtained for TRA (1.00) and ACP (0.68) were smaller than the $t_{critical}$ values (12.71), these results revealed that there are no significant differences in the two analytical methods. The electrochemical setup for the flow injection electrochemical sensor is presented in Fig. 8.5.

Santos et al. employed BDDE for the simultaneous quantification of hydrochlorothiazide (hypertension drug) and losartan (cardiovascular drug) [48]. The electrode was anodically pretreated to change the diamond phase of the BDDE from hydrophobic to hydrophilic by introducing oxygen functional groups to the surface. DPV was used for co-detection of the analytes at concentration range 3.0 $\times$ 10^{-6} to 7.4 $\times$ 10^{-5} mol/L with detection limits of 1.2 $\times$ 10^{-6} and 9.5 $\times$ 10^{-7} mol/L. The sensor was successfully used for simultaneous detection of the analytes in pharmaceutical formulations with satisfactory results.

Generally, most reports in the literature pretreated boron-doped diamond electrodes either anodically or cathodically, the former was done to change the morphology of the diamond surface from hydrophobic to hydrophilic by incorporating oxygen functional groups into the surface of the BDDE while cathodic

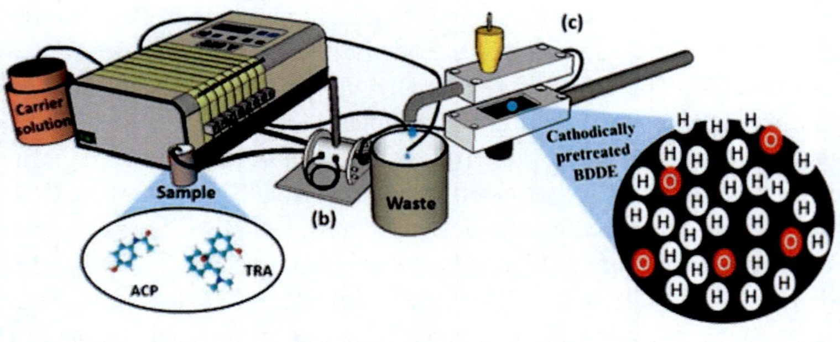

Fig. 8.5 The flow injection setup for co-detection of acetaminophen and tramadol using BDDE [47]. (Copyright (2015), with permission from Elsevier)

pre-treatment changed the diamond surface of the BDDE from hydrophilic to hydrophobic by incorporating hydrogen functional groups. Beyond that, most reports in the literature for the application of BDDE for sensing of pharmaceutical are not modified with nanomaterials based on the reports that they are more conductive and display higher current signal than other carbon-based electrodes. However, lower detection limits and better sensitivity would have been achieved or reported in the literature, if nanomaterials were used to modify BDDE for electrochemical sensors. The application of arguably the noblest carbon-based electrode will be discussed in the next section.

8.1.3 Glassy Carbon Electrode (GCE)

This electrode is arguably the most utilised form of sp^2 hybridised carbon electrode owing to its importance analytical properties including low density, high-temperature resistance, low friction, extreme resistance to chemical attack, hardness, impermeability to gases and liquids [49, 50]. These properties have attracted researchers and industrial communities to employ this noble electrode for different analytical applications [51–53]. For example, Bagheri et al. used bare glassy carbon electrodes to study the electrochemistry of raloxifene in pharmaceutical and human plasma [54]. Before the application of GCE for any electrochemical analysis, it was thoroughly cleaned by polishing with Al slurry on a polishing pad and gently rinsed with distilled water. To obtain a reproducible current–potential peak for the GCE, a cyclic voltammetry experiment was carried out in the presence of 0.1 M H_2SO_4 at a scan rate of 100 mV/s and applying a potential from 0.0 to 1.6 V for ten cycles. The effective surface area and the surface coverage of the GCE were calculated using the Randles–Sevcik equation and coverage equation as depicted in (8.1) and (8.2).

$$I_\mathrm{p} = 0.4463\ (F^3/RT)^{1/2}\ An^{3/2}D_\mathrm{o}{}^{1/2}C_\mathrm{o}\upsilon^{1/2} \tag{8.1}$$

where I_p is the peak current, F is the Faraday's constant (95,000 C/mol), R is the universal gas constant (8314 J/mol K), T is the room temperature, A is the surface area of the electrode (cm^2) constant, n is the number of electron transfers, D_0 (cm^2/s) is the diffusion coefficient, C_o (mol/cm^3) is the concentration of the redox probe and υ is the scan rate measured in (V/s). ^{2}ace area of the GCE was calculated using (8.1) to be 0.0207 cm. The effective surf.

The surface coverage of the GCE was calculated using (8.2).

$$I_\mathrm{p} = (\alpha n^2\ F^3\Gamma\upsilon A)/2.718RT \tag{8.2}$$

where α is the transfer coefficient (0.57), other symbols have their usual meanings and Γ (mol/cm^2) is the surface coverage with a calculated value of 4.7×10^{-10} mol/cm^2 for the GCE. The results obtained from the effective surface area and the surface coverage spurred the investigation of the electrochemical behaviour of raloxifene on GCE surface using CV. The CV measurements revealed that the reaction at the electrode surface was quasi-reversible, adsorption-controlled and two protons/electrons are involved in the process. Finally, DPV was used for the determination of raloxifene in human plasma and tablet samples.

A new voltammetry protocol for the electrochemical determination of fentanyl (analgesic drug) was assembled by modifying GCE with carbon nanoonions (CNO) [55]. The electrochemical report revealed that the GCE/CNO quantified fentanyl at a working potential of 0.85 (vs. Ag/AgCl). Various parameters including supporting electrolytes, pH, accumulation potential and time were interrogated before quantifying the analytes with DPV and chronoamperometry techniques. The chronoamperometry behaviour of the analyte on the sensor was analysed by applying a working potential of 1.0 V and the enhancement in the current signal was directly proportional to the increase in concentration from 10 to 100 μM. The sensor was responsive over a concentration range of the analyte from 1 to 60 μM with a detection limit of 300 nM. The synthesis and fabrication of fentanyl sensor are depicted in Fig. 8.6.

In another report, multiwalled carbon nanotubes (MWCNTs) is dispersed on a GCE for the quantification of cefpirome in pharmaceutical samples [56]. The sensor was prepared by pre-conditioning the GCE in a BR buffer solution and applying a potential (−0.1 to −1.2 V) for twenty cycles until a stable voltammogram was obtained. Thereafter, 10 μL MWCNTs suspension was drop cast on the electrode and allowed to dry at room temperature. The impact/effect of MWCNTs on GCE was interrogated with CV and DPV. The results obtained in the electrochemical techniques gave one well-defined reduction peak in cetyltrimethylammonium bromide, which was ascribed to the reduction of the exocyclic bond (−C=N) present in the analyte. In addition, there was a tremendous improvement in the reduction current response of the modified electrode (GCE/MWCNTs) than the bare GCE due to the rich electrochemical properties of MWCNTs such as fast electron-mediators, large surface area, ability to promote electron transfer reactions and high chemical stability. These properties were exploited in the detection of cefpirome in bulk and pharmaceutical formulation using DPV and SWV.

Rezaei et al. used carboxylated multiwalled carbon nanotubes (CMWCNTs) to modify GCE for adsorptive stripping voltammetry in the quantification of methyldopa (MTD) [57]. The adsorption properties of MWCNT were improved by incorporating some functional groups (-OH, COOH and CHO) into its matrix, which aid in enhancing the electrochemical properties of MWCNTs. The sensor was prepared by a sonicating mixture of (8 mg CMCNTs and 4 mL acetonitrile) until a stable suspension was formed and 50 μL of the suspension was carefully cast on the GCE surface to form a uniform film ascribed as GCE/CMWCNTs. The Randle-Sevcik equation was used to calculate the surface area of the GCE and GCE/MWCNTs. The microscopic surface areas of both electrodes were 0.0314 and 0.1688 cm^2, which simply means the surface area of GCE/MWCNTs was 5.4 times greater than the bare GCE. Different optimisation parameters including scan rate, pH, supporting electrolytes, effects of the amount of CMWCNTs on the peak currents were investigated. Under

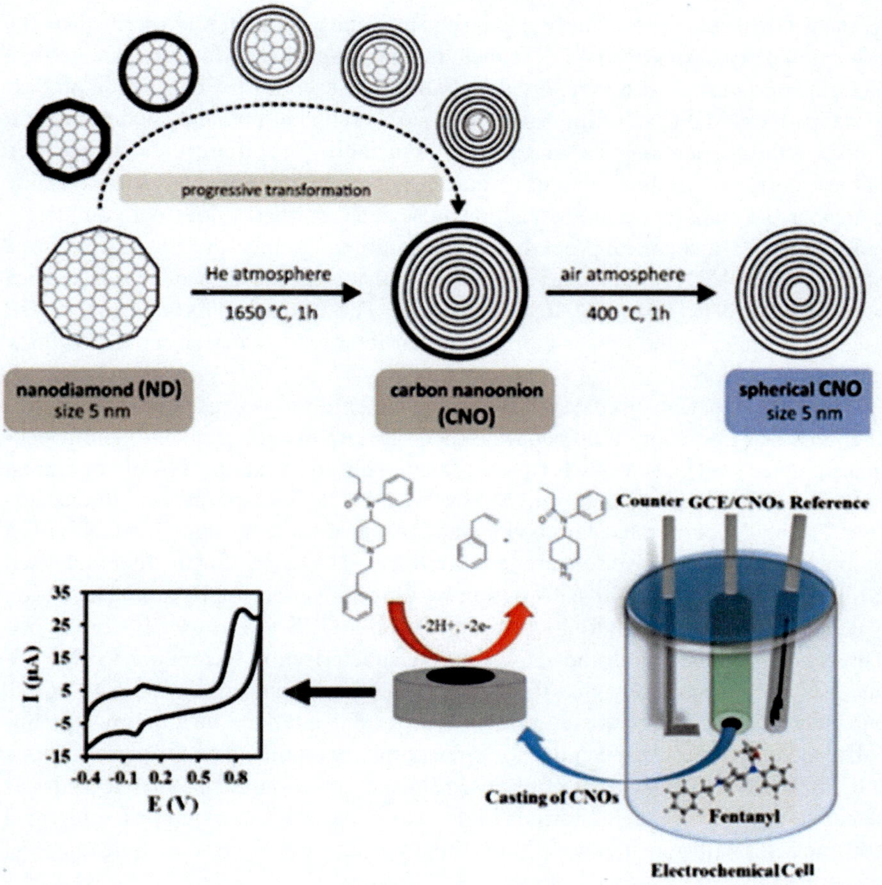

Fig. 8.6 The synthesis, fabrication and characterisation of the fentanyl sensor [55]. (Copyright (2020), with permission from Elsevier)

the optimised experimental conditions, the calibration curves were linear in different concentration ranges (0.1–30 and 30–300 μM) with a detection limit of 0.08 μM using adsorptive stripping voltammetry.

Furthermore, MWCNT was dispersed in water in the presence of dihexadecyl hydrogen phosphate (DHP) to form MWCNT-DHP, which was drop cast on GCE surface for the voltammetry detection of ethamsylate (antihemorrhagic drug) in pharmaceutical tablet and bulk solution [58]. The electrochemical behaviours of the analyte were studied using different electrodes (bare GCE, GCE/DHP and GCE/MWCNTs-DHP). A poor redox peak was observed for the bare GCE with a separation peak potential of 438 mV but the redox peak potential of the GCE/DHP was poorer than the bare GCE, this was accredited to the insulating properties of the DHP, which hinder electron communication on the bare GCE. The GCE/MWCNTs-DHP gave a well-defined redox peak potential of 40 mV and the peak current signal of the analyte was significantly enhanced, this enhancement was ascribed to the exceptional properties of MWCNTs including large specific surface area and refined electronic properties. It was reported that DHP was used for the fabrication process to stabilise the MWCNTs film on the GCE surface. It is of note that incorporating a binder in the fabrication of a sensor is very crucial because it prevents the leaching of the modifier in the analyte solution during electrochemical analysis. However, the binder or the stabiliser must be cautiously selected so that it does not completely reduce the electrochemical properties of the modifier. For this study, the surface area of the GCE/MWCNTs-DHP was 0.087 cm^2, which is far lower than the surface area of GCE/MWCNTs reported by Rezaei et al. [57]. The reason for this is due to the DHP used as a binder in the sensor fabrication that retards the electron movement/flow rate on the sensor.

Similarly, functionalised MWCNT alongside chitosan was used for the electrochemical determination of epinephrine (neurotransmitter) in urine and pharmaceutical samples [59]. The MWCNTs were functionalised by adding 120 mg of carbon nanotubes (CNT) into 10 mL of (3 M HNO_3) at 70 °C for 24 h under vigorous stirring. The acid incorporated carboxylic functional group into the matrix of CNTs. The biopolymer nanocomposite electrode was fabricated by dispersing the functionalised CNT into 4 mg mL^{-1} chitosan solution by ultrasonication for 10 min. Thereafter, 8 μL of the resulting solution was drop cast on a GCE to form GCE/Chit-*f*CNT. The results obtained from the snowballing characterisation techniques (SEM, CV and EIS) revealed clearly that the hybrid biopolymer nanocomposite Chit-*f*CNT enhanced the electron transfer velocity across the interface and thermodynamically decreased the overpotential for epinephrine oxidation. The fabricated sensor (GCE/Chit-fCNT) was successfully used for the detection of the analyte and was selective in the presence of various biological molecules (serotonin, ascorbic acid and uric acid) with good recovery limits for direct detection from adrenaline injection and urine samples.

Baytak and co-workers decorated functionalised carbon nanotubes on zirconium oxide nanoparticles for the quantification of terbutaline in urine and drug formulation samples [60]. The nanocomposite was synthesised by functionalising CNTs with a mixture of $HClO_4$ and HNO_3 (3:7, v: v) for 5 h in an ultrasonic bath. The *f*-MWCNTs alongside ZrO_2 were dispersed in chloroform and sonicated for 45 min, the resulting solution was dropped on a GCE surface to form the sensor. The sequential step for the preparation of ZrO_2/MWCNTs/GCE is depicted in Fig. 8.7. The electrochemical impacts of the modifiers were investigated using cyclic voltammetry, the ZrO_2/MWCNTs/GCE gave a sharp and well-defined anodic peak at 645 mV followed by the MWCNTs/GCE with oxidation peak at 660 mV and the bare GCE gave a poor and broad oxidation wave at 770 mV. It was difficult to understand and explained the effects of the ZrO_2 on the GCE because it was synthesised in situ with the MWCNTs. The sensor was reported to possess lower overpotential, lower detection limit, excellent reproducibility, precise and accurate determination of terbutaline in pharmaceutical samples. We recommend that in the preparation of nanocomposite for electroanalytical applications, the composite should be synthesised singularly before synergistically to ascertain the effect of each modifier on the electrode surface.

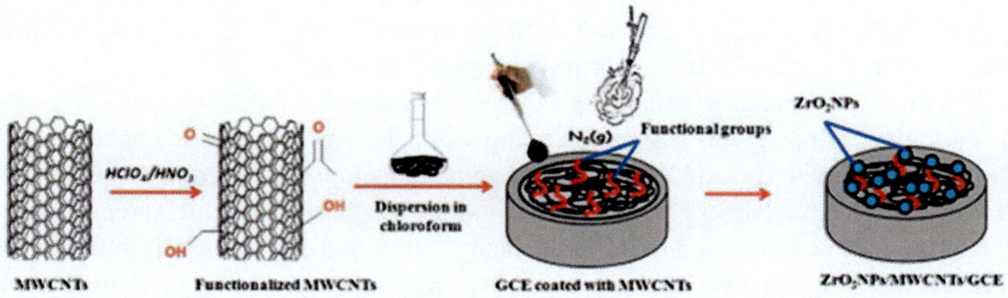

Fig. 8.7 The systematic preparation of ZrO_2/MWCNTs/GCE [60]. (Copyright (2016), with permission from Elsevier)

In another study, adsorptive stripping voltammetry (AdSV) was used for the quantification of thiamine (B_1) in a pharmaceutical sample and commercially available juices [61]. The sensor was constructed by adding 1.25×10^{-4} mol^{-1} $Pb(NO_3)_2$ into 5×10^{-2} mol^{-1} buffer acetate (pH 5.6). The GCE was cleaned as stated earlier and potential of -1.25 V for 120 s was applied to it, during this step, the metallic Pb was electrodeposited/plated on the GCE surface by scanning a potential from -1.25 to -1.55 V under aerated solution forming GCE/Pb film, which is the sensor. The sensor was used for the detection of different concentrations of thiamine from 0.0133 to 0.265 mgL^{-1} and a detection limit of 0.0053 mgL^{-1} was obtained. The pitfall of this sensor is the obvious toxicity of Pb but the author encouraged the use of this sensor on the premise that it is less toxic than Hg and can easily be regenerated by stripping off the Pb film from a preceding measurement and a new Pb film can be regenerated on the GCE for another measurement.

Adsorptive stripping voltammetry was used for the detection of caffeine (CF) using Nafion covered lead film electrode (Nafion/PbFE) [62]. The sensor was prepared by deposition of Pb atoms on GCE, this was achieved by dipping 7.5×10^{-5} molL^{-1} $Pb(NO_3)_2$ solution prepared in 0.1 molL^{-1} HNO_3 into the GCE at a potential of -1.4 V for 30 s. After the deposition process, the electrode was carefully rinsed with deionised water and was left to dry in the air for 120 min. Thereafter, 0.5 µL of a Nafion solution was meticulously dropped on the lead film electrode surface and was left to dry at ambient temperature for a few minutes. This electrode was ascribed as Nafion/PbFE/GCE. The AdSV method was used to study and understand the electrochemical behaviour of caffeine on different electrodes (GCE, PbFE/GCE, Nafion/GCE and Nafion/PbFE). The GCE gave no anodic or cathodic peak signal for 5×10^{-6} mol L^{-1} caffeine solution prepared in 0.1 mol L^{-1} unlike the PbFE/GCE, which gave a small oxidation peak of CF at 1.44 V while the Nafion/GCE and Nafion/PbFE displayed two well-resolved and measurable anodic oxidation peaks, but no cathodic peak was observed. However, the latter offer higher current signal than the former due to the larger surface area, which aids in the accumulation of more caffeine solution on the electrode surface. It is of note that the efficacy of using Nafion for sensor development is due to its ability to pre-concentrate analyte in the polymer layer. This sensor (Nafion/PbFE) was successfully used for the detection of

CF in coffee, tea, energy drink, soft drink and pharmaceutical formulation and the results obtained agreed with a spectrophotometric method.

An innovative electrochemical sensor for the quantification of clozapine (CLZ) in biological samples was assembled using magnetic nanocomposite consisting of Fe_3O_4, β-Alanine (Ala) and Pd [63]. The magnetic nanoparticles (MNPs) were used for the sensor development owing to their unique magnetic, catalytic and electrical properties. More so, other MNPs were incorporated into Fe_3O_4 NPs to prevent its aggregation or agglomeration due to their large surface to volume ratio, limited functional groups for selective binding and strong dipole-dipole attraction between particles. The magnetic nanocomposite was prepared by co-precipitation and reduction method. Briefly, different concentrations of $FeCl_3.6H_2O$, $FeCl_2.4H_2O$ and β-Alanine were dissolved in deionised water and 2 M NaOH was used to adjust the pH to 11. Thereafter, the solution was stirred with a magnetic stirrer and reflux for 12 h. Finally, the nanocomposite was decanted to separate the nanocomposite from the aqueous solution and dried overnight to form Fe_3O_4/Ala. The Pd nanoparticles were incorporated into Fe_3O_4/Ala by the reduction of $PdCl_2$ using hydrazine hydrate in ethanol and refluxed for 10 h, the resulting MNPs formed was ascribed as Fe_3O_4/Ala/Pd. The sensor (GCE/Fe_3O_4/Ala/Pd) improved the oxidation peak current signal of CLZ and was detected at nanomolar level, which is 3–70 nM with a detection limit of 1.53 nM using DPV. The merits of the sensor are long term stability, low cost, fast electrochemical response and low detection limit.

A core-shell nanostructure was used to modify GCE for the electrochemical quantification of diclofenac sodium in human urine and pharmaceutical samples [64]. The core-shell nanostructure (ZnO core@Cu shell nanoparticles) was synthesised by a wet chemical method. The zinc oxide nanoparticles (ZnONPs) were synthesised by dropping 0.02 M NaOH into 0.01 M $ZnCl_2$ solution with continuous stirring, the resulting slurry was centrifuged at 6000 rpm and washed copiously with distilled water, dried and calcined at 500 °C for 3 h. The ZnONPs were powdered by ball milling technique and was activated with $PdCl_2$ and $SnCl_2$. Thereafter, it was coated with Cu shell by electroless plating method using formaldehyde as a reducing agent. The average particle size was 10 to 15 nm as displayed by the atomic force microscopy (AFM). The synthesised ZnO core@Cu shell nanoparticles were used to modify GCE by dispersing in 10 mL methanol followed by sonication, thereafter, 0.5 μL aliquot of the solution was gently drop dried on the electrode surface in an artificial heat chamber. The GCE/core-shell nanostructure displayed strong electrocatalytic and electrooxidation towards the analyte with significant enhancement in peak current signal in comparison to the bare GCE as depicted in the cyclic voltammograms. The modified electrode was reported to possess exceptional qualities such as fast electron transfer, repeatability, constancy and reproducibility.

In another report, the electrochemical detection and effective photocatalytic degradation of paracetamol (PCM) were reported using MoS_2/TiO_2 nanocomposite [65]. The synthesised nanocomposite was drop cast on a glassy carbon electrode and dried in a furnace at 40 °C for 15 min. Thereafter, the electrochemical behaviours of PCM were interrogated on different electrodes (GCE, GCE/TiO_2, GCE/MoS_2 and GCE//MoS_2/TiO_2). The MoS_2/TiO_2 had the highest peak current signal followed by GCE/MoS_2, GCE/TiO_2 and GCE. The highest peak current obtained for the MoS_2/TiO_2 is due to the synergistic effects of the nanocomposite that enhanced and facilitated electron transport on the electrode surface. It is noteworthy that the electrochemical behaviour of paracetamol involves two protons and two-electron process, this accounts for its electrochemical behaviour. The redox mechanism of paracetamol is presented in Scheme 8.2. The was used to measure the oxidation peak current of PCM at a different concentration from 0.5 to 750 μM with a detection limit of 0.01 μM [65]. In addition, the nanocomposite was used for the degradation of PCM under visible light and a degradation rate of 40% was achieved in 25 min.

A comparative study between sp^3 (boron-doped diamond), sp^2 (glassy carbon and pyrolytic graphite) carbon-based electrode was reported for the electrochemical detection of terconazole (an antifungal drug) [66]. The electrodes were electrochemically characterised using 1 mmol L^{-1} $Fe(CN)_6{}^{3-/4-}$ redox probe prepared in 1 mmol L^{-1} KCl a supporting electrolyte at a scan rate of 0.1 mV/s. The peak potential difference between the anodic and cathodic peaks for the electrodes are 410±20 mV, 550±25 mV and 65±4 mV for the boron-doped diamond electrode (BDDE), pyrolytic graphite electrode (PGE) and glassy carbon electrode (GCE), respectively. These results clearly revealed the differences in the chemical composition of moieties in various electrode surfaces. The scan rate study showed that the anodic peak current is directly proportional to the square root of scan rates for BDDE and GCE revealing that the electrochemical kinetics is diffusion-controlled but the PGE was an adsorption-controlled process. For all the electrodes, the electrochemical oxidation of terconazole involved one proton and two electrons transfer. Finally, the electrodes were used in the detection of the analyte and detection limits of 0.4, 0.5 and 1.24 μM were obtained for BDDE, GCE and PGE. The advantage of using BDD over GCE and PGE is the signal reproducibility due to in situ anodic activation, but GCE and PGE are activated via *ex-situ* activation, which affects the reproducibility signal of the electrodes.

Different nanomaterials played significant roles in the fabrication of various sensors constructed with glassy carbon electrodes but the use of GCE is discouraged for electroanalysis owing to the tedious pre-treatment procedure. More importantly, the modifiers employed to resolve the high potential and sluggish kinetics of GCE leached into the analyte solution during electrochemical analysis, which greatly affects the reproducibility and repeatability features of the electrode. However, these

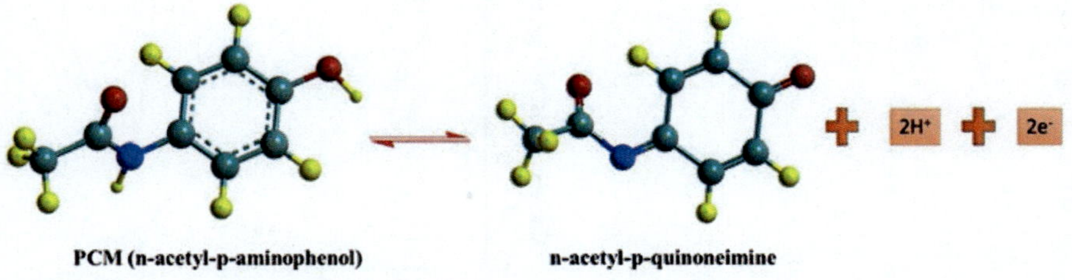

Scheme 8.2 The proposed electrochemical redox reaction of paracetamol [65]

problems can be tackled or resolved using binders such as Nafion, dihexadecyl hydrogen phosphate (DHP) or drying the modified electrode in a furnace at 40 °C for 15 min. The electrochemical application of other carbon-based electrodes is discussed in the next section.

8.1.4 Pencil Graphite Electrode (PGE)

The pencil graphite electrode is arguably the least utilised form of carbon electrode that has been utilised for the quantification of a few varieties of analytes by voltammetry techniques. The analytical merits of PGE over other carbon-based electrodes are ease of modification, disposability, highly selective, wide linear range, low cost, surface polishing is not required, high electrochemical reactivity and large surface area [67].

Although few reports have been reported on the application of PGE for pharmaceutical applications. For instance, an unmodified disposable pencil graphite electrode was used for the detection of amlodipine (AML) [68]. The conductivity and surface area of PGE were obtained from EIS and cyclic voltammetry in 3 mm $Fe(CN)_6^{3-/4-}$. The electron charge transfer resistance (R_{ct}) and surface area of PGE were 76.56 Ω and 0.165 cm^2. The R_{ct} revealed that PGE showed better electrical conductivity and fast electron transfer rate, while the surface area of the PGE was far higher than the surface area of the GCE reported by Bagheri et al. [54]. The PGE was successfully used for the detection of AML using SWV and DPV with detection limits of 0.02 and 0.04 pM.

A comparative study between PGE and GCE for the voltammetric quantification of acyclovir (ACV) was reported [69]. The ACV response of GCE was broad, weak and shifted to the positive potential around 1200 mV, this is attributed to the sluggish electron transfer as well as higher adsorption of the analyte on GCE surface. Moreover, the peak current intensity of the PGE was higher than GCE revealing that the electroactive surface area of PGE was higher than GCE as shown in Fig. 8.8. The

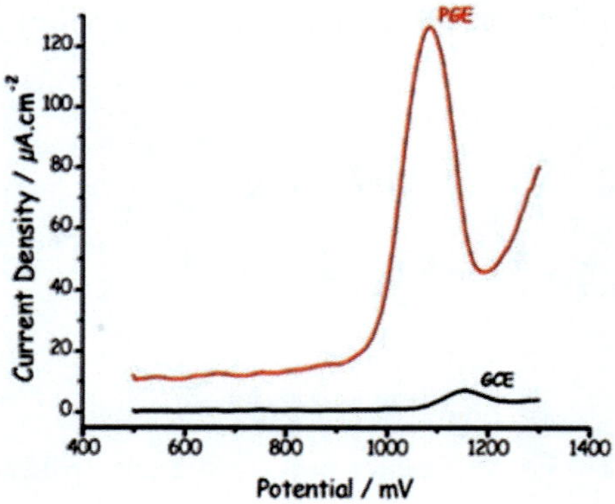

Fig. 8.8 The DPV of 0.14 mM ACV in BR buffer solution prepared in 0.1 M KCl [69]. (Copyright (2016), with permission from Elsevier)

reason for the higher peak current signal for PGE was due to its porosity and high conductivity. Thus, the PGE was employed in the detection of different concentrations of ACV from 1 to 100 μM with a DL of 0.3 μM.

A yellow PX4R monomer was polymerised on PGE for the electrochemical quantification of dopamine (DP) [70]. The PX4R monomer was polymerised on the PGE surface by cycling potential from −0.6 to 1.6 V at a scan rate of 100 mV/s. During multiple cycles, there was a gradual reduction in the voltammogram and enhancement in the potential cycling, this observation confirmed that the PX4R membrane was successfully deposited on the PGE. The poly(yellow PX4R)/PGE gave lower peak potential and better redox peak current signal than the bare PGE. Hence, it was used for detecting dopamine in the presence of Serotonin (5-HT).

Phosphotungstic acid (PWA) alongside with reduced graphene oxide (rGO) was used to modify PGE for the electrochemical detection of paroxetine (PRX) in pharmaceutical and biological media [71]. The chronoamperometric measurement was used to obtain information about the diffusion coefficient (D) of PRX using a constant potential of 950 mV. The diffusion coefficient of different PRX concentrations (0.1, 0.46 and 0.82 mM) was calculated using Cottrel equation:

$$I = \mathrm{NFACD}^{1/2}\pi^{-1/2}t^{-1/2} = Kt^{-1/2} \tag{8.3}$$

where all parameters remain the same and D is the diffusion coefficient (cm^2s^{-1}), which was calculated to be 5.3×10^{-5} cm^2 s^{-1} for PGE/rGO/PWA. The D value of PGE and PGE/rGO was not calculated, which makes it difficult to compare the D values of all the working electrodes. The DPV for various concentrations of PRX (0.008–1.0 μM) was not stable and at higher concentrations, there was a cathodic shift in the peak current signal, this showed that the sensor was unstable and this might be due to the leaching of the modifier in the analyte solution during electrochemical analysis.

In another study, MWCNT was networked on PGE surface for the determination of aceclofenac (an anti-inflammatory drug) in pharmaceutical and biological samples [72]. The sensor was prepared by dispersing 1 mg/mL MWCNTs in distilled water and ultrasonicated for 120 min to obtain a homogenous solution. Thereafter, 10 μL of the dispersing solution was carefully drop dried on PGE surface for 12–16 h to form PGE/MWCNTs. The CV of aceclofenac (ACF) displayed three oxidation peaks at (+0.12, +0.32 and +0.51 V) and a reduction peak at −0.26 V for PGE/MWCNTs, with a four-fold enhancement in peak current signal than bare PGE. This revealed that MWCNTs increased the electrochemically active surface area and the rate of charge transfer reaction on the PGE. The oxidation peak currents of ACF were responsive in the concentration range of 1×10^{-6} to 60×10^{-6} M with a detection limit of 2.6 $\times 10^{-9}$ M.

Adsorptive stripping voltammetry (AdsDPV) technique was used for co-detection of ascorbic acid (AA) and acetylsalicylic acid (ACA) using nano-smectite and MWCNTs as nanowires [73]. The sequential step for sensor development is shown in Scheme 8.3. Various parameters including immobilisation time, pH, the percentage

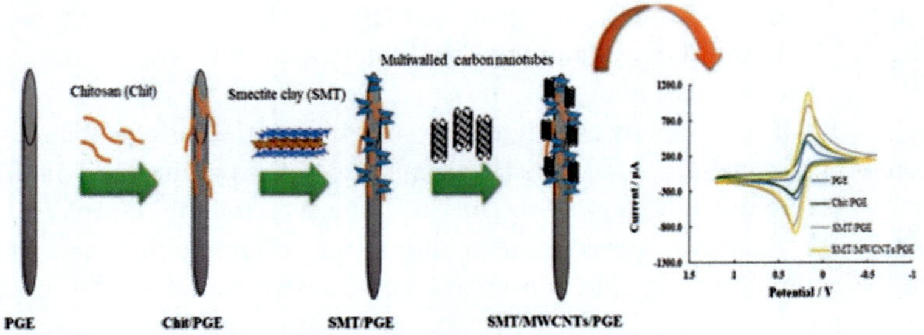

Scheme 8.3 The scheme for the development of SMT/MWCNTs/PGE nanowires [73]

composition of the modifiers, accumulation time and potential were optimised before carrying out the electrochemical analysis. Various electrodes (PGE, Chit/PGE, SMT/PGE and SMT/MWCNTs/PGE) were developed for the electrochemical analysis, but the MWCNTs/PGE sensor was not constructed (Fig. 8.9a and b), this makes it difficult to compare the various thermodynamic parameters (surface area, diffusion coefficient, charge transfer resistance and Warburg impedance) of MWCNTs/PGE to other electrodes. The fabricated sensor (SMT/MWCNTs/PGE) was successfully used in the simultaneous detection of AA and ACA with DLs of 0.096 and 0.241 μM.

The comparative study between PGE and GCE revealed that it possesses high surface area and excellent conductivity than GCE. More importantly, the conductivity and surface area of PGE are enhanced by modifying it with nanomaterials. We strongly recommend this electrode for various electrochemical analyses. The next section discussed the electrochemical application of disposable screen-printed carbon electrodes.

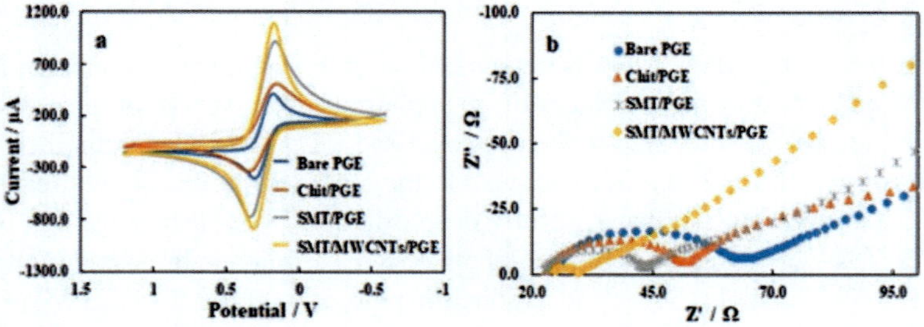

Fig. 8.9 The CV and Nyquist plot of different electrodes in 5 mM $Fe(CN)_6^{3-/4-}$ [73]. (Copyright (2019), with permission from Elsevier)

8.1.5 Screen-Printed Carbon Electrode

The screen-printed electrode (SPCE) consists of a three-electrode component including reference, counter and the working electrode [74]. The carbon ink in the SPCE is painted on the conductive tracks to form the counter and working electrode, while the reference electrode is based on silver electroactive tracts. The SPCE comprises binding components, additives, carbon conductive ink and organic solvents. These electrodes can be pretreated thermally, chemically, mechanically and electrochemically [75].

The screen-printed electrodes are transducer substrates employed for various electrochemical applications owing to their significant analytical advantages such as large surface area, disposability, low cost, portability, mass production and ease of modification [76, 77]. These properties have allured scientists and industrial communities in employing this electrode for various electrochemical applications. For instance, Sasal et al. employed unmodified SPCE for the quantification of sildenafil citrate (erectile dysfunction drugs) in biological and pharmaceutical samples [78]. The active surface area (ASA) and geometric area (GA) of the SPCE were calculated to be 0.05891 and 0.1256 cm^2 using the Randles–Sevcik equation. It is of note that the effective ASA is less than the GA due to the insulating region in the interior part of the SPCE. The scan rates study revealed that the anodic and cathodic peak current is directly proportional to the square root of the scan rates, this result displayed reversible and diffusion-controlled system for the faradaic reaction on $Fe(CN)_6^{3-/4-}$ redox probe at the SPCE. The linear concentration range of sildenafil citrate detected by the SPCE is within the range of 2×10^{-7} to 2×10^{-9} mol/L and a DL of 5.9×10^{-10} mol/L was obtained.

A bare SPCE was used for the voltammetric detection of meclizine antihistamine drug using sodium dodecyl sulphate (SDS) as the surfactant [79]. The electrochemical oxidation peak of the analyte was observed at 1.0 V and a threefold enhancement in the peak current signal was obtained in the presence of SDS, this enhancement was due to the uniformity of SDS molecules on the SPCE, which assisted and facilitated electrostatic attractions of the analyte on the electrode surface. Thus, the electrode was effectively used in detecting the analyte in human fluids and pharmaceutical tablets with satisfactory precision and accuracy.

Adsorptive stripping voltammetry was used in the quantification of synthetic (17α-ethinylestradiol) and natural (17β-estradiol) hormones in urine and pharmaceutical formulations [80]. The hanging mercury drop electrode (HMDE) was used to compare the electrochemical reduction of the analytes and the oxidation reaction that occurs on an SPCE. Different optimisation parameters such as pH 10, adsorption potential ($E_{ads} = -0.6$ V) and time ($t_{ads} = 30$ s) were used for the detection of both analytes. The detection limits (DLs) obtained using HMDE for the 17α-ethinylestradiol and 17β-estradiol were 14.8 and 0.3 μg/L. For SPCE, the DLs of both analytes were 277 and 242 μg/L, respectively. However, the DLs obtained from HDME is far lower than the SPCE but the toxicity of HDME discouraged its applications for environmental and electrochemical analysis.

Mesoporous WS_2 nanorods were decorated on nitrogen-doped reduced graphene oxide (N-doped rGO) for the electrochemical detection of Roxarsone (antibiotic drug) in pharmaceuticals and biological samples [81]. The scheme for the synthesis and the electrochemical detection of Roxarsone (ROX) is depicted in Scheme 8.4. The sensor was fabricated by dispersing the synthesised WS_2 NRs/N-rGOs nanocomposite in ethanol and sonicated to form a homogeneous suspension (HS). Thereafter, 7 μL of the HS was carefully drop cast on the SPCE and dried at ambient temperature to form WS_2 NRs/N-rGO/SPCE. Similarly, the same procedure was used in the fabrication of other electrodes (WS_2 NRs/SPCE, N-rGO/SPCE) as a control experiment to understand the impact/influence of each modifier on the SPCE. The interfacial electron migration behaviour of the various modified electrodes was interrogated on 5 mM $Fe(CN)_6^{3-/4-}$ using CV technique. The WS_2 NRs/N-rGO/SPCE displayed low resistance and high current signal than other electrodes, this is due to the synergy of the conductivity of N-rGO and excellent catalytic property of the mesoporous WS_2 NRs. Furthermore, the electrochemical reduction of 100 μM ROX was carried out on SPCE, N-rGO/SPCE, WS_2 NRs/SPCE and WS_2 NRs/N-rGOs/SPCE, respectively. The WS_2 NRs/N-rGOs/SPCE displayed catalytic reduction peak for the analyte at −0.68 V, highest peak current and lowest peak potential, followed by WS_2 NRs/SPCE, N-rGO/SPCE and SPCE. For this reason, the fabricated sensor (WS_2 NRs/N-rGOs/SPCE) was successfully used in the quantification of ROX in urine, pharmaceutical samples and human serum. It is important to note that the concentration of ROX was analysed in urine and human serum samples by diluting with buffer solution and spiking with standard solution of ROX.

Simultaneous detection of nifepidine and atenolol (antihypertensive drugs) was done by modifying SPCE with MgO nanoplatelet [82]. The MgO nanoplatelet was prepared by hydrothermal synthesis. Briefly, 2.033 g of $MgCl_2$ was added to water and the resulting solution was transferred into 100 mL Teflon-lined stainless-steel autoclave at 160 °C for 12 h to form a white precipitate, which was collected and

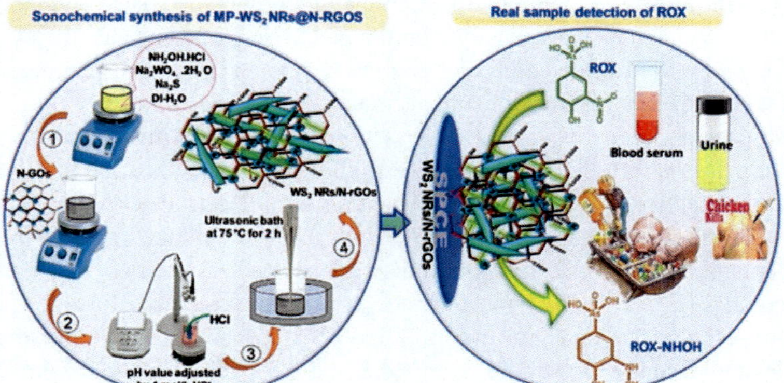

Scheme 8.4 The synthesis of WS_2 NRs/N-rGOs and electrochemical detection of ROX in biological samples [81]. (Copyright (2019), with permission from Elsevier)

washed copiously with water/ethanol. The white precipitate was dried overnight and calcined at 300 °C for 3 h to form MgO nanoplatelet (MgO NPLs). The sensor was constructed by dispersing 5 mg MgO NPLs and 100 μL of 1% polytetrafluoroethylene (PTFE) in deionised water for 30 min. Thereafter, 5 μL of the suspension was dropped on the SPCE for 60 min at 50 °C in the oven to form the sensor MgO NPLs/SPCE. The influence of MgO NPLs on the heterogeneous electron transfer properties of SPCE was evaluated using Nicholson's formula depicted in (8.4).

$$\Psi = \mathrm{k}^{\mathrm{o}} \ \{\pi D n \upsilon F/RT\}^{-1/2} \tag{8.4}$$

where Ψ is the kinetic parameter, k^{o} is the standard heterogeneous rate constant, D is the diffusion coefficient, υ is the scan rate, n is the number of electron transfers in the reaction, F is the Faraday constant, R is the gas constant and T is the temperature. The heterogeneous rate constants for SPCE and MgO NPLs/SPCE were calculated to be 8.21×10^{-4} and 1.72×10^{-3} cms^{-1}, respectively. This clearly showed that the modifier enhanced the electrochemical performance and promoted electron transfer rate of SPCE. Thus, it was efficiently used for the simultaneous detection of nifedipine and atenolol in human urine and pharmaceutical tablets with good precision and accuracy. Generally, incorporating nanomaterials into the matrix of electrodes help to increase the surface area, facilitate electron transport, improve stability, selectivity and sensitivity. It is noteworthy to point out that various electrodes have a different electroactive surface areas and this plays a significant role in the detection limit obtained during electrochemical quantification of analytes.

8.1.6 Conclusion

This chapter focuses on the application of various electrodes for the electrochemical detection of different pharmaceuticals in the environment. The limitations of using the unmodified electrode for electrochemical analysis include electrode fouling, sluggish electron communication, high background current, non-reproducibility of signal response after surface pre-treatment and time-consuming pre-treatment procedure. These limitations are overcome using smart and intelligent nanomaterials owing to their analytical features such as large surface area, high catalytic efficiency, mechanical, electrical, optical, chemical and magnetic properties. More so, the nanomaterials employed in modifying various electrodes leached in the analyte solution during electrochemical analysis, this problem can be resolved using stabilising agents such as Nafion, dihexadecyl hydrogen phosphate (DHP) or heating the modified electrode in an oven at 40 °C for few minutes. More importantly, the sensitivity and detection limit obtained during electrochemical examination depends on the type of electrode used, the nanomaterials and the optimisation parameters that are investigated during electrochemical study.

References

1. K. Balakrishna, A. Rath, Y. Praveenkumarreddy, K. Siri, A review of the occurrence of pharmaceuticals and personal care products in Indian water bodies. Ecotoxicol. Environ. Saf. **137**, 113–120 (2017). https://doi.org/10.1016/j.ecoenv.2016.11.014
2. S. Mangala, S. Norashikin, M. Shaifuddin, S. Sukiman, F. Adzima, M. Nasir, Z. Hana, N. Kamarudin, T. Hanidza, T. Ismail, A. Zaharin, Pharmaceuticals residues in selected tropical surface water bodies from Selangor (Malaysia): occurrence and potential risk assessments. Sci. Total Environ. **642**, 230–240 (2018). https://doi.org/10.1016/j.scitotenv.2018.06.058
3. S. Fekadu, E. Alemayehu, R. Dewil, B. Van Der Bruggen, Pharmaceuticals in freshwater aquatic environments: a comparison of the African and European challenge. Sci. Total Environ. **654**, 324–337 (2019). https://doi.org/10.1016/j.scitotenv.2018.11.072
4. J. Jose, J.S. Pinto, B. Kotian, A.M. Thomas, R.N. Charyulu, Comparison of the regulatory outline of ecopharmacovigilance of pharmaceuticals in Europe, USA, Japan and Australia, Sci. Total Environ. **709**, 134815 (2020). https://doi.org/10.1016/j.scitotenv.2019.134815
5. P.K. Thai, L. Xuan, V. Ngan, P. Hong, P. Thi, N. Quang, N.T.T. Dang, N. Kieu, B. Tam, N. Thi, K. Anh, Occurrence of antibiotic residues and antibiotic-resistant bacteria in effluents of pharmaceutical manufacturers and other sources around. Sci. Total Environ. **645**, 393–400 (2018). https://doi.org/10.1016/j.scitotenv.2018.07.126
6. E.T. Furlong, A.L. Batt, S.T. Glassmeyer, M.C. Noriega, D.W. Kolpin, H. Mash, K.M. Schenck, Nationwide reconnaissance of contaminants of emerging concern in source and treated drinking waters of the United States: pharmaceuticals. Sci. Total Environ. **579**, 1629–1642 (2017). https://doi.org/10.1016/j.scitotenv.2016.03.128
7. B. Hong, S. Yu, Y. Niu, J. Ding, Q. Lin, X. Lin, W. Hu, Spectrum and environmental risks of residual pharmaceuticals in stream water with emphasis on its relation to epidemic infectious disease and anthropogenic activity in watershed. J. Hazard. Mater. **385**, 121594 (2020). https://doi.org/10.1016/j.jhazmat.2019.121594
8. J. Chul, D. Yoon, E. Byeon, J. Soo, U. Hwang, J. Han, J. Lee, Adverse effects of two pharmaceuticals acetaminophen and oxytetracycline on life cycle parameters, oxidative stress, and defensome system in the marine rotifer Brachionus rotundiformis. Aquat. Toxicol. **204**, 70–79 (2018). https://doi.org/10.1016/j.aquatox.2018.08.018
9. B. Doğan, A. Elik, N. Altunay, Determination of paracetamol in synthetic urea and pharmaceutical samples by shaker-assisted deep eutectic solvent microextraction and spectrophotometry, Microchem. J. **154**, 104645 (2020). https://doi.org/10.1016/j.microc.2020.104645
10. K. Basavaiah, P. Nagegowda, Determination of ranitidine hydrochloride in pharmaceutical preparations by titrimetry and visible spectrophotometry using bromate and acid dyes. Farmaco **59**, 147–153 (2004). https://doi.org/10.1016/j.farmac.2003.11.012
11. Y. Zhang, R. Zhang, X. Yang, H. Qi, C. Zhang, Recent advances in electrogenerated chemiluminescence biosensing methods for pharmaceuticals. J. Pharm. Anal. **9**, 9–19 (2019). https://doi.org/10.1016/j.jpha.2018.11.004
12. C. Miossec, T. Mille, L. Lanceleur, M. Monperrus, Simultaneous determination of 42 pharmaceuticals in seafood samples by solvent extraction coupled to liquid chromatography—tandem mass spectrometry. Food Chem. **322**, 126765 (2020). https://doi.org/10.1016/j.foodchem.2020.126765
13. S.A. Ozkan, B. Uslu, From mercury to nanosensors: past, present and the future perspective of electrochemistry in pharmaceutical and biomedical analysis. J. Pharm. Biomed. Anal. **130**, 126–140 (2016). https://doi.org/10.1016/j.jpba.2016.05.006
14. A.O. Idris, J.P. Mafa, N. Mabuba, O.A. Arotiba, Nanogold modified glassy carbon electrode for the electrochemical detection of arsenic in water. Russ. J. Electrochem. **53**, 190–197 (2017). https://doi.org/10.1134/S1023193517020082
15. A.O. Idris, J.P. Mafa, N. Mabuba, O.A. Arotiba, Dealing with interference challenge in the electrochemical detection of As(III) —A complexometric masking approach. Electrochem. Commun. **64**, 18–20 (2016). https://doi.org/10.1016/j.elecom.2016.01.003

16. M.M. Ghoneim, H.S. El-desoky, M.M. Abdel-galeil, Electrochemistry of the antibacterial and antifungal drug nitroxoline and its determination in bulk form, pharmaceutical formulation and human blood. Bioelectrochemistry **80**, 162–168 (2011). https://doi.org/10.1016/j.bioelechem.2010.08.003
17. T. Ha, T. Nguyen, J. Lee, H. Kim, K. Min, B. Kim, Current research on single-entity electrochemistry for soft nanoparticle detection : introduction to detection methods and applications, Biosens. Bioelectron. **151**, 111999 (2020). https://doi.org/10.1016/j.bios.2019.111999
18. N. Karimian, P. Hashemi, A. Afkhami, H. Bagheri, The principles of bipolar electrochemistry and its electroanalysis applications. Curr. Opin. Electrochem. **17**, 30–37 (2019). https://doi.org/10.1016/j.coelec.2019.04.015
19. T.G. Rygar, F.M. Arken, U.S. Chröder, F.S. Cholz, Electrochemical analysis of solids. A review. Chem. Comm. **67**, 163–203 (2002). https://doi.org/10.1135/cccc20020163
20. A. Pardakhty, S. Ahmadzadeh, S. Avazpour, V. Kumar, Highly sensitive and efficient voltammetric determination of ascorbic acid in food and pharmaceutical samples from aqueous solutions based on nanostructure carbon paste electrode as a sensor. J. Mol. Liq. **216**, 387–391 (2016). https://doi.org/10.1016/j.molliq.2016.01.010
21. S.A. Atty, G.A. Sedik, F.A. Morsy, D.M. Naguib, H.E. Zaazaa, A novel sensor aluminum silicate modified carbon paste electrode for determination of anti-depressant dothiepin HCl in pharmaceutical formulation and biological fluids. Microchem. J. **148**, 725–734 (2019). https://doi.org/10.1016/j.microc.2019.05.007
22. A. Motaharian, K. Naseri, O. Mehrpour, S. Shoeibi, Electrochemical determination of atypical antipsychotic drug quetiapine using nano-molecularly imprinted polymer modified carbon paste electrode. Anal. Chim. Acta. **1097**, 214–221 (2020). https://doi.org/10.1016/j.aca.2019.11.020
23. M.M. Khalil, G.M.A. El-aziz, Multiwall carbon nanotubes chemically modified carbon paste electrodes for determination of gentamicin sulfate in pharmaceutical preparations and biological fl uids. Mater. Sci. Eng. C. **59**, 838–846 (2016). https://doi.org/10.1016/j.msec.2015.10.095
24. M. Keyvanfard, K. Alizad, Determination of isoproterenol in pharmaceutical and biological samples using a pyrogallol red multiwalled carbon nanotube paste electrode as a sensor. Chin. J. Catal. **37**, 579–583 (2016). https://doi.org/10.1016/S1872-2067(15)61036-1
25. M. Ibrahim, H. Ibrahim, N. Almandil, A. Kawde, Gold nanoparticles/f-MWCNT nanocomposites modified glassy carbon paste electrode as a novel voltammetric sensor for the determination of cyproterone acetate in pharmaceutical and human body fluids. Sens. Actuators B. Chem. **274**, 123–132 (2018). https://doi.org/10.1016/j.snb.2018.07.105
26. Š Sanja, V. Guzsvány, J. Anoj, T. Heged, M. Mikov, K. Kalcher, Imidazolium-based ionic liquids as modifiers of carbon paste electrodes for trace-level voltammetric determination of dopamine in pharmaceutical preparations. J. Mol. Liq. **306**, 1–8 (2020). https://doi.org/10.1016/j.molliq.2020.112900
27. Z. Rezayati, S. Saeed, H. Davarani, A. Taheri, Y. Bide, A yolk shell Fe3O4@PA-Ni @ Pd/Chitosan nanocomposite -modified carbon ionic liquid electrode as a new sensor for the sensitive determination of fl uconazole in pharmaceutical preparations and biological fl uids. J. Mol. Liq. **253**, 233–240 (2018). https://doi.org/10.1016/j.molliq.2018.01.019
28. M. Afzali, A. Mostafavi, T. Shamspur, Square wave voltammetric determination of anticancer drug flutamide using carbon paste electrode modified by CuO/GO/ PANI nanocomposite. Arab. J. Chem. **13**, 3255–3265 (2020). https://doi.org/10.1016/j.arabjc.2018.11.001
29. N.P. Shetti, M.M. Shanbhag, S.J. Malode, R.K. Srivastava, Amberlite XAD-4 modified electrodes for highly sensitive electrochemical determination of nimesulide in human urine. Microchem. J. **153**, 1–9 (2020). https://doi.org/10.1016/j.microc.2019.104389
30. P.P. Talay, Y. Yard, Z. Şentürk, Electrochemical oxidation of ranitidine at poly (dopamine) modified carbon paste electrode : its voltammetric determination in pharmaceutical and biological samples based on the enhancement effect of anionic surfactant. Sens. Actuators B Chem. **273**, 1463–1473 (2018). https://doi.org/10.1016/j.snb.2018.07.068.

31. C. Stefanov, C.C. Negut, L. Alexandra, D. Gugoasa, J. Koos, F. Van Staden, Sensitive voltammetric determination of riboflavin in pharmaceutical and biological samples using FSN-Zonyl-Nafion modified carbon paste electrode. Microchem. J. **155**, 1–9 (2020). https://doi.org/10.1016/j.microc.2020.104729
32. N. Setoudeh, S. Jahani, M. Kazemipour, Zeolitic imidazolate frameworks and cobalt-tannic acid nanocomposite modified carbon paste electrode for simultaneous determination of dopamine, uric acid, acetaminophen and tryptophan: investigation of kinetic parameters of surface electrode and its a. J. Electroanal. Chem. **863**, 1–14 (2020). https://doi.org/10.1016/j.jelechem.2020.114045
33. D. Shi, N. Huang, L. Liu, B. Yang, Z. Zhai, Y. Wang, Nanostructured boron-doped diamond electrode for seawater salinity detection. Appl. Surf. Sci. **512**, 1–8 (2020). https://doi.org/10.1016/j.apsusc.2020.145652
34. V. Rehacek, I. Hotovy, M. Marton, M. Mikolasek, P. Michniak, A. Vincze, A. Kromka, M. Vojs, Voltammetric characterization of boron-doped diamond electrodes for electroanalytical applications. J. Electroanal. Chem. **862**, 1–9 (2020). https://doi.org/10.1016/j.jelechem.2020.114020
35. J. Sochr, M. Rievaj, Rapid and sensitive electrochemical determination of codeine in pharmaceutical formulations and human urine using a boron-doped diamond film electrode. Electrochim. Acta. **87**, 503–510 (2013). https://doi.org/10.1016/j.electacta.2012.09.111
36. Š Renáta, B. Kränková, Š Michaela, P. Martinková, L. Janíková, J. Chýlková, M. Vojs, Influence of boron content on electrochemical properties of boron-doped diamond electrodes and their utilization for leucovorin determination. J. Electroanal. Chem. **821**, 2–9 (2018). https://doi.org/10.1016/j.jelechem.2018.02.007
37. R. Bogdanowicz, M. Ficek, N. Malinowska, S. Gupta, R. Meek, Electrochemical performance of thin free-standing boron-doped diamond nanosheet electrodes. J. Electroanal. Chem. **862**, 1–7 (2020). https://doi.org/10.1016/j.jelechem.2020.114016
38. H. Dejmkov, V. Ostatn, K. Schwarzov, Recent progress in the applications of boron doped diamond electrodes in electroanalysis of organic compounds and biomolecules e a review a, Ale Simona Baluchov a. Anal. Chim. Acta. **1077**, 30–66 (2019). https://doi.org/10.1016/j.aca.2019.05.041
39. D. Kuzmanovi, M. Khan, E. Mehmeti, R. Nazir, N. Ramdan, R. Amaizah, D.M. Stankovi, Determination of pyridoxine (vitamin B6) in pharmaceuticals and urine samples using unmodified boron-doped diamond electrode. Diam. Relat. Mater. **64**, 184–189 (2016). https://doi.org/10.1016/j.diamond.2016.02.018
40. Ľ. Švorc, I. Strežova, K. Kiani, D.M. Stankovi, P. Ot, An advanced approach for electrochemical sensing of ibuprofen in pharmaceuticals and human urine samples using a bare boron-doped diamond electrode. J. Electroanal. Chem. **822**, 144–152 (2018). https://doi.org/10.1016/j.jelechem.2018.05.026
41. S. Pysarevska, L. Dubenska, S. Plotycya, A state-of-the-art approach for facile and reliable determination of benzocaine in pharmaceuticals and biological samples based on the use of miniaturized boron-doped diamond electrochemical sensor. Sens. Actuators B Chem. **270**, 9–17 (2018). https://doi.org/10.1016/j.snb.2018.05.012
42. L. Švorc, K. Cinková, J. Sochr, M. Vojs, P. Michniak, M. Marton, Sensitive electrochemical determination of amlodipine in pharmaceutical tablets and human urine using a boron-doped diamond electrode. J. Electroanal. Chem. **728**, 86–93 (2014). https://doi.org/10.1016/j.jelechem.2014.06.038
43. J. Armando, E. Rom, O. Fatibello-filho, Square-wave voltammetric determination of bezafibrate in pharmaceutical formulations using a cathodically pretreated boron-doped diamond electrode. Talanta **103**, 201–206 (2013). https://doi.org/10.1016/j.talanta.2012.10.033
44. Š Renáta, F. Hlobe, J. Skopalová, P. Canka, L. Janíková, J. Chýlková, Electrochemical oxidation of anti-inflammatory drug meloxicam and its determination using boron doped diamond electrode. J. Electroanal. Chem. **858**, 1–10 (2020). https://doi.org/10.1016/j.jelechem.2019.113758

45. B. Uslu, B.D. Topal, S.A. Ozkan, Electroanalytical investigation and determination of pefloxacin in pharmaceuticals and serum at boron-doped diamond and glassy carbon electrodes. Talanta **74**, 1191–1200 (2008). https://doi.org/10.1016/j.talanta.2007.08.023
46. P. Talay, Y. Yardım, Z. Şentürk, Individual and simultaneous electroanalytical sensing of epinephrine and lidocaine using an anodically pretreated boron-doped diamond electrode by square-wave voltammetry. Diam. Relat. Mater. **101**, 1–10 (2020). https://doi.org/10.1016/j.diamond.2019.107649
47. A.M. Santos, F.C. Vicentini, L.C.S. Figueiredo-filho, P.B. Deroco, O. Fatibello-filho, Flow injection simultaneous determination of acetaminophen and tramadol in pharmaceutical and biological samples using multiple pulse amperometric detection with a boron-doped diamond electrode. Diam. Relat. Mater. **60**, 1–8 (2015). https://doi.org/10.1016/j.diamond.2015.10.005
48. M. Cristina, G. Santos, C. Ricardo, T. Tarley, L. Henrique, D. Antonia, E. Romão, Evaluation of boron-doped diamond electrode for simultaneous voltammetric determination of hydrochlorothiazide and losartan in pharmaceutical formulations. Sens. Actuators B. Chem. **188**, 263–270 (2013). https://doi.org/10.1016/j.snb.2013.07.025
49. L. Coustan, B. Daniel, Electrochemical activity of platinum, gold and glassy carbon electrodes in water-in-salt electrolyte. J. Electroanal. Chem. **854**, 1–8 (2019). https://doi.org/10.1016/j.jelechem.2019.113538
50. T.S.S.K. Naik, B.E.K. Swamy, Pre-treated glassy carbon electrode sensor for catechol: a voltammetric study. J. Electroanal. Chem. **826**, 23–28 (2018). https://doi.org/10.1016/j.jelechem.2018.08.022
51. T.S.S.K. Naik, S. Saravanan, K.N.S. Saravana, U. Pratiush, P.C. Ramamurthy, A non-enzymatic urea sensor based on the nickel sulfide/graphene oxide modified glassy carbon electrode. Mater. Chem. Phys. **245**, 1–7 (2020). https://doi.org/10.1016/j.matchemphys.2020.122798
52. B. Sanna, C.D. Mruthyunjayachari, P. Malathesh, Electrochemical sensing based MWCNT-Cobalt tetra substituted sorbaamide phthalocyanine onto the glassy carbon electrode towards the determination of 2-Amino phenol: a voltammetric study. Sens. Actuators B Chem. **301**, 1–8 (2019). https://doi.org/10.1016/j.snb.2019.127078
53. Y. Xu, K. Huang, Z. Zhu, T. Xia, Effect of glassy carbon, gold, and nickel electrodes on nickel electrocrystallization in an industrial electrolyte. Surf. Coat. Technol. **370**, 1–10 (2019). https://doi.org/10.1016/j.surfcoat.2019.04.072
54. A. Bagheri, H. Hosseini, Electrochemistry of raloxifene on glassy carbon electrode and its determination in pharmaceutical formulations and human plasma. Bioelectrochemistry **88**, 164–170 (2012). https://doi.org/10.1016/j.bioelechem.2012.03.007
55. E. Sohouli, A. Homayoun, F. Shahdost-fard, E. Naghian, A glassy carbon electrode modified with carbon nanoonions for electrochemical determination of fentanyl. Mater. Sci. Eng. C. **110**, 1–10 (2020). https://doi.org/10.1016/j.msec.2020.110684
56. R. Jain, Voltammetric determination of cefpirome at multiwalled carbon nanotube modified glassy carbon sensor based electrode in bulk form and pharmaceutical formulation. Colloids Surf. B Biointerfaces **87**, 423–426 (2011). https://doi.org/10.1016/j.colsurfb.2011.06.001
57. B. Rezaei, N. Askarpour, A.A. Ensafi, Adsorptive stripping voltammetry determination of methyldopa on the surface of a carboxylated multiwall carbon nanotubes modified glassy carbon electrode in biological and pharmaceutical samples. Colloids Surf. B Biointerfaces. **109**, 253–258 (2013). https://doi.org/10.1016/j.colsurfb.2013.04.004
58. Y. Wei, C. Shao, J. Liu, Voltammetric determination of ethamsylate in bulk solution and pharmaceutical tablet by nano-material composite-film coated electrode. Mater. Sci. Eng. C. **29**, 2442–2447 (2009). https://doi.org/10.1016/j.msec.2009.07.007
59. K.K. Reddy, M. Satyanarayana, K.Y. Goud, K.V. Gobi, H. Kim, Carbon nanotube ensembled hybrid nanocomposite electrode for direct electrochemical detection of epinephrine in pharmaceutical tablets and urine. Mater. Sci. Eng. C. **79**, 93–99 (2017). https://doi.org/10.1016/j.msec.2017.05.012
60. A.K. Baytak, T. Teker, S. Duzmen, M. Aslanoglu, A sensitive determination of terbutaline in pharmaceuticals and urine samples using a composite electrode based on zirconium oxide nanoparticles. Mater. Sci. Eng. C. **67**, 125–131 (2016). https://doi.org/10.1016/j.msec.2016.05.008

61. K. Tyszczuk-rotko, New voltammetric procedure for determination of thiamine in commercially available juices and pharmaceutical formulation using a lead film electrode. Food Chem. **134**, 1239–1243 (2012). https://doi.org/10.1016/j.foodchem.2012.03.017
62. K. Tyszczuk-rotko, I. Be, Nafion covered lead film electrode for the voltammetric determination of caffeine in beverage samples and pharmaceutical formulations. Food Chem. **172**, 24–29 (2015). https://doi.org/10.1016/j.foodchem.2014.09.056
63. E. Tammari, A. Nezhadali, S. Lotfi, H. Veisi, Fabrication of an electrochemical sensor based on magnetic nanocomposite Fe 3O4/B-alanine/Pd modified glassy carbon electrode for determination of nanomolar level of clozapine in biological model and pharmaceutical samples. Sens. Actuators B. Chem. **241**, 879–886 (2017). https://doi.org/10.1016/j.snb.2016.11.014
64. N.C. Honakeri, S.J. Malode, R.M. Kulkarni, N.P. Shetti, Electrochemical behavior of diclofenac sodium at coreshell nanostructure modified electrode and its analysis in human urine and pharmaceutical samples. Sens. Int. **1**, 1–8 (2020). https://doi.org/10.1016/j.sintl.2020.100002
65. N. Kumar, A. Singh, B. Mizaikoff, S. Singh, C. Kranz, Electrochemical detection and photocatalytic performance of MoS2/TiO2 nanocomposite against pharmaceutical contaminant: paracetamol. Sens. Bio-Sens. Res. **24**, 2–10 (2019). https://doi.org/10.1016/j.sbsr.2019.100288
66. J. Fischer, J. González-martín, H. Dejmková, Voltammetric study of triazole antifungal agent terconazole on sp3 and sp2 carbon-based electrode materials. J. Electroanal. Chem. **863**, 1–10 (2020). https://doi.org/10.1016/j.jelechem.2020.114054
67. Á. Torrinha, C.G. Amorim, M.C.B.S.M. Montenegro, A.N. Araújo, Biosensing based on pencil graphite electrodes. Talanta. **190**, 235–247 (2018). https://doi.org/10.1016/j.talanta.2018.07.086
68. N. Jadon, R. Jain, A. Pandey, Electrochemical analysis of amlodipine in some pharmaceutical formulations and biological fluid using disposable pencil graphite electrode. J. Electroanal. Chem. **788**, 7–13 (2017). https://doi.org/10.1016/j.jelechem.2017.01.055
69. D. Giray, S. Karakaya, Differential pulse voltammetric determination of acyclovir in pharmaceutical preparations using a pencil graphite electrode. Mater. Sci. Eng. C. **63**, 570–576 (2016). https://doi.org/10.1016/j.msec.2016.02.079
70. J.K. Shashikumara, B.E.K. Swamy, A simple sensing approach for the determination of dopamine by poly (Yellow PX4R) pencil graphite electrode. Chem. Data Collect. **1**, 1–10 (2020). https://doi.org/10.1016/j.cdc.2020.100366
71. A.H. Oghli, A. Soleymanpour, Polyoxometalate/reduced graphene oxide modified pencil graphite sensor for the electrochemical trace determination of paroxetine in biological and pharmaceutical media. Mater. Sci. Eng. C. **108**, 1–10 (2020). https://doi.org/10.1016/j.msec.2019.110407
72. P. Manjunatha, Y.A. Nayaka, B.K. Chethana, C.C. Vidyasagar, R.O. Yathisha, Development of multi-walled carbon nanotubes modified pencil graphite electrode for the electrochemical investigation of aceclofenac present in pharmaceutical and biological samples. Sens. Bio-Sens. Res. **17**, 7–17 (2018). https://doi.org/10.1016/j.sbsr.2017.12.001
73. D.E. Bayraktepe, Z. Yazan, M. Önal, Sensitive and cost effective disposable composite electrode based on graphite, nano-smectite and multiwall carbon nanotubes for the simultaneous trace level detection of ascorbic acid and acetylsalicylic acid in pharmaceuticals. Talanta **203**, 131–139 (2019). https://doi.org/10.1016/j.talanta.2019.05.063
74. A. Smart, A. Crew, R. Pemberton, G. Hughes, O. Doran, J.P. Hart, Screen-printed carbon based biosensors and their applications in agri-food safety. Trends Anal. Chem. **127**, 1–16 (2020). https://doi.org/10.1016/j.trac.2020.115898
75. R. Adriano, D. De Faria, L. Guilherme, D. Heneine, Application of screen-printed carbon electrode as an electrochemical transducer in biosensors. Int. J. Biosens. Bioelectron. Mini. **5**, 9–10 (2019). https://doi.org/10.15406/ijbsbe.2019.05.00143
76. S. Eissa, A comparison of the performance of voltammetric aptasensors for glycated haemoglobin on different carbon nanomaterials-modified screen printed electrodes. Mater. Sci. Eng. C. **101**, 423–430 (2019). https://doi.org/10.1016/j.msec.2019.04.001
77. A. Smart, A. Crew, R. Pemberton, G. Hughes, O. Doran, J.P. Hart, Screen-printed carbon based biosensors and their applications in agri- food safety. Trends Anal. Chem. **127**, 1–16 (2020). https://doi.org/10.1016/j.trac.2020.115898

78. A. Sasal, K. Tyszczuk-rotko, Screen-printed sensor for determination of sildenafil citrate in pharmaceutical preparations and biological samples, Microchem. J. **149**, 104065 (2019).https://doi.org/10.1016/j.microc.2019.104065
79. A.A. Khorshed, M. Khairy, C.E. Banks, Voltammetric determination of meclizine antihistamine drug utilizing graphite screen-printed electrodes in physiological medium. J. Electroanal. Chem. **824**, 39–44 (2018). https://doi.org/10.1016/j.jelechem.2018.07.029
80. J. José, M. Gómez, J. Valenzuela, A. Vera, V. Arancibia, Determination of a natural (17 β-estradiol) and a synthetic (17α-ethinylestradiol) hormones in pharmaceutical formulations and urine by adsorptive stripping voltammetry. Sens. Actuators B. Chem. **297**, 1–10 (2019). https://doi.org/10.1016/j.snb.2019.126728
81. T. Chen, U. Rajaji, S. Chen, S. Chinnapaiyan, R.J. Ramalingam, Ultrasonics—Sonochemistry facile synthesis of mesoporous WS2 nanorods decorated N-doped RGO network modified electrode as portable electrochemical sensing platform for sensitive detection of toxic antibiotic in biological and pharmaceutical samples. Ultrason. Sonochem. **56**, 430–436 (2019). https://doi.org/10.1016/j.ultsonch.2019.04.008
82. M. Khairy, A.A. Khorshed, F.A. Rashwan, G.A. Salah, H.M. Abdel-wadood, C.E. Banks, Simultaneous voltammetric determination of antihypertensive drugs nifedipine and atenolol utilizing MgO nanoplatelet modified screen-printed electrodes in pharmaceuticals and human fluids. Sens. Actuators B. Chem. **252**, 1045–1054 (2017). https://doi.org/10.1016/j.snb.2017.06.105

Chapter 9
Biosensing Applications of Electrode Materials

Kingsley Eghonghon Ukhurebor, Uyiosa Osagie Aigbe, Robert Birundu Onyancha, Onoyivwe Monday Ama, Can-voro Osemwengie Amadasun, Joseph Onyeka Emegha, Otolorin Adelaja Osibote, Samuel Ogochukwu Azi, Azeez Olayiwola Idris, and Kabir Opeyemi Otun

Abstract The application of biosensors (BSs) in various fields of our endeavours such as medical science and agriculture as well as environmental management and monitoring has attained great importance. Consequently, this has resulted in more discoveries of comprehensive and influential investigative tools using organic or biological sensing components as BSs. Topical developments in organic or biological procedures and instrumentation have improved the sensitive extent of BSs. Several BSs such as nanoparticles/nanomaterials, polymers, microbes built BSs,

K. E. Ukhurebor (✉)
Department of Physics, Faculty of Science, Edo State University Uzairue, PMB 04, Auchi, Edo State, Nigeria
e-mail: ukhurebor.kingsley@edouniversity.edu.ng

U. O. Aigbe · O. A. Osibote
Department of Mathematics and Physics, Faculty of Applied Sciences, Cape Peninsula University of Technology, Cape Town, South Africa

R. B. Onyancha
Department of Physics and Space Science, School of Physical Sciences and Technology, Technical University of Kenya, Nairobi, Kenya

O. M. Ama · C. O. Amadasun
Department of Chemical Sciences, University of Johannesburg, Doornfontein, Johannesburg 2028, South Africa

J. O. Emegha · S. O. Azi
Department of Physics, Faculty of Physical Sciences, Ambrose Alli University, PMB 14, Ekpoma, Edo State, Nigeria

A. O. Idris
Department of Physics, Faculty of Physical Sciences, University of Benin, PMB 1154, Benin City, Edo State, Nigeria

K. O. Otun
Institutes for the Development of Energy for African Sustainability, University of South Africa, Private Bag, Johannesburg 1709, South Africa

Department of Chemical, Geological and Physical Sciences, College of Pure and Applied Sciences, Kwara State University, PMB 1530, MaleteIlorin, Nigeria

O. M. Ama et al. (eds.), *Modified Nanomaterials for Environmental Applications*, Engineering Materials, https://doi.org/10.1007/978-3-030-85555-0_9

etc. have widespread potential applications. However, it is significant to assimilate multifaceted methods to develop BSs that can possibly be used for more diverse applications. Agreeably, nanotechnology has been of great assistance in the transformation of BSs, biosensing-based mechanisms and their applications not just in electrode materials but other aspects of human endeavours. Advancement in the field of nano-sciences has contributed significantly to a wide range of innovative technologies that have allowed us to considerably improve and even reconsidered procedures and products or to generate wholly innovative functionalities. Hence, in this chapter, we will present a comprehensive appraisal of the manufacture of diverse nanostructured conducting polymers as well as their composites with various forms. Our major emphasis will primarily be on conducting polymer-integrated electrochemical biosensors (ECBs) that are used for addressing extensive biological areas as well as the construction and applications of the molecularly reproduced polymer-based BSs. The applications of graphene-based ECBs and biosensing-based electrode systems will also be discussed.

Highlights

In this chapter, the following are included:

- The meaning and historical development of biosensors
- Operational and electronic principles of biosensors
- Applications of nanostructured conducting polymers for electrochemical biosensors
- Graphene and its applications for biosensors and biosensing-based electrode systems
- Research trends, challenges, limitations and recommendations for contemporary future of biosensor technology.

9.1 Introduction

Biosensors (BSs) are fundamental components in intelligent systems that are useful for the monitoring and management of almost if not all aspects of our doings; medically, environmentally and otherwise. The recent advancements in the field of nanoscience have been of great assistance in the development of innovative technologies which have considerably led to enormous improvement, procedures, products and the creation of absolute innovative functionalities [1–3]. Nanotechnology has transformed BSs, biosensing-based systems as well as their numerous applications in such systems [1–3].

There are several types of BSs that are used for various fields ranging from environmental to medical sciences and their usages depends on the sensing rudiments or transducers. Ever since the first discovering of the glucose BSs in 1956, by Clark Jr [4], which came into limelight commercially in 1975 [5], there are now several BSs discovered for several commercial purposes. These contemporary BSs have a wider range of applications which offers additional specifications, sensitivities, fastness, tangibility and multiplicative results compared to the previous conventional ones like the chemical sensors or BSs [3, 6].

BSs are broadly categorized into two classes which are based on sensing component and transduction modes. The sensing components consist of enzymes, antibodies (immunosensors), microorganisms (cell BSs), biological tissues and organelles. While the transduction modes hinge on the physiochemical variation resulting from the sensing component. Accordingly, different transducer BSs can be piezoelectric, electrochemical, calorimetric and optical [7]. According to Reyes De Corcuera and Cavalieri [7], the common types of piezoelectric transducer BSs are "acoustic and ultrasonic", the common types of electrochemical transducer BSs are "amperometric, conductometric and potentiometric", while the common types of optical transducer BSs are "absorbance, fluorescence and chemiluminense".

According to Arora [6], BSs can also be categorized based on the period/order they were discovered. In these categories, we have "first-generation BSs, second-generation BSs and third-generation BSs". The "first-generation BSs" are the modest approach involving the unswerving discovery of either increase of an enzymatically produced product or decrease of a substrate of redox enzymes using a natural mediator for electron transfer. Examples are glucose BSs which uses enzyme glucose oxidase and oxygen in detecting the decrease in oxygen level or increase in hydrogen peroxide corresponding to the level of glucose [5]. The "second-generation BSs" use non-natural redox mediators like "ferricynide, quinones and ferrocene" for the movement of the electron which increases the reproducibility and sensitivity. Examples are self-monitoring amperometric glucose BSs [6]. The "third-generation BSs" in which the redox enzymes are immobilized on the electrode surface in such a manner that direct electron transfer is possible between the enzyme and transducer [6]. According to Borgmann et al. [8], it uses organic conducting material like "*Tetrathiafulvalnetetracynoquinodi Methane (TTF-TCQN)*".

Recent progressions in integrated electronics have led to the reduction of complicated electronic systems into nano sizes (mm^2 sizes); this possibility occurs in the implementation of complicated and sophisticated instrumentation in inexpensive and portable devices for quick discovery of detrimental and poisonous agents [1, 2]. Consequently, nanotechnology is becoming significantly useful in the production of BSs. The development of the various nanomaterials (NMs) for BSs and biosensing-based systems regarding the applications of nanotechnology have contributed significantly to the advancement of food security and safety, environmental analysis as well as in healthcare diagnostic applications [3, 9]. Now, the development in the field of nanoscience has led to the deployment of nanotechnology for inexpensive, numerous-produced, portable BSs that can unswervingly monitor environmental contaminants as well as to assimilate these sensors into mechanisms and systems for numerous market applications in areas of food security and safety, environmental analysis as well as in healthcare diagnostic applications [3, 9].

Several studies have shown that NMs are useful in the manufacture of BSs and applications in sensing-based mechanisms. The advancements in nanotechnological have led to improved performances in the aspects of discernment and sensitivity, producing lesser detection ranges of analysis [1–3, 10, 11]. NMs are beneficial in their high definite surface; this enables the immobilization of improved quantities of bioreceptor components. NMs are normally furnished with appropriate roles, through direct functionalization or through coating with efficient polymers, without disturbing their precise features; however, this attribute depends on their chemical configuration and composition [1, 2].

The recent advancements and development in electrochemical biosensors (ECBs) and biosensing-based electrode machinery as well as their significant applications from nanostructured constituents are becoming of great importance to various researchers in the field of nanoscience. Hence, in this chapter, we will present an appraisal of the manufacture of diverse nanostructured conducting polymers as well as their composites and various forms. The various means which could be used for the development of nanostructured conducting polymers and their investigative effects both chemically, physically as well as topographically will be exhibited. However, our attention will primarily be on conducting polymer-integrated ECBs that are used for addressing extensive biological areas such as genetic factors like biopolymer of deoxyribonucleic acid (DNA), peptides, proteins and other biological bioindicators, together with their expediency as cell-built chips and the construction and applications of the molecularly reproduced polymer-based BSs. Momentarily, the applications of graphene-based ECBs and biosensing-based electrode technology will also be discussed.

9.2 Biosensors Operational Principles

As stated earlier in the introduction section, BSs appeared first in the 1950s following its invention by a Biochemist Jr [4], whose intention was to develop a device that will efficiently detect the quantity of oxygen that is circulating in the blood. The developed BSs was then known as the Clark electrode and was virtuously just an oxygen detector. Several BSs were subsequently developed, with the first being the enhancement on Clark electrode BSs (oxygen detector) with the overlaying of a glucose-sensitive enzyme film over the existing electrode and this further improved the functionality of the Clarke electrode BSs towards acting as a blood sugar sensor. The blood sugar sensor was used for the detection and measurement of the level of blood glucose. In furtherance to this development was the use of a layer of urease enzyme which was able to detect urea in blood, urine, etc.

Habitually, a BS could function inside or outside a living organism. Hitherto, the target is the distinctive variable to which the device is applied to detect or measure frequently originate from the living organism or ecosystem. Consequently, BSs could be categorized according to the following features:

Targeted Analyte: This is the target substance to be sensed, examples are glucose BSs, oxygen BSs, nitrogen BSs, DNA BSs, etc. Note that the analyte must have an affinity for the sensing surface to which it may react chemically, electrochemically, etc. [12].

Bioreceptor: This is the deactivated medium (an immobilized hard or soft biocomponent) often selectivity specific, that is to be exposed to the targeted substance or analyte [12]. The common bioreceptor categories are enzymes (such as "oxidoreductases, peroxidases, polyphenol oxidases aminooxidases", etc. [13], tissues, antibody, DNA, microbes, biomimetics [14], nanoparticles. The bioreceptor is often interfaced with the transducer.

Generation: The period during when the BSs was developed and released. This implies that BSs manufactured, developed and released into the market over a given period with similar characteristics belong to the same or a particular generation. There are "first generation, second generation, third generation", etc. [6] as earlier discussed. Like other devices such as computers and other electronic devices, the latest generation often shows more sophistication than the earlier generation. The sophistication may be expressly defined in terms of precision, energy consumption, multifunctionality, accuracy, size, the scale of integration, refinement, etc.

Sensing or Transduction Technique: This reflects the kind of detection BSs performs, that is, the chemical, biochemical, organic, structural or physical principle adopted [10, 11, 15–17]. Ordinarily, the strategy is based on the behaviour of the transducer. Modern strategies include electrical, optical, piezoelectric, photoelectric, thermometric, photoelectrochemical, nano and electrochemical sensing strategies.

Nature of Organic Constituent Used: The biological or organic material; is often in a deactivated state which is layered over the transducer or electrode.

BSs are not just detectors; they also serve analytical functions depending on the domain of their applications. For example, as an analytical tool, a BS may be deployed to detect the precise existence of specific biomolecules, microorganisms, biological/organic structures or any other biological/organic analytes. BSs operate on the principle of cell signalling or signal transduction; a process involving the relay of physical or biochemical signals that reflects molecular events from a cell or substance in such a manner that is measurable [5, 13, 14].

The major recognized components in BSs are shown in Fig. 9.1. The figure illustrates a block model of a typical BS with processor and display unit.

Detecting/Sensing Element: A biorecognition component; for instance, in a glucose sensor, the biorecognition element is a deactivated glucose-sensitive enzyme.

The Transducer: The chemical, biochemical, organic, structural or physical device that translates variations in the target biophysical parameter such as oxygen, glucose, etc. to a physically measurable output signal and/or vice versa.

Signal Processor: This could be an electrical/electronic device with/without a display system, a processor and an amplifier [13].

BSs are more advanced vis-à-vis its selectivity and sensitivity when compared to other existing types of diagnostic tools. Some of the significant benefits and advantages of BSs include fast and continuous measurement, reduced use of reagents, elimination of calibration and re-calibration, high precision, quick response time, ability to measure non-polar molecules that other conventional or traditional devices can hardly estimate [13].

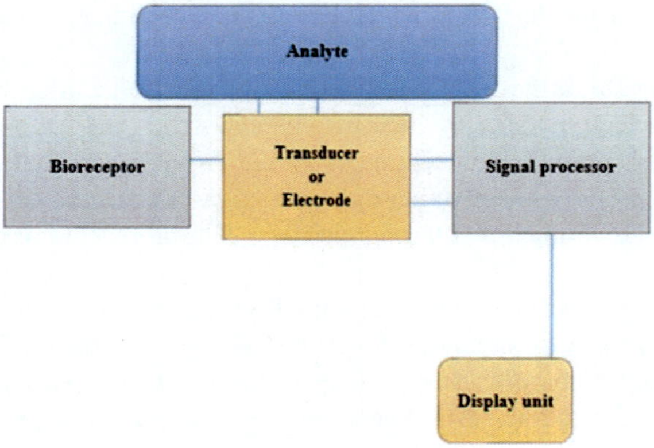

Fig. 9.1 Block model of a typical BS

9.3 Electronic Principles of Biosensor

BSs are known to operate on cell signalling principle, whose foremost components are; the element of biorecognition, biotransducer and electronic components that integrate, microprocessor, amplifier and display/monitor respectively. The biorecognition or sensing component is essentially a bioreceptor designed to detect or act on a particular analyte (the target substance whose relevant characteristics is to be detected or measured). The transducer receives input from the bioreceptor and depending on the transducer employed, a signal is relayed to the signal processor. The signal output amplitude is directly proportional to an analyte's concentration [18]. The signal is then amplified by the electronic device and processed accordingly. For example, in an amperometric sensor, the bioreceptor comprises a deactivated enzyme or specific "biomaterial," and the said deactivated "biomaterial" is kept in affinity with the transducer. The analyte reacts with the "biomaterial" [18], in this way, a new analyte may be formed which in effect gives the calculable electronic reaction. Sometimes, the analyte is transferred to a system that can be attached to the gas, heat, electron ions or hydrogen ions discharge, in this, the transducer can adjust the device linked, then converts to electrical signals that could be altered and computed [18].

BSs could be distinguished by way of the transduction sensing strategy employed in its construction. Common families are electrochemical, calorimetric, optical, etc. In some cases, one family may dovetail into the other in which case there may not be very clear boundaries of distinction [18]. Illustratively, Fig. 9.2 shows some types and subtypes of BSs based on sensing strategy employed in the design. Seven major types are recognized. The eight types are the nano-BSs which is based on NMs. As illustrated, nano-BSs could be a hybrid and may utilize the sensing strategies applied in other major types.

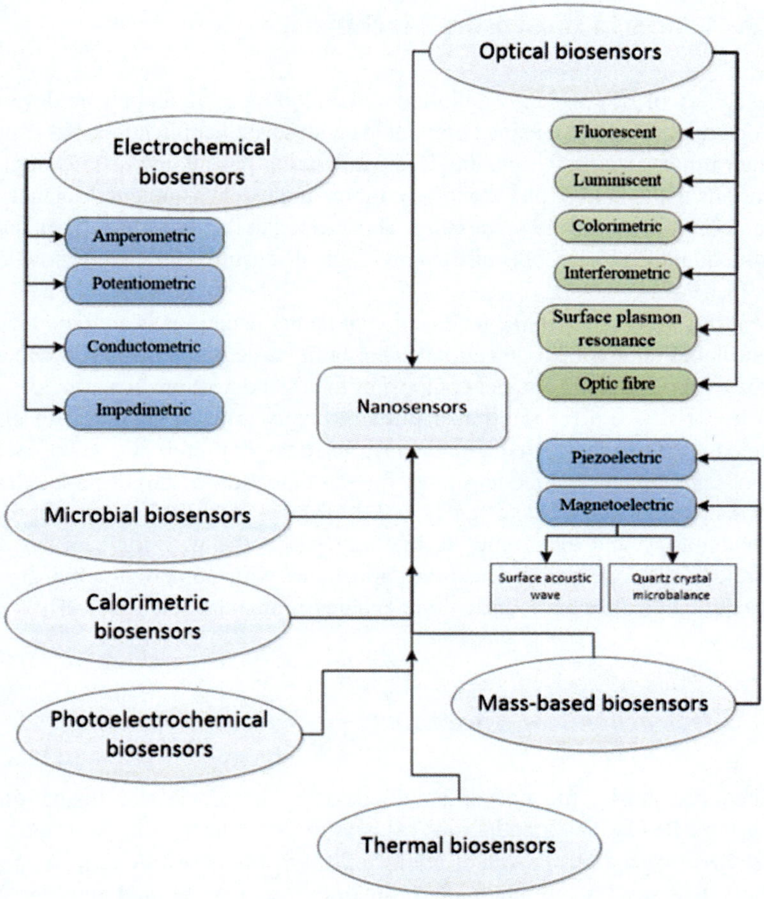

Fig. 9.2 Diagrammatical illustration of some types of BSs

9.4 Biosensors Mode of Operations

The components of BSs are instrumental to its operation. First, every BSs are designed to perform specific functions. The way and manner BSs operate are dependent on the bioreceptor (enzyme, DNA, phage, antibody, etc.) and the sensing strategy (that is, the workings of the transducer). The transducer's electrical signal is mostly small and overlays on a high baseline. Usually, the processing of the signal requires the deduction of a baseline location signal, collected from a similar transducer without covering any biocatalyst. The BSs reaction's comparatively slow character significantly eases the issue of electrical noise filtration. The direct output is an analogue signal at this point. However, the signal may be digitized and relayed to the microprocessor unit where the information is processed, guided by preferred units and exported to a data store.

9.5 Advances in Biosensors Machineries

The discovery of BSs which is an influential and pioneering diagnostic device that has to do with biological sensing components with several applications has espoused dominant importance in various fields. Its utilization has attained some significant applications in the field of pharmacology, biomedicine, environmental science, agriculture, food protection and processing. This has led to the development of accurate and influential diagnostic tools utilizing biological sensing components now known as BSs [3, 19–21].

According to Turner [5], the technical approaches used in BSs are built on label-built and label-unrestricted detection. Label-built detection is primarily dependent upon the explicit features of label composites to target detection. Nevertheless, these categories of BSs are reliable but habitually involve the combination of explicit sensing components fabricated with the restrained target protein. On the other hand, the label-unrestricted technique allows for the detection of target molecules that are not categorized or hard to tag [22, 23]. Topical interdisciplinary approaches of biotechnology and electronics technology paved the way for evolving label-unrestricted BSs for several detection approaches with numerous applications in the areas of medical science, agriculture and environmental science [1–3].

9.5.1 Electrochemical Biosensors

As stated previously, the conventional discovery of glucometer using glucose oxidase-built BSs by Clark and Lyons [4], was recorded as the first invented ECBs. Glucose BSs are generally prevalent among infirmaries as these are crucial for persons with diabetic issues for sporadic monitoring of the amount of blood glucose. Nevertheless, according to Harris et al. [24], glucose BSs are usually having problems as a result of the unbalanced enzyme action (inhomogeneity) for which additional calibration is required. Essentially, these possible problems lead to the discovery of a variety of biomolecules that are having discrepancies in their electrochemical features [5, 25, 26], which according to Wang et al. [27], paved way for the discovery of more practical glucose BSs.

Presently, ECBs are characteristically equipped with adjustable metal surface and carbon electrodes employing biomaterials, like enzymes, antibody or DNA. BSs' productivity signal is habitually produced via explicit binding or catalytic reactions of biomaterials on the electrode surface [27]. Accordingly, the necessity for the development of electrochemical sensors became essential for medical diagnosis of diseases where the initial discovery or monitoring appears indispensable [28]. In this situation, as rightly reported by Vigneshvar et al. [20], the invention of non-enzymatic BSs is frequently considered utilizing synthetic constituents using proteins. Remarkably, several forms of biomolecules have differential electrodes that are stable and selective which eventually contribute to the discovery of novel kinds of ECBs for several purposes [24].

In this perception, Wang et al. [27] studied the advancement of "ferroceneboronic acid" and "ferrocene-adapted boronic acids" for BSs discovery as a result of the existence of binding site known as "boronic acid moiety" and an electrochemically dynamic fragment known as "ferrocene residue". "Ferroceneboronic acid" and its by-products possess an exceptional characteristic of binding to 1,2- or 1,3-diol deposits of sugars forming cyclic boronate ester bonds. The redox characteristics of "ferroceneboronic acid-sugar adduct" vary from free "ferroceneboronic acid" forming a basis for electrochemical discovery. Furthermore, "boronic acids" have an affinity for binding to Fe–ions, which gives an added benefit for the invention of non-conventional ion-selective electrodes using Fe–ion. According to Wang et al. [27], the main limitation in applying this technique is the necessity for immobilization of "ferroceneboronic acid" by-products on the surface of the electrode as these by-products serve as additional sample solutions (reagents). The use of polymers in addition to or with silver electrodes together with appropriate adjustment of "ferroceneboronic acid" by-products could lead to enhancement of "ferroceneboronic acid electrochemical sensor" for biomedical research like in the diagnosis of diabetes in which glucose quantity will remain supplemental [20].

According to Mello et al. [29], another contemporary invention is in the aspect where ECBs are used to measure the extents of antioxidants and reactive oxygen species in physiological systems. As precisely reported by Erden and Kilic [30], the major problem with this perception is the discovery of uric acid as a main culmination product of body fluid purine metabolism, which accounts for some diagnostic means for several medical abnormalities, diseases or infections.

Albeit there was the need to develop an inexpensive as well as sensitive technique. Electrochemical-built method for the quantification of uric acid oxidation as well as for the measurement of glucose appears superlative. However, the similarity of uric acid in terms of oxidation by ascorbic acid poses a main experimental obstacle to invent exceedingly sensitive ECBs. In overcoming this experimental obstacle, researchers have established "an amperometric detection- built BSs" that possesses the capability to quantify both reduction and oxidation possibilities [20, 30].

Given the financial implications and reproducibility for this technique, it becomes imperative to restrain or screen print the enzyme on electrodes, otherwise on the nanomaterial-built electrodes, that are ideal for evolving the disposable, selective, expensive and sensitive uric acid BSs for humdrum investigation. Incidentally, topical advances in "three dimensional (3D) bioprinting" which is aimed at producing BSs with live cells condensed in 3D micro-environments [25]. Accordingly, a novel wireless mouth-protector BSs for detecting salivary uric acid quantity in real-time and incessant fashion was established [31] and the technique could be extended to customizable monitoring devices for various medical and suitability applications [20].

According to Bahadir and Sezginturk [32], ECBs have been effectively used for hormone quantifications; hitherto, its perspective need to be detailly studied [20]. An additional possible aspect of technological growth in BSs depends on aiming nucleic acids. Obviously, cellular "micro ribonucleic acid (miRNA)" expression is a superlative bioindicator for onset diagnosis of medical abnormalities, diseases or infections and for aiming those improves the efficiency of gene treatment for genetic syndromes.

According to Hamidi-Asl et al. [33], in the normal sense, the detection of miRNAs, are mostly by microarray, northern blotting as well as a polymerase chain reaction. However, contemporary technology offers superlative ECBs for miRNA detection built on label-unrestricted detection having guanine oxidation consequent to the hybrid development that has miRNA and its inosine ancillary capture the investigation. All these discoveries according to Vigneshvar et al. [20], are a result of the contemporary approaches in "biofabrification" for the advancement of electrochemical-built BSs technology in biomedical sciences.

9.5.2 Biosensors Used in Environmental Monitoring and Management

Another crucial area where BSs technology is needed in the field of environmental management and monitoring. This is another important aspect wherein BSs technology would undoubtedly assist in the swift identification of pesticidal and other environmentally harmful chemical deposits in order to avert the corresponding health dangers [34, 35].

According to Verma and Bhardwaj [35], the traditional or conventional means, such as "high-performance liquid chromatography, capillary electrophoresis and mass spectrometry" are efficient for the investigation of environmental pesticides; however, there are some restrictions such as intricacy, time-intense measures, necessity of high-end devices and operative proficiencies. Therefore, even if it is believed that modest BSs have great advantages; yet, it is not easy to invent integrated BSs that can analyze several categories of pesticides. Hence, steady enzyme-built BSs have been invented for understanding the physiological vis-à-vis the biochemical, chemical, biological, structural and physical influence of pesticides in the environment, food security and quality management [35, 36]. The appropriate understanding of the influence of pesticides and other influential issues on the environment will definitely be of assistance in the mitigation of the incessant change in the climate system which is inferably contributing to environmental hazards and other essential functionalities in the ecosystem such as irregular radio waves propagation in the atmosphere [37–43]. In affirmation of this, it has been reported that "acetylcholinesterase inhibition-built BSs" have been invented. Over the years, for the purpose of swift analysis, this method has received great improvement with additional topical developments in "acetylcholinesterase inhibition-based BSs" including immobilization means as well as other diverse approaches for fabrication [20, 44]. In the same way, "piezo-electric BSs" have been established for sensing the organophosphate and carbamate on the environmental influence pesticides [45]. "Organochlorine pesticides" are recognized for affecting the ecosystem where pesticides such as "endosulfan" cause substantial environmental impairments [46]. Undoubtedly, "organochlorine pesticides" have been reported to cause alteration in the reproductive system of in both male and female fish disparately [46] and given these facts, discovery of BSs for detecting aquatic ecosystem would have more consequences as a result of biomagnification [20]. In handling this quest, ECBs have experienced a revolution with swift advances in fabrication as well as the use of constituents like NMs [3, 25, 35].

At this juncture, it is of great significance to place distinct prominence for the selection or collection of receptors for BSs advancement, the use of diverse transduction procedures and fast screening approaches for applications of BSs in medical sciences and agricultural activities (food production and safety) as well as environmental protection, monitoring and management. To aid this, BSs fabrication appears to be vital; hence, the improvements in this aspect have been absolutely elucidated by several researchers.

9.5.3 Optical/Visual Biosensors

As described previously, biomedical or/and environmental applications require the advancement of unpretentious, instantaneous as well as extreme-sensitive BSs. This according to several researchers would be actualized with immobilizers such as gold, carbon-built constituents, silica, glass or quartz [20, 25, 34, 47–51].

Actually, the integration of gold NMs or quantum dots with the use of microfabrication affords novel technology for the advancement of extremely sensitive and transferrable cytochrome P450 enzyme BSs for specific purposes [20, 52, 53]. Additionally, chemical fiber-optic sensors have several significances in several fields, such as medication discovery, biosensing and other aspects of biomedical sciences.

Presently, hydrogels are now used as DNA-built sensors, with chemical fiber-optic used for developing elements for immobilization procedures [20, 54]. Immobilization in hydrogels occurs in 3D when compared to other constituents, which permits high loading measurements of sensing particles [20]. Hydrogels which are also known as "polyacrylamide are hydrophilic cross-related polymers" [55] and could be formed into diverse procedures for immobilization such as thin films, NMs, etc. According to Vigneshvar et al. [20], hydrogels have seemed as a modest substrate for "DNA immobilization" with other benefits which include entrapment, meticulous release, analyte enrichment as well as DNA fortification. These physiognomies as reported by Khimji et al. [55], are exceptional to hydrogels when compared to other constituents which give biomolecular immobilization. Also, the virtuous optical transparent nature of hydrogels gives an appropriate approach for visual discovery. As such Khimji et al. [55], reported that comprehensive procedures for the "immobilization of DNA BSs" in "monolithic polyacrylamide gels and gel micro-substances/microparticles" are frequently considered as a technical improvement in the science of BSs. However, according to Kwon and Bard [56], single/sole molecule discovery has also been established employing electrochemical oxidation of hydrazine for DNA discovery and detection.

9.5.4 Silica, Quartz/Crystal and Glass Biosensors

Topical approaches in the advancement of BSs have aided the use of silica, crystal or quartz and glass constituents as a result of their exceptional characteristics. According to Vigneshvar et al. [20], silicon NMs have a superior potential for the developments for BSs applications technologically as a result of their biocompatibility, copiousness as well as physiognomies electronically, optically and mechanically. Also, silicon NMs have been reported to have non-toxicity features which is a vital requirement of their applications biomedically and environmentally. These applications according to Peng et al. [47] as well as Shen et al. [57], range from bioimaging, biosensing and tumour or cancer treatment. Additionally, fluorescent silicon NMs have promising benefits and applications in bioimaging [20].

Fascinatingly, Shen et al. [57] reported that silicon nanowires together with gold NMs give hybrids that are useful as innovative silicon-built nano-reagents for efficient tumour or cancer treatments. Covalent bonding of "thiol-improved DNA oligomers" on silica or glass aids DNA films for improved procedure in "ultraviolet (UV) spectroscopy and hybridization techniques" [55]. Notwithstanding, these benefits of silicon NMs, there are also some prevalent challenges which include the advancement of comprehensive inexpensive production procedures and the biocompatibility after the biomolecular interaction, require evaluation. According to Vigneshvar et al. [20], the answer to these issues is to pave way for silicon NMs to be the present-day BSs constituents.

The "wire and non-electrode quartz crystal micro-equilibrium BSs" need an alternative platform for investigating the connections between biomolecules/bioparticles/biomaterials with lofty sensitivity. It has been reported that the pulsations of quartz oscillators were excited and spotted by antennas employing electromagnetic waves shorn of any wire connections. This non-communicating specific quantification is a crucial property for evolving ultrahigh-sensitive discovery of proteins in liquids by means of "quartz crystal BSs -built instrumentation" [50]. Given the exceptional properties of silica or quartz or glass constituents, numerous innovative BSs have been established with high-end technology for advancing bioinstrumentation, biomedical technology and some other aspects of human endeavours [20, 25, 47, 50, 57]; hitherto, inexpensive and eco-friendly/biosafety needs adequate attention.

9.5.5 Nanomaterials-Built Biosensors

Several NMs such as silicon, gold, silver as well as copper NMs and carbon-built constituents like graphene, graphite and carbon nanotubes have been used for evolving BSs immobilization [10, 20, 23, 34, 51, 52, 57–61]. Furthermore, nanomaterial/nanoparticle-built constituents give enormous sensitivity and specificity for evolving electrochemical and other categories of BSs. Gold NMs have more potentials when compared to other metallic NMs use; this is because of their permanency against oxidation and practically gold NMs are not noxious, whereas other NMs such as silver oxidizes and exhibits noxiousness if used for within the field of medical science such as medication delivery. Basically, the procedure of NMs for BSs has possible challenges, which need to be adequately addressed when been used for biomedical issues [62, 63]. Also, according to Ding et al. [64], NMs-built signal amplification approaches have potential benefits and complications. However, NMs are considered as critical constituents in bio-diagnostic instruments, essentially for improving the sensitivity and discovery extents for sole or single particle discovery [20, 25].

Accordingly, it is noteworthy to say that the discovery of platinum built NMs for electrochemical amplification with sole/single tag response for the discovery of low or truncated concentration of DNA [20, 56]. In the same way, Niu et al. [16] reported that semiconductor quantum dots and iron oxide nanocrystals that have some optical and magnetic characteristics that could be efficiently connected with tumour or cancer aiming ligands, like monoclonal antibodies, peptides, or small/minor particles that are targeted for tumour or cancer antigens with lofty affinity and specificity [20]. Accordingly, quantum dots technologies are useful for comprehending the tumour or cancer micro-environment for their treatment as well as for the conveyance of nano-medical issues [10]. It was reported by Vigneshvar et al. [20], that the use of cantilever dimensions BSs such as "milli-cantilever, micro-cantilever and nano-cantilevers BSs" are uniformly scrutinized disparagingly as a result of their enormous application in several fields.

9.5.6 Genetically Programmed/Synthetic Fluorescent Biosensors

The advancements of identified BSs using genetically programmed or what is also known as synthetic fluorescence have been of great assistance in the understanding of the biological procedure as well as several molecular trails within the cells [2, 3, 10, 11, 20, 65–67]. According to Oldach and Zhang [66], the discovery of fluorescent-identified antibodies was first established to image immobile cells. This approach certainly provided innovative means that assisted in the advancement of some sensors that use biological/organic proteins, minor particle binding to analytes and second messengers [2, 3, 10, 11, 20].

Presently, synthetic fluorescent BSs are established for the analysis of motor proteins via single or sole particle discovery with definite analyte concentration [65]. Notwithstanding, these advantages, Vigneshvar et al. [20] reported that are some implications such that the procedure of investigating the discovery and analysis appears to be challenging. According to Oldach and Zhang [66], the discovery of the various fluorescent proteins especially green fluorescent protein brought about several benefits in the aspect of optical investigation design and effectiveness.

Recently, genetically programmed BSs aiming particles linked with energy generation, reactive oxygen classes and cAMP gives more improved clarity of mitochondrial physiology [68]. Similarly, cGMP is a significant signalling particle and a medication aim for the cardiovascular system. Consequently, "Förster resonance energy transfer (FRET)-built BSs" have been established for envisaging cGMP, cAMP and Ca^{2+} in cells [69].

Quite a lot of these BSs work competently in primary culture and live-cell in vivo imaging [2, 3, 10, 11, 17, 66, 67]. A number of the main aspects have now been suited out for evolving BSs for live-animal imaging [20]. Enhancing such methods, Oldach and Zhang [66] reported that minor-angle X-ray scattering for evolving calcium sensors and fluorescence resonance energy transmission-investigations for kinase sensing are being mentioned as superlative BSs procedures in contemporary physiology. However, some microbial and cell organelle-built BSs were established with precise targets [70].

As previously discussed, optical BSs, electrochemical and electromechanical are established for sensing miRNA considerably more effective than other molecular procedures [71]. In view of the invention of in vivo imaging with minor particle BSs, an improved understanding of cellular activity and several other particles such as RNA, DNA and miRNA have been known [2, 3, 10, 11, 20, 55, 71].

According to Vigneshvar et al. [20], the present renovation in this aspect needs a complete genome method using improved optical-built genetic BSs. Presently, it is now well known that optical-built BSs together with fluorescence and NMs have attained tremendous accomplishment in terms of applications and sensitivity [2, 3, 10, 11, 20].

9.5.7 Microbial-Built Biosensors

In order to achieve more trends in the area of environmental management and monitoring as well as for bioremediation, the utilization of advanced innovative technology built on synthetic biology and genetic/protein engineering for programming microbes through precise signal outputs, sensitivity and selectivity [3, 20]. For instance, living cells having enzyme action to degrade or reduce xenobiotic composites would have extensive applications in bioremediation [72]. Also, microbial fuel-built BSs have been established to monitor or manage biochemical oxygen requests and environmental toxicity [3, 20].

Some microorganisms specifically bacteria have all it takes to degrade or reduce the organic substrate and producing electrical current for fermentation. Essentially, the technology encompasses the application of a bio-electrochemical means that manages the power of the respiratory system of microorganisms in converting organic substrates unswervingly into electrical energy [20]. Notwithstanding these possibilities, Vigneshvar et al. [20] reported that there are still some limitations in microbial BSs as a result of truncated power density in the aspect of production and operational costs.

There are being tremendous efforts by researchers to meaningfully improve the performance and cost effectiveness through innovative systemic methods, in which technologies have been of great assistance in evolving self-mechanical engineered microbial BSs [20, 73, 74]. Microbial BSs have also made known possible applications in the management and monitoring of pesticides and heavy metal discovery wherein eukaryotic microorganisms have are better off when compared to prokaryotic microorganisms; this is mainly as a result of the benefit of evolving complete cell BSs with selective and sensitive usefulness in relation to the discovery of heavy metal and the harmfulness of pesticides [20, 75]. Also, the more advanced eukaryotic microorganisms could possess extensive sensitivity to diverse noxious particles and have consequences to advanced living organisms.

Interestingly, the applications of microorganisms-built BSs are diverse extending from environmental management and monitoring to energy generation. However, innovative approaches will provide better novel BSs with higher sensitivity than selectivity from microbial basis from eukaryotic microorganisms to engineered prokaryotic microorganisms. In future, these microorganisms-built BSs would provide more beneficial applications for monitoring and managing environmental metal effluence and sustainable energy generation [20, 73, 74].

9.6 Benefits and Technological Comparison of Some Biosensors

Some of the different kinds of BSs as well as their applications have been previously discussed. However, this section will discuss the technological comparison and specificity as well as the detection extent, linear extent, investigation time and compactness of these BSs.

Advancements in electrochemical sensors with high-quantity approaches concentrating on detection extent, investigation time and compactness provided huge-scale user markets for low-cost BSs designed for glucose and pregnancy assessments via anti-human chorionic gonadotropin immobilization strip with lateral-flow technique [25]. According to Vigneshvar et al. [20], the control of analytes using polymers and NMs are vital for the improvement of the sensitivity and detection extent. As a result of this, the lateral-flow technique permits the direct conveyance of samples to anticipated spots to generate precise interactions as an alternative to random interactions. Some of the above-mentioned BSs have applied this technique [20]; also, it has paved the way for bio-fabrication using either contact or non-contact- built modelling.

The use of NMs such as gold, silver and silicon-built bio-fabrication produced innovative procedures. Also, coating/covering of polymers on these NMs brings huge advancements in contact-built electrochemical sensing. One of the foremost benefits of these categories of electrochemical sensors is their sensitive and specific nature with real-time investigation. Though, there are some limitations such as the recovering capability or long-term procedure of polymers or/and other constituents [20]; hitherto, decrease in cost makes such electrochemical sensors more within one means.

Single or sole analyte discovery using contact-built sensing has great benefits, for instance, real-time measurement of particles with lofty specificity. To aid this, FRET, bioluminescent resonance energy transmission, fluorescent-built as well as surface plasmon resonance-built transducers was developed for improving the sensitive and specific nature in terms of sole or single particle discovery [20, 54]. These procedures also have some limitations in terms of multiple analyte discovery as a result of signal emission intersection/overlap [20]; hitherto, resonance energy transmission procedures habitually established for multiple analyte discovery, which is extremely rewarding in clinical/medication diagnosis as a result of transformation in biomarkers between patients and interrelated infections.

The application of microcantilevers or nanocantilevers as transducers in electrochemical sensor bio-fabrication is similarly having widespread potential in multiple analytes discovery [20]. Also, non-contact-built sensors by means of 3D bioprinting utilizing inkjet or laser direct inscription afforded improved consequences. However, the cost involved and customizable capability has great limitations in these procedures.

Remarkably, some of these high-quantity BSs have been combined with electrochemical sensing for precise applications. Most of the prominent sensitive, real-time, in addition to portable amperometric ECBs, have been established for infection diagnosis using body fluids. Generally, ECBs combined with bio-fabrication have truncated discovery extent for sole or single analyte discovery specific nature employing real-time investigation and at an inexpensive price given the portable nature of the device [20, 31].

Presently, optic-built BSs are also one of the foremost technologies in biosensing that has to do with fiber-optic chemistry. Single or sole particle discoveries, such as DNA or peptide, are done appropriately using hydrogel-built cross-linking as a result of the benefit of having lofty loading capability and hydrophilic nature. More so, optical BSs which had widespread applications in biomedical and forensic science for DNA measurement was developed [56]. The combination of organic or biological constituents, such as "enzyme/substrate, antibody/antigen and nucleic acids" brought great version to optical BSs technology. Also, it has now been possible to integrate microorganisms, other living organisms (animal or plant) cells and tissue segments in the biosensing system. The advancements in molecular optoelectronics presented optical biometric recognition systems. Integrated optics technology gave room for the incorporation of both passive and active optical mechanisms on the same substrate for evolving minimized or curtailed compact sensing strategies through the fabrication of multiple sensors on a sole or single chip. In this situation, high class polymers provided hybrid systems for optical BSs. In effect, optic-built BSs technology was enhanced as a result of recent innovations in surface morphology investigation employing advanced electron and atomic force microscopy.

Regardless of these, the discovery extent of optical BSs at no time would have been close to femto-level given the instrumentation cost and device non-portability. As a result of this, recent optic technology that has to do with nanomechanical BSs built on microcantilevers and surface resonance technology assisted in the provision for advanced DNA chips at best for real-time precise and sensitive investigation [76–78]. Basically, the benefits of optical BSs comprise fast instant investigation with the resistance of signal to electrical/magnetic intrusion and the budding for a spectrum of data. However, the main disadvantage is its expensive nature as a result of certain instrumentation necessities. Other procedural difficulties include complexity in immobilization particularly for bio-fabrication and the need to disparagingly resolved to decontaminate the environment to take complete advantage of optical BSs.

Bio-fabrication of mechanical devices needs improved outcomes for mass-built BSs. Irrefutably, both electrochemical and optical BSs make the furthermost of this technology for devising superior BSs. According to Arlett et al. [79], the major developments in micro-fabrication and nanofabrication technologies made it possible for the development of mechanical devices with nanosized moving fragments. The capability to fabricate such devices using semiconductor processing measures are linked to biophysics and bioengineering moralities in the direction of the advancement of practical micro-electromechanical and nano-electromechanical BSs that were produced in large quantities [79].

Glass, silicon and quartz constituents have been applied either after tagging with fluorescence or gold NMs efficaciously. Although, these BSs have superior precision in terms of sole or single particle discovery, cost effectiveness, mass production is hardly achievable [10, 47, 57]. Several challenges are however still noticed in mass-built sensors in terms of improved capture agents to produce at nanoscale utilizing microelectronic fabrication for high-throughput investigation. As such, the great application possibilities of semiconductor materials and quantum dot technology are worth mentioning.

However, there are present technologies in BSs that are capable of simultaneously executing, real-time and quantitative analyses for bulky arrays. It is the use of micro-scale and nanoscale cantilevers fabrication that has made this possible [11, 57]. Another key technical uprising in BSs is the detection of genetically encoded or synthetic fluorescent BSs for analyzing molecular devices of biological or organic procedures [3, 10, 48, 49, 65–67]. Nevertheless, these sensors have a great prospect for sole or single particle detection with precise analyte measurement, the procedure and investigative preparation and discovery are hard and also need advanced instrumentation.

As stated previously, biomaterials and microbial fuel-built BSs also have discrepancies as a result of the highly sensitive and selective nature; hitherto, mass production and genetic engineering procedures to develop a microbial strain need intricate measures and high cost. Nonetheless, one benefit of microbial BSs is the capability to use them as a means for bioremediation, which has superior significance in environmental management and monitoring. Nevertheless, the development and release of such modified genetic strain need to be strictly analyzed with appropriate governing rules and regulations as well as the ethical clearance (that was the issue of bioethics comes in) and production cost management. Hence, it has been reported that the development of diverse micro- BSs and nano- BSs frameworks that as to do with combined technologies that utilize electrochemical or optic bioelectronic principles with a combination of bioparticles or biological materials, polymers and NMs are needed to attain exceedingly sensitive and miniaturized devices [3, 10, 17, 48, 49, 76].

9.7 Research Trends, Future Challenges and Limitations of Biosensor Machineries

Integrated approaches utilizing multiple technologies such as electrochemical, electromechanical and fluorescence-cum-optical-built BSs and genetically engineered microorganisms are recent approaches for BSs detections. Shown in Table 9.1 is a summary of the operational principle and applications of some BSs revised from Vigneshvar et al. [20].

Most of these BSs have great application for medical possibilities in disease/infection diagnosis. Table 9.2 shows the summary of the medical applications of some BSs in disease/infection diagnosis.

As the quest and need for applying BSs for swift investigation with cost-efficiency entail bio-fabrication that now pave way for the identification of cellular to complete animal action with a discovery extent of high precision for sole or single particles. According to Dias et al. [54], BSs ought to be designed to work under multiplex and complex circumstances. In that condition, both 2D and 3D discovery are needed using sophisticated transducers for detecting and measuring small or minor analytes of interest. Some breakthrough discoveries have been made with contact or non-contact-built modelling at diverse stages. We now have BSs are developed for discovering additional robust recovering long-term usage. These diagnostic b BSs are developed by means of therapeutics, which assist both the medical staff/team and their patients for a more assimilative understanding of infections/diseases and the treatment procedures (therapy). Hence, Fracchiolla et al. [81] reported that fluorescence resonance energy transfer-built BSs provides an exceptional diagnostic technique for evaluating the effectiveness of imatinib management and treatment of continuing or chronic myeloid leukemia. The present use of affibodies, aptamers, peptide arrays

Table 9.1 The operational principle and applications of some BSs

BSs type	Operational principle	Applications
Glucose oxidase electrode-built BSs [14]	Function on the electrochemistry principle using glucose oxidation	For analyzing the amount of glucose in biological sample
HbA1c BSs [27]	Function on the Electrochemistry principle by means of ferroceneboronic acid	For vigorous analytical technique for the measurement of glycated haemoglobin
Uric acid BSs [30, 31]	Function on the electrochemistry principle	For detecting clinical infections, abnormalities or diseases
Acetylcholinesterase inhibition-built BSs [44]	Function on the electrochemistry principle	For understanding the influence or impact of pesticides
Piezoelectric BSs [45]	Function on the electrochemistry principle	For the detection of organophosphate and carbamate
Microfabricated BSs [76]	Function on the optical/visual BSs principle using cytochrome P450 enzyme	For developing medications/drugs
Hydrogel-built BSs r/polyacrylamide-built BSs [55]	Function on the optical/visual BSs principle	For the immobilization of biomolecular
Silicon BSs [47, 57]	Function on the optical/visual/fluorescence principle	For bioimaging, biosensing and the treatment (therapy) of tumour/cancer
Quartz–crystal BSs [50]	Function on the electromagnetic principle	For the development of ultrahigh-sensitive detection of proteins in liquids
NMs-built BSs [23, 51, 52, 56–61, 63, 80]	Function on the electrochemical or optical/ visual/fluorescence principle	Used for multifaceted purposes such as biomedicine as well as diagnostic tools
Genetic-programmed or fluorescence- built BSs [27, 65–67]	Function on the fluorescence principle	For the understanding of the biological procedure as well as several molecular systems within cells
Microbial fuel cell- built BSs [73, 75]	Function on the optical principle	For monitoring and management, the biochemical oxygen demand and environmental contamination from heavy metals, pesticides as well as other chemicals

Table 9.2 Medical applications of BSs in disease/infection diagnosis

BSs	Disease/infection diagnosis
Glucose oxidase electrode-built BSs and HbA1c BSs [2, 3, 20, 21]	For diabetes diagnosis
Uric acid BSs [2, 3, 16, 17, 20]	For cardiovascular and general/widespread diseases diagnosis
Microfabricated BSs [2, 16, 20]	For visual/optical corrections
Hydrogel-built b BSs /polyacrylamide-built BSs [3, 13, 20]	For recovering/regenerative medicine
Silicon BSs [13, 18, 20]	Development and applications in cancer/tumour biomarker
NMs-built BSs [1–3, 13, 17, 20]	For healing therapeutic applications for disease/infection

as well as imprinted molecular polymers are conventional illustrations for potential research methods in this aspect [20, 22, 35, 82].

However, some achievement has also been reported with limited possible particles for advanced therapeutic, antimicrobial and medication delivery. According to Grabowska et al. [83], the development in this aspect leads to discovery of ECBs as reliable and dependable diagnostic means for pathogen discovery of avian influenza virus in the intricate matrices. However, additional new reports showed possible applications of affinity-built BSs in medication for sports and doping control investigation [84].

Also, there are now a multiplicity of even or wearable ECBs for real-time non-hostile screening of electrolytes and metabolites in body fluid as indicators or determinants of a wearer's medical status [85]. Additional attractive application is in the assessment of meat and fish quality utilizing hypoxanthine BSs through fabrication [86]. According to Kumar and Rani [87], the advancement in BSs for the discovery of biological or organic warfare agents such as virus, bacteria and contaminants (toxins) are now sometimes used for various means of BSs such as nucleic acid, electrochemical, piezoelectric and optical, which now have massive benefits in the military operations (defence, protection, safety and security), environmental and medical science.

The combination of NMs and polymers with different categories of BSs have assisted in the provision of hybrid devices that have now improved the usage in the above-mentioned applications [3, 10, 11, 17, 22, 25, 48, 49]. Moreover, scientific advancement has been of great assistance in the development of microbial BSs utilizing biological or organic synthesis approaches that are now contributing to energy and environmental management, monitoring and sustainability [20, 73]. Several other reports have shown the significance of using microbial fuel cells for evolving techniques for the treatment/remediation of water as power sources for environmental sensors [20].

9.8 Applications of Nanostructured Conducting Polymers for Electrochemical Biosensors

Nanostructured conducting polymers have given great consideration in recent times as a result of their numerous applications in diverse fields [3]. It has been reported that conducting polymers have arisen as one of the furthermost auspicious materials in numerous biological and biomedical applications, such as BSs, tissue engineering applications, energy conversion, microelectronics, actuators, polymer batteries and several other biological applications as a result of their exceptional conductivity, stability, their simple means of preparation and exceptional electrical characteristics that could translate the biochemical data into electrical signals [3, 11, 87–90].

According to El-Said et al. [3], nanostructured conducting polymers have great applications in the field of bioengineering; because nanostructured conducting polymers are brilliant matrixes for the effectiveness of several biological particles as such, they improved their efficiencies as BSs. Furthermore, some studies have shown that the mixtures of the nanostructures of metals/metal oxides with conducting polymers are to improve the stability and sensitivity as the BSs [11]. Consequently, some techniques have been described for evolving homogeneous nanostructures thin or reedy layer metal/metal oxide on the nanostructured conducting polymer exteriors [3].

It has been reported that conducting polymers have arisen as one of the furthermost auspicious materials in numerous biological and biomedical applications, such as BSs and tissue engineering applications [88–90]. This range of applications is a result of their biocompatibility and their exceptional electrical physiognomies that could translate the biochemical data into electrical signals. According to Soylemez et al. [91], conducting polymers also have numerous functional groups, that provide all-out enzyme loading via the interface between the enzyme particles and the polymers' functional groups. Hence, a regimented framework BSs could be realized.

Of recent nanostructured conducting polymers are now representing exceptional building blocks for evolving vastly sensitive BSs as a result of their exceptional properties that combine with those of the conducting polymers (like biocompatibility, direct electrochemical synthesis) and the NMs such as large surface area, flexibility for the immobilization of biomolecules and quantum effect [92]. Some synthetic approaches were described for the nanostructured conducting polymers synthesis, together with template-built means (that are either rigid or lenient template), template-permitted synthesis and physical approaches like electrospinning [93–97].

Basically, the "polyaniline nanostructures" is one of the outstanding "nanostructured conducting polymers" that have been developed with the assistance of prototype-guided polymerization within networks of microporous zeolites, permeable membranes and chemical path together with self-organized supra-molecules or stabilizers [98–100]. According to Che et al. [98], "polyaniline nanostructures" possesses advanced sensitivity and the response time is awesome when compared to conventional ones, according to them is a result of the advanced dynamic surface area as well as the reduced depth of the penetration target of the particles or molecules. As a result of this advanced dynamic surface area and permeable structure, there is a vast diffusion of particles or molecules into the outline and this makes them applicable as BSs.

The combination of metal NMs and "conjugated polymers" to produce nano constituents is envisioned to intensify the electrical conductivity [101, 102]. One of the utmost recent important deliberations has evidently proved that gold and silver NMs could be implanted into a polymeric film, which basically amplified the surface area for adjustment of various bioparticles/biomolecules. Recently, there has been substantial attention to the system of electrode adjustment using NMs and conducting polymers [3, 36, 102–107].

9.8.1 Nanostructured Conducting Polymers-Integrated Electrochemical Biosensors

Some of the most recent developed "nanostructured conducting polymers-integrated ECBs" and their various applications as summarized from the study of El-Said et al. [3] are briefly discussed below.

9.8.1.1 Nanostructured Conducting Polymers for Electrochemical Detection of Glucose and Hydrogen Peroxide (H_2O_2)

Due to the present extensive nature of diabetes mellitus and the severe health issues globally; the complications arising from diabetes mellitus have been medically reported to be more detrimental than diabetes itself [3, 108]. Consequently, evolving a precise and fast assess for early diagnosis of diabetes illness is evidently an urgent issue that should be given the appropriate attention.

Diagnostic assessments that have been reported for the management of diabetes which is built on the quantity of glucose level present in the blood [109–111]. According to Zhu et al. [109], Elkholy et al. [101], Lee et al. [111], Salazar et al. [112], the glucose sensors can be grouped as enzymatic and enzyme-free sensors. However, we would concentrate briefly on the uses of conducting polymers for evolving extremely delicate glucose sensors.

In the study carried out by Saini et al. [100], where they reportedly prepared "polyaniline nanostructures" employing "sodium dodecylsulphate" as glucose and hydrogen peroxide BSs. The "sodium dodecylsulphate" serves as a perfect structure-guiding agent for the synthesis of well-ordered nanostructured polymer made up of context protonated amine like "polyaniline nanostructures". Using these "nanostructured polymers" having huge precise surface range can increase the conductivity of "polyaniline" and brings about the control with high filling of "horseradish peroxidase" and "glucose oxidase" without many difficulties. Additionally, these nanostructures increase the amount of electrons convey and the up-to-date feedback. These improved "polyaniline nanostructures" served as optical and electrical BSs with decent performances, fast feedback time, extensive linear range as well as virtuous selectivity, firmness and reproducibility [3]. Similarly, in the work of Yang et al. [108], where they reportedly fabricate and applied poly(3,4-ethylenedioxythiophene) (PEDOT) nanofibers for electrochemical discovery of glucose built on ensnared of the "glucose oxidase enzyme" into the PEDOT nanofibers through the "galvanostatic polymerization procedure" at platinum (Pt) electrode. This sensor has been reported to have lofty sensitivity, lofty electrochemical stability and lesser limit of detection than the glucose oxidase enzyme combined PEDOT film sensors in connection to their massive surface area.

In the study carried out by Soganci et al. [113], they attempt to change the "graphite rod electrode" with a "super-structured conducting polymers" comprised of "amine substituted thienyl-pyrrole derivative" that is built on the "electro-polymerization" procedure [113]. This "conducting polymers" is categorized by the existence of free "amine groups" that allows the "covalent bonding" of the electrode and the "biorecognition elements" such as "glucose oxidase". This sensor has been proven to be efficient in the detection of glucose in beverages [3]. It has also been demonstrated by Munteanu et al. [114], that a double electrochemical sensor together with "optical-microscopy" as an opto-electrochemical sensor could be effective for the detection of glucose as well as hydrogen peroxide [114]. The working principle of this sensor is built on applications of "osmium complex-built redox polymer"; thus, its oxidation state can be altered when the glucose and hydrogen peroxide are interacting with the enzyme, such interaction depends on time [3].

9.8.1.2 Nanostructured Conducting Polymers for Cell-Built Chip

According to Yea et al. [115], it is habitually challenging to comprehend cell activities based on the quantity of only the nucleic acid or protein appearance levels because the cells are much more intricate than the entirety of its mechanisms. There are presently several electrically conductive frameworks that have been applied for manufacturing cell-built chips for improving the bond, propagation as well as the distinction of some cell kinds like neurons [48, 49, 115–118]. However, we will briefly concentrate on the uses of conducting polymers improved electrodes in evolving cell-built chips and their applications.

The study carried out by Lee et al. [117], where they built an electrochemical conducting scaffold made up of "pyrrole N-hydroxyl succinimidyl ester and pyrrole copolymer", at that point they enhanced the "copolymer" with a "nerve growth factor" and which was beneficial for the immobilizations of "pheochromocytoma cells (PC12 cells)". They believed that the cells have protracted neurites correspondingly to that for cells that are cultured using "nerve growth factor (NGF)".

In the study carried out by El-Said et al. [116], where they use a "thin layer of polyaniline emeraldine base" covered "indium–tin oxide electrode" as a cell-built chip. Conversely, the metal electrodes, "polyaniline nanostructures emeraldine base/indium–tin oxide electrode" exhibited an exceptional "electrochemical activity" at a neutral pH level in the absence of "co-depositing acidic counterion". The developed electrode was also used as a cell-built chip for rapid and easy means for the determination of the cell electrochemical characteristics, the cell capability, the cell bond, the cell development and for the monitoring of the consequences of various anticancer medications/treatments on the cell capability. They also worked on in-situ electrochemical synthesis of "polypyrrole nanowires" together with a "nanopermeable alumina" prototype [119]. They also show the formation of exceedingly ordered permeable "alumina substrate" and the development of the "ester and pyrrole (PPy) nanowires" inside the "nanopermeable structures" by means of "direct electrochemical oxidation" of the "pyrrole monomer". The actions, morphology, bonding and the development of the cell as well as the "biocompatibility of PPy nanowires/nanoporous alumina substrate" in the direction of both the investigation of tumorous and cancerous issues are better off when compared to other substrates. This according to them is a justification that "PPy nanowires/nanoporous alumina substrate" presented an improved cell bonding and development when compared to the other control substrates. These results from their study have further demonstrated the capability of "PPy nanowires/nanoporous alumina substrate" as "biocompatibility electroactive polymer substrate" that could be beneficial for both dynamic and cancer or tumour cell cultures.

Another useful study was conducted by Strover et al. [120], where they combined "pyrrole" and "thiophene moieties" in its monomer for the fabrication of "PolyPyThon-based molecular brushes". A film from this "conducting polymers" was placed on the "gold substrate" that was used as the scaffold for the "electrical stimuli-responsive surfaces of human fibroblasts cells" [120].

From all these studies, we could therefore say that the uses of the cell-built chips are promising alternative to animal experiments, as a result of the shortcomings of the uses of animal models since they are sacrilegious to animals' privileges, expensive, laborious and also poor significance to human environmental/natural science. Furthermore, the uses of cell-built chips as BSs in monitoring the effects of anticancer medications or for monitoring the distinction of the neural cells disclosed several benefits, that includes:

- Progressively additional sophisticated representation of absorption, circulation, metabolism, excretion and noxiousness procedure.
- Improved understanding of the medication interface mechanisms in the human system.
- Exhibited a tremendous potential to improved prediction of drug efficiency and safety.

9.8.1.3 Nanostructured Conducting Polymers for Some Various Contemporary Biosensors

There are presently some reports that conducting polymers are now extensively used for the preparation of sensor platforms and have several beneficial impacts as a result of the combination of their functional groups during the fabrication [3]. In this section, the fabrication and applications of various nanostructured conducting polymers as well as their compounds and their electrochemical BSs benefits for the different organic targets like neurotransmitter, nucleic acids (DNA/RNA), adenosine triphosphate (ATP), etc. The foremost purine bases which contribute to the development of the nucleic acids are "guanine and adenine". They are the essential composites in the various organic systems [3]. The irregular concentration of "guanine and adenine" in the fluid content of the body is a result of the low immunity system [3]. Therefore, it is essential and necessary to monitor the "guanine and adenine" concentration in living creatures.

In the study carried out by El-Said et al. [121], where they developed "poly (4–aminothiophenol) nanostructures" coated with "gold nanodots patterned indium tin oxide electrode" employing a simple technique. The improved "gold (Au) nanodots patterned indium tin oxide electrode" was developed by means of an evaporated heat of pure Au metal on top of an "indium–tin oxide surface" via "polystyrene monolayer". Thereafter, they use these "Au nanodots" as a prototype for self–assembly immobilization of "ATP" particles trailed using the "electrochemical polymerization" of "ATP to generate poly (4-amino thiophenol) (PATP)". This improved electrode was used to check and manage the concentration of the combination of "adenine and guanine" by means of "level of detail (LOD) of 500 and 250 nM", correspondingly. Also, the improved electrode was expansively used for sensing "adenine and guanine" in hominoid serum.

Aksoy et al. [122], in their study, established a selective "electrochemical dopamine BSs" using "polyimide" and "polyimide-boron nitride" compounds as a selective membrane for dopamine discovery. The application of "boron nitride" constituents into the "polyimide matrix" led to the improvement of the selectivity and sensitivity as well as the swift transfer of the electron (reversibility), together with a "LOD" of approximately 4×10^{-8} m.

According to El-Said et al. [116], the hybridization of "polyaniline nanostructures" and NMs can provide enormous potential in the study of sensors as a result of the improvement of its electron conductivity as well as its ability to serve as a scaffold for the immobilization of the organic species. "Pseudomonas aeruginosa" is one of the furthermost pathogenic gram-negative microorganisms which can result to "corneal ulcers" as well as loss of sight under 48 h. "Pyocyanin (PYO)" is the biomarker which has been proven for the detection, monitoring and management of "pseudomonas aeruginosa" [3].

As reported by Elkhawaga et al. [110], in their study where they organized "polyaniline/gold nanostructures/indium–tin oxide electrode as "PYO" sensor in a culture of "polyaniline nanostructures". From the results of this study, it was shown that polyaniline/gold nanostructures/ indium–tin oxide electrode sensitivity nature towards PYO biomarker is higher than that either bare indium–tin oxide electrode or Au nanostructures/indium–tin oxide electrode with LOD of 500 nm.

The improvement of the activity of electrochemical of "polyaniline/gold nanostructures/indium–tin oxide improved electrode" in the direction of the PYO connected to the existence of positive charges on its surface which increased the rate of transfer of the negative charged PYO built on the electrostatic force of attractive massively.

In furtherance, Khalifa et al. [123], applied polyaniline/gold nanostructures/indium–tin oxide electrode for the diagnosis of "pseudomonas aeruginosa" from fifty samples they collected from persons diagnosed of "corneal ulcers"; the results obtained were used to compared other results such as that from the screen-printed electrode, conventional or traditional methods, automatic identification technique as well as the magnification of the 16 s rRNA gene employing the "polymerase chain reaction" as a typical assessment for "pseudomonas aeruginosa" identification. According to them, the electrochemical discovery of "PYO" using the "square wave voltammetry" method via "polyaniline/gold nanostructures" improved "ITO electrode" was the only procedure that could conform completely with the molecular technique in terms of the sensitive and specific attributes as well as showing positive and negative analytical values when likened with the "solid-phase extraction (SPE), conventional as well as colony morphology, pigment production and biochemical tests and automated, as well as the automated system for antibiotic susceptibility testing (AST) and microbiology identification (ID) and the polymerase chain reaction (PCR)" procedures. Hence, "polyaniline/gold nanostructures/indium–tin oxide electrode" was endorsed as one of the fastest, cheapest, most precise and most selective "PYO biomarker sensor" in "pseudomonas aeruginosa" for the "corneal ulcer" issues as a result of the "square wave voltammetry (SWV) method" [3, 123].

The use of "zika virus" which is a "flavivirus" has recently been given significant awareness for the development of a swift "zika virus" identification assessment as a result of the presence of the virus-related infection in little children (infants). Tancharoen et al. [124], in their study, reported the development of a novel kind of "zika virus ECBs". The BSs comprised of "surface imprinted polymers/graphene oxide compounds". The "limit of detention (LOD)" of these "zika virus ECBs" was in conformity with that of the "reverse transcription polymerase chain reaction (RT-PCR) technique". Additionally, they have the potentials to sense the "zika virus" in the presence of the "dengue virus and serum samples" [124].

According to reports, another cellular metabolite that has the potential to associate with some of the life-threatening health maintenance circumstances is the "lactate" [125]. As a result of this, Pappa et al. [125], in their study attempt to fabricate a "micrometer-scale polymer-built transistor prototype", that would be used for the discovery of "lactate". The application of electron-conveying (which is an n-type) biological polymer, integrate hydrophilic lateral chains which have improved the ion transport/injection, facilitated the enzyme conjugation and acted as a series redox center. This their developed "micrometer-scale polymer-built transistor prototype sensor" demonstrated swift, selective and sensitive metabolite potentials.

9.8.1.4 Molecularly Embossed Polymer-Built Biosensors

According to Rico-Yuste et al. [126], the molecularly embossed polymers known as "molecularly imprinted polymers" is one of the furthermost sensitive and selective constituents use for the development of contemporary BSs. The production procedure of "the molecularly embossed polymers" has to do with the application of three main machineries:

- The prototype molecules (embossed molecules)
- The functional monomers
- The cross-connector molecules.

In the production procedure of the molecularly embossed polymers and its applications wherein the embossed molecules are joined to the "functional monomer" to produce several reaction sites. The polymerization procedure was carried out incidentally with a cross-connector. The prototype molecules were detached from the ensuing polymers to produce embossed cavities which could rebind with the prototype molecules all through the recognition procedure. The selectivity machinery of the molecularly embossed polymers is a result of the fact that these molecularly embossed polymers presented physically improved recognition positions for the rebinding of intended molecules like the medications in intricate mixtures. The selection mechanisms of the molecularly embossed polymers to particles are linked to their corresponding chemical groups and the relations between the prototype particles and embossed groups that are being subjected to on the form and stiffness of the prototype particles.

Consequently, these cavities can be applied for enantiomeric determination. Molecularly embossed polymers were used for "electrochemical recognition" of diverse targets [127–130]. "Cyanocobalamin" is among the eight "B-complex vitamins" that can influence the function of the brain as well as the complete nervous system [3].

According to Samari et al. [131], Nakos et al. [132], the reduction of the "cyanocobalamin levels replicate some health issues such as "haematological and neurological disorders". Singh et al. [133], in their study, described the production of biomimetic embossed polymer comprised of a sheet of abridged "graphene oxide" adorned with "multi-walled carbon nanotubes and methyl blue compound" and functionalized with "poly acryloylurea". This BS was used for ultra-trace discovery of "cyanocobalamin" built on "differential-pulse voltammetry" procedure, which proved a direct proportionality with the concentration extent from 0.021 to 1.44 ng/mL with LOD of 0.0056 ng/mL and categorically no interference in actual samples. According to them, the increased electrochemical conductivity of the BS is a result of some interactions. In the study carried out by Liu et al. [128], they also reported the production of molecularly embossed polymer-improved glossy carbon electrode as an "electrochemical nonylphenol BSs". The molecularly embossed polymer electrode comprised of the surface of "acrylamide-functionalized multi-walled carbon nanotubes" as a carrier surface, "nonylphenol" as a prototype particle, "methacrylic acid" as "functional monomer" as well as "ethylene glycol dimethacrylate" as a cross connecting agent and the polymerization was carried out based on the thermal or heat polymerization procedure. This produced sensor exhibited a translational range within the concentration range from 0.1 to 30 μM with a limit of detention (LOD) of 0.02 μM.

9.9 Graphene and Its Applications for Biosensors and Biosensing-Based Electrode Systems

Of recent graphene, evolving as an exact 2D substance (sp^2-hybridized carbon), has attracted robust scientific and technological attention as a result of its exceptional physicochemical features such as lofty surface area, outstanding conductivity, lofty mechanical strength as well as affluence of functionalization and numerous production [1, 134]. Several studies have shown that graphene has great potentials and applications in, electronics, supercapacitors, cells (fuel and solar), batteries and in various area of biotechnologies [134].

Graphene and its various derivative; "graphene oxide, graphene nanoflakes and graphene platelets" are nowadays dominant substances for the fabrication of electrode matrixes for BSs and biosensing-based systems [134]. All graphitic forms belong to the general graphene group which include 0-dimensional fullerenes, 1-dimensional carbon nanotubes and 3-dimensional graphite [134]. According to Nikoleli et al. [134], graphene is normally grouped by the number of slanted layers; one layer, few-layer (that is 2–10 layers) and multilayer (which is sometimes called thin graphite). Ordinarily, for graphene to sustain and preserve its discrete properties, its use must be restricted to one or few-layer morphologies. Graphene is considered as a subsequent generation electronic substance as a result of its exceptional respective properties electronically, optically, mechanically, thermally and electrochemically [134]. Accordingly, graphene is used in the manufacture ECBs and biosensing-based electrode systems as a result of its obvious good low-noise electronic attributes. It is useful for electrochemistry because of its outstanding properties and low environmental influence.

Graphene is useful for electrochemistry as a result of conductive as well as its transparency, with an inexpensive, truncated environmental influence, a large electrochemical potential window, truncated electrical resistance in comparison to glassy carbon, atomic thickness and two well defined redox peaks linearly aligned with the square root of the scan rate magnitude, suggesting that its redox processes are primarily diffusion controlled. As a result of the rate of transmission of the electrons which has been revealed to be surface dependent and could be increased expressively via the formation of definite surface functional groups [134].

The complete volume of graphene is exposed to its backgrounds as a result of its exact 2-D structure, making it effectively use in the detection of adsorbed particles. Graphene-based electrodes also display supercilious enzyme packing as a result of their lofty surface area. This aids the facilitation of lofty sensitivity, exceptional electron transmission promoting capability for some enzymes and exceptional catalytic performance towards several biomolecules [134]. Graphene-based products also possess the essential biocompatibility to be acquiescent for in-situ biosensing. Graphene-based NMs could be categorized by their means of production.

Studies have shown that electrodeposition could be applied for the fabrication nanostructures on electrode materials [134]. They offer several forms of morphologies which include buttressed secluded nanoparticles, consistent structures designed on templates, secluded nanostructures that have high degrees of ramification or adjustable shapes and intricate multicomponent compound structures.

9.9.1 Graphene-Built Enzymatic Electrodes

The "direct electrochemistry of enzymes" comprises direct "ET (DET)" among the electrode and the dynamic heart of the enzymes, in the absence of the involvement of intermediaries or other reagents [1, 135–137]. Innovative enzymatic bioreactors, intermediary-free BSs as well as biomedical devices are required to engage the DET via the immobilization of the enzymes on conducting chemicals (substrates). Agreeably, several recent studies have revealed that graphene can improve DET among enzymes and electrodes [134]. However, there are presently few reports on the application of metal NP utilizing graphene to form outstandingly stable, inexpensive and eco-friendly BSs [138].

9.9.1.1 Enzymatic Biosensors for Glucose Discovery Via Graphene-Built Electrochemical

According to Nikoleli et al. [134], the recent development (third generation) of glucose BSs is based on the DET in enzymatic chemical reaction catalytically, which has been rigorously considered for quite some years now and are still advancing [139]. The "direct electrochemistry of glucose oxidase" is the DET that lies between the electrode surface and the enzyme dynamic location without an arbitrator [134]. Electrons pass through the lengthy tunnelling by means of enzymes, in so doing attaining "direct glucose oxidase electrochemistry" on graphene-improved electrodes [134, 140–144]. The benefits are seen in the prompt kinetics and reasonable sensitivity, which initiate from the good conductivity and huge surface area of graphene that assist the enzyme immobilization.

Razmi and Mohammad-Rezaei [145] reported that "graphene quantum dots" are effective examples of graphene nanomaterial that are presently being applied for the fabrication of innovative electrochemical sensors. This procedure entails the alteration of "graphene quantum dots" as well as the surface immobilization of "glucose oxidase", resulting in an extensive linear glucose response ranging from 5 to 1270 μM, truncated detection extent of about 1.73 μM and reasonable surface exposure of "glucose oxidase" of about 1.8×10^{-9} molcm^{-2}. The ensuing DET is linked to the huge surface-to-volume proportion and the profuse hydrophilic boundaries and planes of "glucose oxidase". These improve immobilized enzyme on the electrode surface the "graphene nanotube/carbon nanotube", the "graphene liquid/ion liquid" and the "graphene/Nafion", offers enormous prospects in the production of more intricate efficient nanostructures [134, 146].

9.9.1.2 Enzymatic Biosensors for Hydrogen Peroxide Discovery Via Graphene-Built Electrochemical

Presently, substantial consideration has been given to the swift discovery of hydrogen peroxide, which has functional applications in chemical investigation, clinical diagnostics as well as environmental safety and sustainability [134].

Several "graphene-built ECBs" have been discovered for hydrogen peroxide discovery. Fan et al. [147], in their study, reported a "graphene capsule" as carrier for "horseradish peroxidase" which was intended to sense hydrogen peroxide in humanoid serum which was an improvement of the study of [148]. The outcomes from this study revealed reasonable stability, reproducibility and better selectivity. This is a great indication that this BS has some capability in medical investigation.

Lu et al. [149], designed a "super-molecular assembly enzymatic functional graphene-built BSs" using "horseradish peroxidase/cyclodextrin" to sense hydrogen peroxide. The "adamantine-improved horseradish peroxidase" networked with the "functionalized graphene nanostructures" via "host-guest supramolecular" interaction for improving the electrochemical act for the hydrogen peroxide discovery. This inventive nanostructure offers exceptional performance in applications catalytically in medical investigations, environmental monitoring and management as well as several other aspects of biosensing. Liu et al. [150], reported on the "enzymatic ECBs" built on "graphene nanoplatelet-titanate nanotube compound" which was fabricated via a hydrothermal technique and displayed sensitive hydrogen peroxide discovery in a comparatively wide investigative extent with a reasonable regaining extent compared to other previous sensors. In the study of Muthurasu et al. [151], a suitable and immensely "high-sensitivity electrochemical enzymatic discovery prototype for hydrogen peroxide was fabricated on "glass carbon electrode" via "horseradish peroxidase peptide amide" bonded on "immobilized graphene quantum dots" onto the electrode surface.

9.9.1.3 Enzymatic Biosensors for Nicotinamide Adenine Dinucleotide (NADH) Discovery Via Graphene-Built Electrochemical

"Nicotinamide adenine dinucleotide" is one of the vital coenzymes presents in the living cells. It is a redox hauler in metabolic procedures and partakes in numerous enzymatic reactions [134]. There are presently "graphene-built ECBs" that are fabricated and attained outstanding electrocatalytic action in the direction of "nicotinamide adenine dinucleotide" [152, 153].

Li et al. [154] reported the use of "graphene-induced self-assembly of peptide nanowires" used for the development of ECBs aimed at "nicotinamide adenine dinucleotide" and improve the "electronic conductivity of the bio-structure". This their "graphene-built NADH BSs" demonstrated reasonable sensitivity, in so doing holding great potential for "nicotinamide adenine dinucleotide" investigation in several aspects. Gasnier et al. [152] fabricated an "electrochemical "nicotinamide adenine dinucleotide BSs" with a graphene adhesive electrode which was used to careful catalyze the "electrooxidation of nicotinamide adenine dinucleotide", in the absence of surface passivation. The resulting BSs revealed several benefits compared to other "electrochemical nicotinamide adenine dinucleotide BSs" since it assimilates the outstanding electrocatalytic actions of graphene, its exceptional physiognomies in networking and integrating with biomolecules as well as its rapid preparation periods. Gai et al. [155] fabricated an "electrochemical nicotinamide adenine dinucleotide BSs" by means of N-doped graphene, which was built and displayed highly similar physiognomies to "nicotinamide adenine dinucleotide dehydrogenase". So, competently catalyzed "adenine dinucleotide BSs oxidation". Furthermore, the "N-doped graphene" was used as a "nicotinamide adenine dinucleotide" conveyance of "nicotinamide adenine dinucleotide" to the electrode surface due to its enormously good conductivity. With truncated circumstantial current, this graphene substance enhanced the selective and sensitive nature of the BSs aimed at "nicotinamide adenine dinucleotide" determination.

9.9.1.4 Enzymatic Biosensors for Cholesterol Discovery Via Graphene-Built Electrochemical

In the study carried out by Dey et al. [156], where they developed extremely sensitive "amperometric BSs" for detecting cholesterol utilizing immobilizing "cholesterol oxidase" and "cholesterol esterase" on the surface of "Pt nanoparticle" adorned chemically with "synthesized graphene". The resulting sensing prototype proved to be sensitive in the discovery of hydrogen peroxide with a limit of detention of 0.5 nm without redox intermediary or enzyme greater than 100 mV with less positive potential regarding the enormous Pt electrode. This their "bioenzyme integrated nanostructured prototype" exhibited a reasonable selective and sensitive discovery of cholesterol ($2.07 \pm 0.1\ \mu A\ \mu M^{-1}\ cm^{-2}$), a limit of detention of 0.2 μM, reasonable stability and quick response periods.

Nguyen et al. [157] developed an "amperometric layer-by-layer BSs for cholesterol", built on an "electrochemical microelectrode" by means of "graphene films synthesized" via "thermal chemical vapour deposition (CVD) technique" covered with "Fe_3O_4-doped polyaniline (PANi) films" as well as immobilization of "cholesterol oxidase" onto the functional electrode using "glutaraldehyde agent". The invented bioelectrodes invented "layer-by-layer" displayed outstanding analytical measurement in the inclusive cholesterol concentration range from 2 to 20 mM, with reasonable sensitivity ($74\ \mu AmM^{-1}\ cm^{-2}$) and quick response period of less than five seconds.

Israr et al. [158] in their study described a "potentiometric and miniaturized cholesterol BSs" built on physical adsorption of "cholesterol oxidase" on "exfoliated glutenin subunits (GSs)" on a "thin copper wire". This their developed "potentiometric BSs" exhibited exceptional "stability, reusability, selectivity as well as sensitivity" of about 82 mV $decade^{-1}$ for the discovery of "cholesterol biomolecules" with logarithmic ranging from 1×10^{-6} M to 1×10^{-3} M, quick response periods of about four seconds and a decent shelf life time of greater than thirty days at room temperature.

Nikoleli et al. [159] fabricated innovative "potentiometric cholesterol BSs" in their study utilizing the immobilization of the "stabilized polymeric lipid membrane" on the graphene electrode. This "stabilized polymeric lipid membrane" is composed of "cholesterol oxidase enzyme and polymerization" combination, which is a dominant influence on the physiognomies of "cholesterol BSs". Their discovered BSs displays significant reproducibility, decent selectivity and reasonable sensing ability with a straight-line slope curve of about 64 mV/decade as well as quick response periods of about five seconds. The robust biocompatibility among the "stabilized polymeric lipid membrane and human biofluids" makes it possible to be used for real blood samples as well as other practical benefits.

9.9.1.5 Enzymatic Biosensors for Urea Discovery Via Graphene-Built Electrochemical

According to Nikoleli et al. [160], there is a designed "miniaturized potentiometric urea lipid film-built BSs" on "graphene nanosheets". The structural characterization of the "graphene nanosheets" for reduction of the "potentiometric urea lipid film-built BSs" was considered by means of an "atomic force microscopy (AFM) and transmission electron microscopy (TEM) quantifications". The "absorption spectroscopy or reflectance spectroscopy is also known as ultraviolet-visible spectroscopy (UV–Vis) and Fourier transform infrared (FTIR) spectroscopy" was applied for studying the pre-conjugated and post-conjugated surfaces of the "graphene nanosheets". These designed "potentiometric urea BSs" displays decent reproducibility, recyclability, selectivity, quick response periods of about four seconds as well as long shelf life and reasonable sensitivity of ca. Also, it was reported to have 70 mV/decade over the "urea logarithmic concentration" ranging from 1×10^{-6} M to 1×10^{-3} M [160].

9.9.2 Graphene-Built Electrochemical DNA Sensors

As reported by Palek and Fojta [161] and Tosar et al. [162], there are studies that have reported the use of "graphene-built BSs" as applied in "electrochemical DNA" detecting. In the electrochemical discovery of nucleic acids, both non-reagent and reagent-built methods have been described. Among the first group that was designed are the "direct or catalyzed oxidation of DNA bases" and the "charge transport reactions" facilitated using the "π-stacked base pairs". While, the reagent-built approaches are generally built on the redox reactions of correspondent particles, that are bonded to the "single-stranded, double-stranded DNA or enzymes" enlisted to the electrode surface via detailed probe-target relations. However, by evading the bring in of a reagent in the procedure that was later require the removal, non-reagent procedures are much easier to systematize than reagent-built procedures; they still demonstrated further consuming time or affluent though they offer lesser sensitivity. Moreover, Artiles et al. [79] reported that even though the conventional carbon constituents, like "graphite and glassy carbon", the "adenine and guanine" ones give systematically beneficial signals but there is done for "cytosine and thymine".

Nevertheless, Zhou et al. [163] in their study revealed that an improved "glassy carbon electrode" affords reasonable signals of all unrestricted bases of DNA and exhibited more approving electron transmission kinetics when compared to graphite improved "glassy carbon" and "glassy carbon electrodes. This significantly improved the reactions of the electrochemical of all the unrestricted bases of DNA, offering the "glassy carbon-improved electrode" an enhanced choice for the biosensing of electrochemical of both the "single-stranded DNA and the double-stranded DNA" at functional pH in the absence of a prehydrolysis stage.

In the study carried out by Lim et al. [164], where they discovered that "anodized epitaxial graphene" with reasonable edge plane flaws was a better choice for evolving biosensing prototype to sense ascorbic acids, dopamine, nucleic acids and uric. The results from this their study confirmed that the oxidation of electrochemical of "pristine epitaxial grown graphene" presented a reasonable quantity of edge plane faults on its surface, this was similar to the previous study of Pumera et al. [165] were they designated "multi-walled carbon nanotubes" that provide prototype with suggestively improved determined responses that are ideal for reasonable determination of electrochemical sensing.

Ambrosi and Pumera [166] in their study demonstrated that "stacked graphene nanofibers" also proved greater electrochemical presentation for the oxidation of DNA bases over "carbon nanotubes" as a result of the outstandingly high quantity of available "graphene sheet edges" on the surface of the "nanofibers" compared to "carbon nanotubes". As reported by them, these "stacked graphene nanofibers" are exact opposites of "carbon nanotubes" since they contain what they referred to as "perpendicularly stacked graphene sheets along the c-axis" which display completely electrochemically active edges and also afford reasonable sensitivity more than the "edge-plane pyrolytic graphite, guanine cytosine or graphite microparticle-based electrodes". They proved the sensitive oxidation of "anti-pandemic influenza/A (H1N1)-interrelated strands" on these "stacked graphene nanofiber-built electrodes", which supposedly can be used for "label-free DNA" investigation.

9.9.3 Graphene-Built Electrochemical Immunosensors

Immunosensors are basically analytical detection protocols that are based on quantifying "antibody-antigen specific conjugation reactions" [134, 167]. Several procedures have been established as a result of these antibody-antigen connections [134]. These immunosensors habitually serve as a typical investigation technique use in clinical and biomedical diagnostics centers, monitoring and management of environmental pollution, as well as in the determination of food security and safety [134]. Presently, "electrochemical immunosensors" have been seen as a significant technology in "immunoassays" as a result of the fact that they are not difficult to establish and apply, they also need minor sample volumes and give swift analyzes, reasonable sensitivity and better selectivity [143].

According to Nikoleli et al. [134], the foremost distinctive electrochemical procedures used in immunosensors are; "amperometry, potentiometry and electrochemical impedance spectroscopy". Also, several kinds of point-of-care testing benefits as a result of the conventional electrochemical procedures have been reported [134, 150], example of such is the" paper- or microfluidic-based detection platforms" which have received substantial consideration in the manufacture of immunosensor [168].

Several studies reviewed summarizes the use of "electrochemical immunosensors for point-of-care diagnosis" [51, 134, 164, 168–170]. Nikoleli et al. [134] in their study reported the fabrication of "miniaturized potentiometric D-dimer BSs on graphene nanosheets" with combined lipid films. The "graphene electrode" was applied for the development of a reasonable selective and sensitive immunosensor for the discovery of "D-dimer" using the immobilization of the "mouse anti-human D-dimer antibody" on "stabilized lipid films". Reportedly, the immunosensor retorted to the elaborate "D-dimer concentrations" with swift response periods of about fifteen seconds. These "potentiometric D-dimer BSs" is not difficult to fabricate and displays better selectivity, reproducibility, recyclability, swift response periods, lengthy shelf life as well as reasonable sensitivity of ca.

9.9.4 Some Commercial Advances in Graphene Biosensors and Biosensing-Based Electrode Systems

There have been some commercial activities in the study of graphene sensors as reported by Nikoleli et al. [134], for example, the "Nokia's Cambridge Research Centre" in November 2013, established a "humidity sensor built on graphene oxide", which is believably quick, thin, transparent and flexible with a reasonable response and recovery periods, before then they also filed for a patent for a "graphene-built photodetector" which is transparent in August 2012. The United States based "Graphene Frontier" declared raising of $1.6 million for the expansion of the growth and production of their "graphene functionalized GFET sensors" in August 2014. These their "Six Sensors" product was "vastly sensitive chemical and biological sensors" that was used for the diagnosis of diseases with reasonable sensitive and efficient attributes with incomparable using traditional sensors. Also, in the same year precisely September 2014, a Germany based "AMO" established a graphene-built photodetector in partnership with "Alcatel Lucent Bell Labs", which was then believed to be one of the fastest photodetectors globally.

Again, in June 2015, a partnership between one of the foremost engineering companies in Germany "Bosch" and researchers from the "Max-Planck Institute for Solid State Research" produced a "graphene-built magnetic sensor" which was believed to be far more sensitive than a comparable device based on silicon.

Also, a team of researchers in Korea from the "Electronics and Telecommunications Research Institute" in June 2015, effectively developed a technology of textile-form gas sensor that is cleanable, flexible and reasonably sensitive. This technology was built on coating graphene utilizing molecular adhesives to textiles such as polyester, nylon or cotton, such textiles could be used for checking the existence of gas in the air. Before then they partner with "Korea's Ulsan" in October 2014 to establish a method that generates "flexible transparent graphene electrodes" which was attached to the skin. This designed was also beneficial for "bio-signal sensors" and allows "electronic tattoo-like stickers".

A "biocompatible BSs" that concurrently sense numerous biomarkers like proteins and DNA, which was produced from carbon constituents, specifically graphene quantum dots and graphene oxide was also established in September 2014 by scholars from "Zhejiang Normal University, China". A month before then, precisely August 2014, scholars from "Trinity College, Ireland, United Kingdom" established wearable sensors employing "coated rubber bands" together with graphene, this was since the rubber band possibly vicissitudes its conductivity during extension; as such minute actions like pulse or breathing could be detected. These established BSs were also believed to be beneficial for medical dealings, automotive and robotics industries, etc. Also, at the same period scientists from the "University of Michigan" established a "graphene-aided wearable vapour sensor" that was used for incessant detection of airborne chemicals and disease monitoring. In May of that same year, scientists from the "Nankai University" also established a "single-cell sensor known as optical refractive index sensor" by means of "graphene field-effect transistors". These innovative BSs were capable of detecting "single cancer cell" as a result of the ultrahigh sensitivity and high-temperature abridged graphene oxide.

However, there are presently several other categories of advanced BSs such as the carbon nanotube-based and microbial fuel cell-based BSs which operate on the optical as well as other recognized principles. These BSs are fundamentally considered and utilized for the detecting, monitoring and management of heavy metals and other chemicals or biochemicals in the environment as well as in the biomedical/medical domain [171–179].

9.10 Conclusion and Future Contribution to Knowledge

The progress and advancement in BSs technology habitually depend on the sensitive, specific, non-noxiousness, minor/small particle detection and cost-efficiency nature of the b BSs. Bearing in mind these physiognomies would ultimately assist in addressing some of the most critical constraints and limitations faced in BSs technology. Most of the improvements that have been realized and achieved are from the development of electrochemical sensors together with NMs that have consequences in several innovative categories of BSs.

Graphene exhibits explicit strong electronic structure, characteristics and physiochemistry as a result of its exceptional performance in direct electrochemistry of enzymes, the electrochemical discovery of minute biomolecules, environmental analysis, ECBs and biosensing-based electrode machinery. Even though there are several grounds that are yet explore, graphene has greater performance in the development of ECBs and biosensing-based electrode machinery when compared to carbon nanotubes.

Generally, an appropriate combination of biosensing, as well as bio-fabrication with non-natural/ synthetic biological methods applying either/both electrochemical, optical, bio-electronic moralities, would be crucial for efficacious development of comprehensive and influential BSs for contemporary future contribution to knowledge in the field of BSs technology in the area of agricultural, organic/biological, medical and environmental sustainability.

Conclusively, the extent of the development of BSs built on electrodeposited nanostructures has several conceivable applications. Agreeably, graphene is an exceptional electrode material, but there is still considerable room for more scientific research and applicable development of graphene-based systems ECBs and biosensing-based electrode machinery. It is therefore recommended that further scientific research and application of graphene-based ECBs and biosensing-based electrode systems should be advanced to give electrochemical and other BSs exceptional performance and in addition enlarge the realm of electrochemical and other BSs. This can be done by advancing innovative techniques for appropriate synthesis and processing of graphene and other potential materials.

Acknowledgements The authors are grateful to their respective institutions, the editors and authors whose articles were used while writing this chapter.

References

1. Y. Shao, J. Wang, H. Wu, J. Liu, I.A. Aksay, Y. Lin, Graphene based electrochemical sensors and biosensors: a review. Electroanalysis **22**, 1027–1036 (2010)
2. K.J. Stine, Biosensor applications of electrodeposited nanostructures. Appl. Sci. **9**(797), 1–40 (2019)
3. W.A. El-Said, M. Abdelshakour, J.H. Choi, J.W. Choi, Application of conducting polymer nanostructures to electrochemical biosensors. Molecules **25**(307), 1–11 (2020)
4. L.C. Clark Jr., C. Lyons, Electrode systems for continuous monitoring in cardiovascular surgery. Ann. N. Y. Acad. Sci. **102**, 29–45 (1962)
5. A.P.F. Turner, I. Karubeand, G.S. Wilson, *Biosensors Fundamentals and Applications* (Oxford University Press, Oxford, 1987)
6. N. Arora, Recent advances in biosensors technology: a review. Octa J. Biosci. **1**(2), 147–150 (2013)
7. J.I. Reyes De Corcuera, J.R. Cavalieri, Prototype instruments for laboratory and on-line measurement of lipoxygenase activity. Food Sci. Technol. Int. **9**(1), 5–9 (2003)
8. S. Borgmann, A. Schulte, S. Neugebauer, W. Schuhmann, Amperometric biosensors. Adv. Electrochem. Sci. Eng. (2011)

9. W. Nwankwo, A.S. Olayinka, K.E. Ukhurebor, Nanoinformatics: why design of projects on nanomedicine development and clinical applications may fail? in *Proceeding of the 2020 International Conference in Mathematics, Computer Engineering and Computer Science (ICMCECS)*, Lagos, Nigeria (IEEE Xplore, 2020), pp. 1–7
10. R. Long, Y. Guo, L. Xie, S. Shi, J. Xu, C. Tong, et al., White pepper-derived ratiometric carbon dots for highly selective detection and imaging of coenzyme A. Food Chem. **315** (2020)
11. Q. Zhou, D. Tang, Recent advances in photoelectrochemical biosensors for analysis of mycotoxins in food. TrAC - Trends n Anal. Chem. **124** (2020)
12. K. Kahn, K.W. Plaxco, Principles of biomolecular recognition, in *9 Recognition Receptors in Biosensors*. ed. by M. Zourob (Springer, New York, 2010), pp. 3–46
13. P. Mehrotra, Biosensors and their applications—A review. J. Oral Biol. Craniofacial Res. **6**(2), 153–159 (2016)
14. B. Bharat, Biomimetics: lessons from nature—An overview. Phil. Trans. R. Soc. A **367**, 1445–1486 (2009)
15. R. Monosik, M. Streansky, E. Surdik, Biosensors-classification, characterization and new trends. Acta Chimica Slovaca **5**, 109–120 (2012)
16. X. Niu, Y. Zhong, R. Chen, F. Wang, Y. Liu, D. Luo, A "turn-on" fluorescence sensor for Pb^{2+} detection based on graphene quantum dots and gold nanoparticles. Sens. Actuators, B Chem. **255**, 1577–1581 (2018)
17. L. Farzin, M. Shamsipur, L. Samandari, S. Sadjadi, S. Sheibani, Biosensing strategies based on organic-scaffolded metal nanoclusters for ultrasensitive detection of tumor markers. Talanta **214** (2020)
18. K. Sode, T. Yamazaki, I. Lee, T. Hanashi, W. Tsugawa, Biocapacitor: a novel principle for biosensors. Biosens. Bioelectron. **76**, 20–28 (2016)
19. S. Rodriguez-Mozaz, M.J. Alda, M.P. Marco, Biosensors for environmental monitoring: A global perspective. Talanta **65**(2), 291–297 (2005)
20. S. Vigneshvar, C.C. Sudhakumari, B. Senthilkumaran, H. Prakash, Recent advances in biosensor technology for potential applications—An overview. Front. Bioeng. Biotechnol. **4**(11), 1–9 (2016)
21. A. Salek-Maghsoud, F. Vakhshiteh, R. Torabia, S. Hassani, M.R. Ganjali, P. Norouzi, M. Hosseini, M. Abdollahi, Recent advances in biosensor technology in assessment of early diabetes biomarkers. Biosens. Bioelectron. **99**, 122–135 (2018)
22. M. Citartan, S.C. Gopinath, J. Tominaga, T.H. Tang, Label-free methods of reporting biomolecular interactions by optical biosensors. Analyst **138**, 3576–3592 (2013)
23. S. Sang, Y. Wang, Q. Feng, Y. Wei, J. Ji, W. Zhang, Progress of new label-free techniques for biosensors: a review. Crit. Rev. Biotechnol. **15**, 1–17 (2015)
24. J.M. Harris, C. Reyes, G.P. Lopez, Common causes of glucose oxidase instability in in vivo biosensing: a brief review. J. Diabetes Sci. Technol. **7**, 1030–1038 (2013)
25. A.P. Turner, Biosensors: sense and sensibility. Chem. Soc. Rev. **42**, 3184–3196 (2013)
26. J. Wang, G. Chen, H. Jiang, Z. Li, X. Wang, Advances in nano-scaled biosensors for biomedical applications. Analyst **138**, 4427–4435 (2013)
27. B. Wang, S. Takahashi, X. Du, J. Anzai, Electrochemical biosensors based on ferroceneboronic acid and its derivatives: a review. Biosensors (Basel) **4**, 243–256 (2014)
28. F.J. Gruhl, B.E. Rapp, K. Lange, Biosensors for diagnostic applications. Adv. Biochem. Eng. Biotechnol. **133**, 115–148 (2013)
29. L.D. Mello, A. Kisner, M.O. Goulart, L.T. Kubota, Biosensors for antioxidant evaluation in biological systems. Comb. Chem. High Throughput Screen. **16**, 109–120 (2013)
30. P.E. Erden, E. Kilic, A review of enzymatic uric acid biosensors based on amperometric detection. Talanta **107**, 312–323 (2013)
31. J. Kim, S. Imani, W.R. de Araujo, J. Warchall, G. Valdes-Ramirez, T.R. Paixao et al., Wearable salivary uric acid mouthguard biosensor with integrated wireless electronics. Biosens. Bioelectron. **74**, 1061–1068 (2015)
32. E.B. Bahadir, M.K. Sezginturk, Electrochemical biosensors for hormone analyses. Biosens. Bioelectron. **68**, 62–71 (2015)

33. E. Hamidi-Asl, I. Palchetti, E. Hasheminejad, M. Mascini, A review on the electrochemical biosensors for determination of microRNAs. Talanta **115**, 74–83 (2013)
34. F. Long, A. Zhu, H. Shi, Recent advances in optical biosensors for environmental monitoring and early warning. Sensors (Basel) **13**, 13928–13948 (2013)
35. N. Verma, A. Bhardwaj, Biosensor technology for pesticides—A review. Appl. Biochem. Biotechnol. **175**, 3093–3119 (2015)
36. P. Paulraj, N. Janaki, S. Sandhya, K. Pandian, Single pot synthesis of polyaniline protected silver nanoparticles by interfacial polymerization and study its application on electrochemical oxidation of hydrazine. Colloids Surf. A. **377**, 28–34 (2011)
37. W. Nwankwo, S.A. Olayinka, K.E. Ukhurebor, Green computing policies and regulations: a necessity? Int. J. Sci. Technol. Res. **9**(1), 4378–4383 (2020)
38. W. Nwankwo, K.E. Ukhurebor, An x-ray of connectivity between climate change and particulate pollutions. J. Adv. Res. Dyn. Control Syst. **11**(8) Special Issue, 3002–3011 (2019)
39. K.E. Ukhurebor, W. Nwankwo, Estimation of the refractivity gradient from measured essential climate variables in Iyamho-Auchi, Edo State, South-South region of Nigeria. Indonesian J. Electr. Eng. Comput. Sci. **19**(1), 276–284 (2020)
40. K.E. Ukhurebor, U.O. Aigbe, A.S. Olayinka, W. Nwankwo, J.O. Emegha, Climatic change and pesticides usage: a brief review of their implicative relationship. Assumption Univ. eJ. Interdiscip. Res. **5**(1), 44–49 (2020)
41. K.E. Ukhurebor, S.O. Azi, Review of methodology to obtain parameters for radio wave propagation at low altitudes from meteorological data: new results for Auchi area in Edo State, Nigeria. J. King Saud Univ. – Sci. **31**(4), 1445–1451 (2019)
42. K.E. Ukhurebor, S.A. Olayinka, W. Nwankwo, C. Alhasan, Evaluation of the effects of some weather variables on UHF and VHF receivers within Benin City, South-South region of Nigeria. J. Phys.: IOP Conf. Ser. **1299**, 012052 (2019)
43. K.E. Ukhurebor, O.J. Umukoro, Influence of meteorological variables on UHF radio signal: recent findings for EBS, Benin City, South-South, Nigeria. IOP Conf. Ser.: Earth Environ. Sci. **173**, 012017 (2018)
44. C.S. Pundir, N. Chauhan, Acetylcholinesterase inhibition-based biosensors for pesticide determination: a review. Anal. Biochem. **429**, 19–31 (2012)
45. G. Marrazza, Piezoelectric biosensors for organophosphate and carbamate pesticides: a review. Biosensors (Basel) **4**, 301–317 (2014)
46. B. Senthilkumaran, Pesticide and sex steroid analogue-induced endocrine disruption differentially targets hypothalamo-hypophyseal-gonadal system during gametogenesis in teleosts—A review. Gen. Comp. Endocrinol. **219**, 136–142 (2015)
47. F. Peng, Y. Su, Y. Zhong, C. Fan, S.T. Lee, Y. He, Silicon nanomaterials platform for bioimaging, biosensing, and cancer therapy. Acc. Chem. Res. **47**, 612–623 (2014)
48. W.A. El-Said, C.H. Yea, Il-K. Kwon, J.W. Choi, Fabrication of electrical cell chip for the detection of anticancer drugs and environmental toxicants effect. Biochip. J. **3**, 105–112 (2009)
49. W.A. El-Said, C.H. Yea, H. Kim, J.W. Choi, Fabrication of self-assembled RGD layer for cell chip to detect anticancer drug effect on HepG2 cells. Curr. Appl. Phys. **9**, 76–80 (2009)
50. H. Ogi, Wireless-electrodeless quartz-crystal-microbalance biosensors for studying interactions among biomolecules: a review. Proc. Jpn. Acad. Ser. B Phys. Biol. Sci. **89**, 401–417 (2013)
51. S.U. Senveli, O. Tigli, Biosensors in the small scale: methods and technology trends. IET Nanobiotechnol. **7**, 7–21 (2013)
52. X. Guo, Single-molecule electrical biosensors based on single-walled carbon nanotubes. Adv. Mater. **25**, 3397–3408 (2013)
53. E. Schneider, D.S. Clark, Cytochrome P450 (CYP) enzymes and the development of CYP biosensors. Biosens. Bioelectron. **39**, 1–13 (2013)
54. A.D. Dias, D.M. Kingsley, D.T. Corr, Recent advances in bioprinting and applications for biosensing. Biosensors (Basel) **4**, 111–136 (2014)

55. I. Khimji, E.Y. Kelly, Y. Helwa, M. Hoang, J. Liu, Visual optical biosensors based on DNA-functionalized polyacrylamide hydrogels. Methods **64**, 292–298 (2013)
56. S.J. Kwon, A.J. Bard, DNA analysis by application of Pt nanoparticle electrochemical amplification with single label response. J. Am. Chem. Soc. **134**, 10777–10779 (2012)
57. M.Y. Shen, B.R. Li, Y.K. Li, Silicon nanowire field-effect-transistor based biosensors: from sensitive to ultra-sensitive. Biosens. Bioelectron. **60**, 101–111 (2014)
58. M., Li, R. Li, C.M. Li, N. Wu, Electrochemical and optical biosensors based on nanomaterials and nanostructures: a review. Front. Biosci. (Schol Ed) **3**, 1308–1331 (2011)
59. Y. Zhou, C.W. Chiu, H. Liang, Interfacial structures and properties of organic materials for biosensors: an overview. Sensors (Basel) **12**, 15036–15062 (2012)
60. P.J. Ko, R. Ishikawa, H. Sohn, A. Sandhu, Porous silicon platform for optical detection of functionalized magnetic particles biosensing. J. Nanosci. Nanotechnol. **13**, 2451–2460 (2013)
61. C. Lamprecht, P. Hinterdorfer, A. Ebner, Applications of biosensing atomic force microscopy in monitoring drug and nanoparticle delivery. Expert. Opin. Drug Deliv. **11**, 1237–1253 (2014)
62. L. Su, W. Jia, C. Hou, Y. Lei, Microbial biosensors: a review. Biosens. Bioelectron. **26**, 1788–1799 (2011)
63. E. Hutter, D. Maysinger, Gold-nanoparticle-based biosensors for detection of enzyme activity. Trends Pharmacol. Sci. **34**, 497–507 (2013)
64. L. Ding, A.M. Bond, J. Zhai, J. Zhang, Utilization of nanoparticle labels for signal amplification in ultrasensitive electrochemical affinity biosensors: a review. Anal. Chim. Acta **797**, 1–12 (2013)
65. S. Kunzelmann, C. Solscheid, M.R. Webb, Fluorescent bio-sensors: design and application to motor proteins. EXS **105**, 25–47 (2014)
66. L. Oldach, J. Zhang, Genetically encoded fluorescent biosensors for live-cell visualization of protein phosphorylation. Chem. Biol. **21**, 186–197 (2014)
67. C. Randriamampita, A.C. Lellouch, Imaging early signalling events in T lymphocytes with fluorescent biosensors. Biotechnol. J. **9**, 203–212 (2014)
68. M.R. De, F. Carimi, W.B. Frommer, Mitochondrial biosensors. Int. J. Biochem. Cell Biol. **48**, 39–44 (2014)
69. M. Thunemann, K. Schmidt, W.C. de, X. Han, R.K. Jain, D. Fukumura, et al., Correlative intravital imaging of cGMP signals and vasodilation in mice. Front. Physiol. **5**, 394 (2014)
70. T. Su, Z. Zhang, Q. Luo, Ratiometric fluorescence imaging of dual bio-molecular events in single living cells using a new FRET pair mVenus/ mKOkappa-based biosensor and a single fluorescent protein biosensor. Biosens. Bioelectron. **31**, 292–298 (2012)
71. B.N. Johnson, R. Mutharasan, Biosensor-based microRNA detection: techniques, design, performance, and challenges. Analyst **139**, 1576–1588 (2014)
72. K. Park, J. Jung, J. Son, S.H. Kim, B.H. Chung, Anchoring foreign substances on live cell surfaces using Sortase A specific binding peptide. Chem. Commun. (Camb) **49**, 9585–9587 (2013)
73. J.Z. Sun, K.G. Peter, R.W. Si, D.D. Zhai, Z.H. Liao, D.Z. Sun et al., Microbial fuel cell-based biosensors for environmental monitoring: a review. Water Sci. Technol. **71**, 801–809 (2015)
74. Z. Du, H. Li, T. Gu, A state-of-the-art review on microbial fuel cells: a promising technology for wastewater treatment and bioenergy. Biotechnol. Adv. **25**, 464–482 (2007)
75. J.C. Gutierrez, F. Amaro, A. Martin-Gonzalez, Heavy metal whole- cell biosensors using eukaryotic microorganisms: an updated critical review. Front. Microbiol. **6**, 48 (2015)
76. F.W. Scheller, A. Yarman, T. Bachmann, T. Hirsch, S. Kubick, R. Renneberg et al., Future of biosensors: a personal view. Adv. Biochem. Eng. Biotechnol. **140**, 1–28 (2014)
77. S. Wang, G.M. Poon, W.D. Wilson, Quantitative investigation of protein-nucleic acid interactions by biosensor surface plasmon resonance. Methods Mol. Biol. **1334**, 313–332 (2015)
78. Z. Zhang, J. Liu, Z.M. Qi, D.F. Lu, In situ study of self-assembled nanocomposite films by spectral SPR sensor. Mater. Sci. Eng. C Mater. Biol. Appl. **51**, 242–247 (2015)
79. M.S. Artiles, C.S. Rout, T.S. Fisher, Graphene-based hybrid materials and devices for biosensing. Adv. Drug Deliv. Rev. **63**, 1352–1360 (2011)

80. F. Valentini, F.L. Galache, E. Tamburri, G. Palleschi, Single walled carbon nanotubes/polypyrrole-GOx composite films to modify gold micro-electrodes for glucose biosensors: study of the extended linearity. Biosens. Bioelectron. **43**, 75–78 (2013)
81. N.S. Fracchiolla, S. Artuso, A. Cortelezzi, Biosensors in clinical practice: focus on oncohematology. Sensors (Basel) **13**, 6423–6447 (2013)
82. K. Abe, W. Yoshida, K. Ikebukuro, Electrochemical biosensors using aptamers for theranostics. Adv. Biochem. Eng. Biotechnol. **140**, 183–202 (2014)
83. I. Grabowska, K. Malecka, U. Jarocka, J. Radecki, H. Radecka, Electrochemical biosensors for detection of avian influenza virus – current status and future trends. Acta Biochim. Pol. **61**, 471–478 (2014)
84. F. Mazzei, R. Antiochia, F. Botre, G. Favero, C. Tortolini, Affinity- based biosensors in sport medicine and doping control analysis. Bioanalysis **6**, 225–245 (2014)
85. A.J. Bandodkar, J. Wang, Non-invasive wearable electrochemical sensors: a review. Trends Biotechnol. **32**, 363–371 (2014)
86. A.T. Lawal, S.B. Adeloju, Progress and recent advances in fabrication and utilization of hypoxanthine biosensors for meat and fish quality assessment: a review. Talanta **100**, 217–228 (2012)
87. H. Kumar, R. Rani, Development of biosensors for the detection of biological warfare agents: its issues and challenges. Sci. Prog. **96**, 294–308 (2013)
88. J.L. Arlett, E.B. Myers, M.L. Roukes, Comparative advantages of mechanical biosensors. Nat. Nanotechnol. **6**, 203–215 (2011)
89. P.N. Navya, A. Kaphle, S.P. Srinivas, S.K. Bhargava, V.M. Rotello, H.K. Daima, Current trends and challenges in cancer management and therapy using designer nanomaterials. Nano Converg. **6**, 23 (2019)
90. J.Y. Kim, D. O'Hare, Monolithic nano-porous polymer in microfluidic channels for lab-chip liquid chromatography. Nano Converg. **5**, 19 (2018)
91. S. Soylemez, F.E. Kanik, S.D. Uzun, S.O. Hacioglu, L. Toppare, Development of an efficient immobilization matrix based on a conducting polymer and functionalized multiwall carbon nanotubes: Synthesis and its application to ethanol biosensors. J. Mater. Chem. B **2**, 511–521 (2014)
92. C. Dhand, S.P. Singh, S.K. Arya, M. Datta, B.D. Malhotra, Cholesterol biosensor based on electrophoretically deposited conducting polymer film derived from nano-structured polyaniline colloidal suspension. Anal. Chim. Acta **602**, 244–251 (2007)
93. M.A. Nandi, R.B. Gangopadhyay, A. Bhaumik, Mesoporous polyaniline having high conductivity at room temperature. Microporous Mesoporous Mater. **109**, 239 (2007)
94. S.E. Moulton, P.C. Innis, L.A.P. Kane-Maguire, O. Ngamna, G.G. Wallace, Polymerisation and characterisation of conducting polyaniline nanoparticle dispersions. Curr. Appl. Phys. **4**, 402–406 (2004)
95. M. Mazur, M. Tagowska, B. Pałys, K. Jackowska, Template synthesis of polyaniline and poly (2-methoxyaniline) nanotubes: comparison of the formation mechanisms. Electrochem. Commun. **5**, 403 (2003)
96. K.P. Lee, A. Gopalan, S. Komathi, D. Raghupathy, Polyaniline-based nanocomposites: Preparation, properties and applications, in *Physical Properties and Applications of Polymer Nanocomposites*, ed. by S.C. Tjong, Y.W. Mai 1st ed., (Woodhead Publishing, Cambridge, 2010)
97. J. Huang, R.B. Kaner, A General chemical route to polyaniline nanofibers. J. Am. Chem. Soc. **126**, 851–855 (2004)
98. S. Che, A.E.G. Bennett, T. Yokoi, K. Sakamoto, H. Kunieda, O. Terasaki, T. Tatsumi, A novel anionic surfactant templating route for synthesizing mesoporous silica with unique structure. Nat. Mater. **2**, 801–805 (2003)
99. D.D. Sawall, R.M. Villahermosa, R.A. Lipeles, A.R. Hopkins, Interfacial polymerization of polyaniline nanofibers grafted to Au surfaces. Chem. Mater. **16**, 1606 (2004)
100. D. Saini, R. Chauhan, T. Basu, Fabrication of biosensors based on nanostructured conducting polymer (NSCP). Int. Res. J. Biotechnol. **2**, 145–156 (2011)

101. K.J. Moreno, I. Moggio, E. Arias, I. Llarena, S.E. Moya, R.F. Ziolo, H. Barrientos, Silver nanoparticles functionalized in situ with the conjugated polymer (PEDOT: PSS). J. Nanosci. Nanotechnol. **9**, 3987–3992 (2009)
102. R.G. Melendez, K.J. Moreno, I. Moggio, E. Arias, A. Ponce, I. Llanera, S.E. Moya, On the influence of silver nanoparticles size in the electrical conductivity of PEDOT: PSS. Mater. Sci. Forum **644**, 85–90 (2010)
103. I. Hussain, M. Brust, A.J. Papworth, A.I. Cooper, Preparation of acrylate-stabilized gold and silver hydrosols and gold-polmer composite films. Langmuir **19**, 4831–4835 (2003)
104. H. Cong, M. Radosz, B.F. Towler, Y. Shen, Polymer-inorganic nanocomposite membranes for gas separation. Sep. Purif. Technol. **55**, 281–291 (2007)
105. Y. Feng, T. Yang, W. Zhang, C. Jiang, K. Jiao, Enhanced sensitivity for deoxyribonucleic acid electrochemical impedance sensor: Gold nanoparticle/polyaniline nanotube membranes. Anal. Chim. Acta. **616**, 144–151 (2008)
106. A. Balamurugan, S.M. Chen, Silver nanograins incorporated PEDOT modified electrode for electrocatalytic sensing of hydrogen peroxide. Electroanalysis **21**, 1419–1423 (2009)
107. S.W. Dutse, N.A. Yusof, H. Ahmad, M.Z. Hussein, Z. Zainal, An electrochemical DNA biosensor for Ganodermaboninense pathogen of oil palm utilizing a new ruthenium complex, [Ru(dppz)2(qtpy)] Cl2. Int. J. Electrochem. Sci. **7**, 8105–8115 (2012)
108. G. Yang, K.L. Kampstra, M.R. Abidian, High-performance conducting polymer nanofiber biosensors for detection of biomolecules. Adv. Mater. (2014)
109. H. Zhu, L. Li, W. Zhou, Z. Shao, X. Chen, Advances in non-enzymatic glucose sensors based on metal oxides. J. Mater. Chem. B **4**, 7333–7349 (2016)
110. A.A. Elkhawaga, M.M. Khalifa, O. El-badawy, M.A. Hassan, W.A El-Said, Rapid and highly sensitive detection of pyocyanin biomarker in different Pseudomonas aeruginosa infections using gold nanoparticles modified sensor. PLoS ONE (2019)
111. J.H. Lee, W.A. El-Said, B.K. Oh, J.W. Choi, Enzyme-free glucose sensor based on Au nanobouquet fabricated indium tin oxide electrode. J. Nanosci. Nanotechnol. **14**, 8432–8438 (2014)
112. P. Salazar, V. Rico, A.R. González-Elipe, Nickel-copper bilayer nanoporous electrode prepared by physical vapor deposition at oblique angles for the non-enzymatic determination of glucose. Sens. Actuator B **226**, 436–443 (2016)
113. T. Soganci, H.C. Soyleyici, D.O. Demirko, M. Ak, S. Timur, Use of super-structural conducting polymer as functional immobilization matrix in biosensor design. J. Electrochem. Soc. **165**, 22–26 (2018)
114. R.E. Munteanu, R. Ye, C. Polonschii, A. Ru, M. Gheorghiu, E. Gheorghiu, R. Boukherroub, W. Schuhmann, S. Melinte, S. Gáspár, High spatial resolution electrochemical biosensing using reflected light microscopy. Sci. Rep. **9**, 15196 (2019)
115. C.H. Yea, B. Lee, H.H. Kim, S.U. Kim, W.A. El-Said, J.H. Minc, B.K. Oh, J.W. Choi, The immobilization of animal cells using the cysteine-modified RGD oligopeptide. Ultramicroscopy **108**, 1144–1147 (2008)
116. W.A. El-Said, C.H. Yea, J.W. Choi, Il-K. Kwon, Ultrathin polyaniline film coated on an indium–tin oxide cell-based chip for study of anticancer effect. Thin Solid Film **518**, 661–667 (2009)
117. J.Y. Lee, J.W. Lee, C.E. Schmidt, Neuroactive conducting scaffolds: nerve growth factor conjugation on active ester-functionalized polypyrrole. J. R. Soc. Interface **6**, 801–810 (2009)
118. A.E. Elkholy, F.E.T. Heakal, W.A. El-Said, Improving the electrocatalytic performance of Pd nanoparticles supported on indium/tin oxide substrates towards glucose oxidation. Appl. Catal. A: Gen. **580**, 28–33 (2019)
119. W.A. El-Said, C.H. Yea, M. Jung, H.C. Kim, J.W. Choi, Analysis of effect of nanoporous alumina substrate coated with polypyrrole nanowire on cell morphology based on AFM topography. Ultramicroscopy **110**, 676–681 (2010)
120. L.T. Strover, J. Malmström, O. Laita, J. Reynisson, N. Aydemir, M.K. Nieuwoudt, D.E. Williams, P.R. Dunbar, M.A. Brimble, J. Travas-Sejdic, A new precursor for conducting polymer-based brush interfaces with electroactivity in aqueous solution. Polymer **54**, 1305–1317 (2013)

121. W.A. El-Said, J.W. Choi, Electrochemical Biosensor consisted of conducting polymer layer on gold nanodots patterned Indium Tin Oxide electrode for rapid and simultaneous determination of purine bases. Electrochim. Acta **123**, 51–57 (2014)
122. B. Aksoy, A. Pafsahan, Ö. Güngör, S. Köytepe, T. Seçkin, A novel electrochemical biosensor based on polyimide-boron nitride composite membranes. Int. J. Polym. Mater. Polym. Biomater. **66**, 203–212 (2017)
123. M.M. Khalifa, A.A. Elkhawaga, M.A. Hassan, A.M. Zahran, A.M. Fathalla, W.A. El-Said, O. El-Badawy, Highly specific electrochemical sensing of pseudomonas aeruginosa in patients suffering from corneal ulcers: a comparative study. Sci. Rep. **9**, 18320 (2019)
124. C. Tancharoen, W. Sukjee, C. Thepparit, T. Jaimipuk, P. Auewarakul, A. Thitithanyanont, C. Sangma, Electrochemical biosensor based on surface imprinting for zika virus detection in Serum. Acs Sens. **4**, 69–75 (2019)
125. A.M. Pappa, D. Ohayon, A. Giovannitti, I.P. Maria, A. Savva, I. Uguz, J. Rivnay, I. McCulloch, R.M. Owens, S. Inal, Direct metabolite detection with an n-type accumulation mode organic electrochemical transistor. Sci. Adv. **4**, eaat0911 (2018)
126. A. Rico-Yuste, S. Carrasco, Molecularly imprinted polymer-based hybrid materials for the development of optical sensors. Polymers **11**, 1173 (2019)
127. L. Liu, X. Zhu, Y. Zeng, H. Wang, Y. Lu, J. Zhang, Z. Yin, Z. Chen, Y. Yang, L. Li, An electrochemical sensor for diphenylamine detection based on reduced graphene Oxide/Fe_3O_4-molecularly imprinted polymer with 1,4-Butanediyl-3,30-bis-l vinylimidazolium dihexafluorophosphate Ionic liquid as cross-linker. Polymers **10**, 1329 (2018)
128. B. Liu, J. Yan, M. Wang, X. Wu, Electrochemical sensor based on molecularly imprinted polymer for determination of nonylphenol. Int. J. Electrochem. Sci. **13**, 11953–11960 (2018)
129. H. Essousi, H. Barhoumi, M. Bibani, N. Ktari, F. Wendler, A. Al-Hamry, O. Kanoun, Ion-imprinted electrochemical sensor based on copper nanoparticles-polyaniline matrix for nitrate detection. J. Sens. **2019**, 2019
130. B. Öndes, M. Soysal, Determination of diuron by using electrochemical sensor based on molecularly imprinted polymer film. J. Electrochem. Soc. **166**, 395–401 (2019)
131. F. Samari, B. Hemmateenejad, Z. Rezaei, M. Shamsipur, A novel approach for rapid determination of vitamin B12 in pharmaceutical preparations using BSA-modified gold. Anal. Methods **4**, 4155–4160 (2012)
132. M. Nakos, I. Pepelanova, S. Beutel, U. Krings, R.G. Berger, T. Scheper, Isolation and Analysis of Vitamin B12 from Plant Samples. Food Chem. **216**, 301–308 (2017)
133. R. Singh, S. Jaiswal, K. Singh, S. Fatma, B.B. Prasad, Biomimetic polymer-based electrochemical sensor using methyl blue adsorbed reduced graphene oxide and functionalized multiwalled carbon nanotubes for trace sensing of cyanocobalamin. ACS Appl. Nano Mater. **1**, 4652–4660 (2018)
134. G.-P. Nikoleli, D.P. Nikolelis, C. Siontorou, S. Karapetis, S. Bratakou, N. Tzamtzis, Nanobiosensors based on graphene electrodes: recent trends and future applications, in *Applications of nanomaterials* (Elsevier Ltd. Amsterdam, 2018)
135. C. Karunakaran, K. Bhargava, R. Benjamin, *Biosensors and Bioelectronics*, 1st edn. (Elsevier, Amsterdam, 2015)
136. C. Leger, P. Bertrand, Direct electrochemistry of redox enzymes as a tool for mechanistic studies. Chem. Rev. **108**, 379–2438 (2008)
137. Y. Yao, K. Shiu, Direct electrochemistry of glucose oxidase at carbon nanotube-gold colloid modified electrode with poly (diallyldimethylammonium chloride) coating. Electroanalysis **20**, 1542–1548 (2008)
138. S. Andreescu, L.A. Luck, Studies of the binding and signalling of surface-immobilized periplasmic glucose receptors on gold nanoparticles: a glucose biosensor application. Anal. Biochem. **375**, 282–290 (2008)
139. Y. Wang, Y. Yao, Direct electron transfer of glucose oxidase promoted by carbon nanotubes is without value in certain mediator-free applications. Microchim. Acta **176**, 271–277 (2012)
140. X. Zhang, Q. Liao, M.S. Chu, Liu, Y. Zhang, Structure effect on graphene-modified enzyme electrode glucose sensors. Biosens. Bioelectron. **52**, 281–287 (2014)

141. P. Zhang, X. Zhao, Y. Ji, Z. Ouyang, X. Wen, J. Li, Z. Su, G. Wei, Electrospinning graphene quantum dots into a nanofibrous membrane for dual-purpose fluorescent and electrochemical biosensors. J. Mater. Chem. B **3**, 2487–2496 (2015)
142. S. Liu, J. Tian, L. Wang, Y. Luo, W. Lu, X. Sun, Self-assembled graphene platelet-glucose oxidase nanostructures for glucose biosensing. Biosens. Bioelectron. **26**, 4491–4496 (2011)
143. J. Huang, L. Zhang, R.P. Liang, J.D. Qiu, "On-off" switchable electrochemical affinity nanobiosensor based on graphene oxide for ultrasensitive glucose sensing. Biosens. Bioelectron. **41**, 430–435 (2013)
144. Y. Zhao, A. Kim, G. Wan, B.C. Tee, Design and applications of stretchable and self-healable conductors for soft electronics. Nano Converg. **6**, 25 (2019)
145. H. Razmi, R. Mohammad-Rezaei, Graphene quantum dots as a new substrate for immobilization and direct electrochemistry of glucose oxidase: application to sensitive glucose determination. Biosens. Bioelectron. **41**, 498–504 (2013)
146. Z. Li, C. Xie, J. Wang, A. Meng, F. Zhang, Direct electrochemistry of cholesterol oxidase immobilized on chitosan–graphene and cholesterol sensing. Sens. Actuators B Chem. **208**, 505–511 (2015)
147. Z. Fan, Q. Lin, P. Gong, B. Liu, J. Wang, S. Yang, A new enzymatic immobilization carrier based on graphene capsule for hydrogen peroxide biosensors. Electrochim. Acta **151**, 186–194 (2015)
148. Y. Zhou, H. Yin, X. Meng, Z. Xu, Y. Fu, S. Ai, Direct electrochemistry of sarcosine oxidase on graphene, chitosan and silver nanoparticles modified glassy carbon electrode and its biosensing for hydrogen peroxide. Electrochim. Acta **71**, 294–301 (2012)
149. L.M. Lu, X.L. Qiu, X.B. Zhang, G.L. Shen, W. Tan, R.Q. Yu, Supramolecular assembly of enzyme on functionalized graphene for electrochemical biosensing. Biosens. Bioelectron. **45**, 102–107 (2013)
150. X. Liu, J. Zhang, R. Yan, Q. Zhang, X. Liu, Preparation of graphene nanoplatelet-titanate nanotube composite and its advantages over the two single components as biosensor immobilization materials. Biosens. Bioelectron. **51**, 76–81 (2014)
151. A. Muthurasu, V. Ganesh, Horseradish peroxidase enzyme immobilized graphene quantum dots as electrochemical biosensors. Appl. Biochem. Biotechnol. **174**, 945–959 (2014)
152. A. Gasnier, M. Laura Pedano, M.D. Rubianes, G.A. Rivas, Graphene paste electrode: electrochemical behavior and analytical applications for the quantification of NADH. Sens. Actuators B Chem. **176**, 921–926 (2013)
153. H. Teymourian, A. Salimi, S. Khezrian, Fe_3O_4 magnetic nanoparticles/reduced graphene oxide nanosheets as a novel electrochemical and bioeletrochemical sensing platform. Biosens. Bioelectron. **49**, 1–8 (2013)
154. P. Li, X. Chen, W. Yang, Graphene-induced self-assembly of peptides into macroscopic-scale organized nanowire arrays for electrochemical NADH sensing. Langmuir **29**, 8629–8635 (2013)
155. P.P. Gai, C.E. Zhao, Y. Wang, E.S. Abdel-Halim, J.R. Zhang, J.J. Zhu, NADH dehydrogenase-like behaviour of nitrogen-doped graphene and its application in NAD(+)- dependent dehydrogenase biosensing. Biosens. Bioelectron. **62**, 170–176 (2014)
156. R.S. Dey, C.R. Raj, Development of an amperometric cholesterol biosensor based on graphene-Pt nanoparticle hybrid material. J. Phys. Chem. C **114**, 21427–21433 (2010)
157. H.B. Nguyen, V.C. Nguyen, V.T. Nguyen, H.D. Le, V.Q. Nguyen, T.T. Tam Ngo, Q.P. Do, X.N. Nguyen, N.M. Phan, D. Lam Tran, Development of the layer-by-layer biosensor using graphene films: application for cholesterol determination. Adv. Nat. Sci. Nanosci. Nanotechnol. **4**, 015013 (2013)
158. M.Q. Israr, K. ul Hasan, J.R. Sadaf, I. Engquist, O. Nour, M. Willander, B. Danielsson, Structural characterization and biocompatible applications of graphene nanosheets for miniaturization of potentiometric cholesterol biosensor. J. Biosens. Bioelectron. **2**(3), 109–113 (2011)
159. G.P. Nikoleli, Z. Ibupoto, D. Nikolelis, V. Likodimos, N. Psaroudakis, N. Tzamtzis, M. Willander, T. Hianik, Potentiometric cholesterol biosensing application of graphene electrode with stabilized polymeric lipid membrane. Cent. Eur. J. Chem. **11**, 1554–1561 (2013)

160. G.P. Nikoleli, M.Q. Israr, N. Tzamtzis, D.P. Nikolelis, M. Willander, N. Psaroudakis, Structural characterization of graphene nanosheets for miniaturization of potentiometric urea lipid film-based biosensors. Electroanalysis **24**, 1285–1295 (2012)
161. E. Palek, M. Fojta, Peer reviewed: detecting DNA hybridization and damage. Anal. Chem. **73**, 74–83 (2001)
162. J.P. Tosar, G. Branas, J. Laíz, Electrochemical DNA hybridization sensors applied to real and complex biological samples. Biosens. Bioelectron. **26**, 1205–1217 (2010)
163. M. Zhou, Y. Zhai, S. Dong, Electrochemical sensing and biosensing platform based on chemically reduced graphene oxide. Anal. Chem. **81**, 5603–5613 (2009)
164. C.X. Lim, H.Y. Hoh, P.K. Ang, K.P. Loh, Direct voltammetric detection of DNA and pH sensing on epitaxial graphene: an insight into the role of oxygenated defects. Anal. Chem. **82**, 7387–7393 (2010)
165. M. Pumera, T. Sasaki, H. Iwai, Relationship between carbon nanotube structure and electrochemical behavior: heterogeneous electron transfer at electrochemically activated carbon nanotubes. Chem. Asian. J. **3**, 2046–2055 (2008)
166. A. Ambrosi, M. Pumera, Stacked graphene nanofibers for electrochemical oxidation of DNA bases. Phys. Chem. Chem. Phys. **12**, 8943–8947 (2010)
167. G.P. Nikoleli, D.P. Nikolelis, N. Tzamtzis, N. Psaroudakis, A selective immunosensor for D-dimer based on antibody immobilized on a graphene electrode with incorporated lipid films. Electroanalysis **26**, 1522–1527 (2014)
168. B. Liu, D. Du, X. Hua, X.Y. Yu, Y. Lin, Paper-based electrochemical biosensors: from test strips to paper-based microfluidics. Electroanalysis **26**, 1214–1223 (2014)
169. Y. Wan, Y. Su, X. Zhu, G. Liu, C. Fan, Development of electrochemical immunosensors towards point of care diagnostics. Biosens. Bioelectron. **47**, 1–11 (2013)
170. S. Yang, G. Li, D. Wang, Z. Qiao, L. Qu, Synthesis of nanoneedle-like copper oxide on N-doped reduced graphene oxide: a three-dimensional hybrid for nonenzymatic glucose sensor. Sens. Actuator B **238**, 588–595 (2017)
171. K.E. Ukhurebor, C.O. Adetunji, A.O. Bobadoye, U.O. Aigbe, R.B. Onyancha, I.U. Siloko, J.O. Emegha, G.O. Okocha, I.C. Abiodun. Bionanomaterials for biosensor technology, in *Bionanomaterials: Fundamentals and Biomedical Applications*, ed. by R.P. Singh, K.R.B. Singh (Institute of Physics Publishing, In press, 2021) ISBN: 9780750337656
172. R.G. Kerry, K.E. Ukhurebor, S. Kumari, G.K. Maurya, S. Patra, B. Panigrahi, S. Majhi, J.R. Rout, M.D.P. Rodriguez-Torres, G. Das, H-S. Shin, J.K. Patra, A comprehensive review on the applications of nano-biosensor based approaches for non-communicable and communicable disease detection. Biomater. Sci. https://doi.org/10.1039/D0BM02164D (2021)
173. K.E. Ukhurebor, C.O. Adetunji, Relevance of biosensor in climate smart organic agriculture and their role in environmental sustainability: what has been done and what we need to do, in *Biosensors in Agriculture: Recent Trends and Future Perspectives*, ed. by R.N. Pudake, U. Jain, C. Kole, *Concepts and Strategies in Plant Sciences* (Springer, Cham, pp. 115–136). https://doi.org/10.1007/978-3-030-66165-6_7 (2021)
174. C.O. Adetunji, W. Nwankwo, K.E. Ukhurebor, A.S. Olayinka, A.S. Makinde. Application of biosensor for the identification of pests and diseases mitigating against increase in agricultural production: recent advances, in *Biosensors in Agriculture: Recent Trends and Future Perspectives*, R.N. Pudake, U. Jain, C. Kole, (Springer, Cham, pp. 169–189). https://doi.org/10.1007/978-3-030-66165-6_9 (2021)
175. C.O. Adetunji, K.E. Ukhurebor, Recent trends in utilization of biotechnological tools for environmental sustainability, in *Microbial Rejuvenation of Polluted Environment. Microorganisms for Sustainability*, vol. 27, ed. by C.O. Adetunji, D.G. Panpatte, Y.K. Jhala (Springer, Cham, pp. 239–263, 2021). https://doi.org/10.1007/978-981-15-7459-7_11
176. K.E. Ukhurebor, The role of biosensor in climate smart organic agriculture towards agricultural and environmental sustainability, in *Agrometeorology*, ed. by R.S. Meena (IntechOpen, London). https://doi.org/10.5772/intechopen.93150 (2020)
177. W. Nwankwo, K.E. Ukhurebor, Nanoinformatics: opportunities and challenges in the development and delivery of healthcare products in developing countries. IOP Conf. Ser.: Earth Environ. Sci. **655**, 012018 (2021)

178. K.E. Ukhurebor, U.O. Aigbe, R.B. Onyancha, W. Nwankwo, O.A. Osibote, H.K. Paumo, O.M. Ama, C.O. Adetunji, I.U. Siloko, Effect of hexavalent chromium on the environment and removal techniques: a review. J. Environ. Manage. **280**, 111809 (2021)
179. R.B. Onyancha, U.O. Aigbe, K.E. Ukhurebor, P.W. Muchiri, Facile synthesis and applications of carbon nanotubes in heavy-metal remediation and biomedical fields: a comprehensive review. J. Mol. Struct. **1238**, 1–25 (2021)

Index

A
Absorption, 67
Agglomeration, 172
Anodic stripping voltammetry, 53

B
Bidimensional, 145
Biomaterials, 194
Biosensing-based, 222
Biosensor, 124, 128
Bi-polymeric, 162
Boron-doped, 166

C
Cadmium ions, 21
Carbon-built, 197
Carbon electrode, 52
Chronopotentiometry, 17
Corrosion, 92, 93
Crystallographic, 163
Cyclic voltammetry, 160

D
Degradation processes, 116
Dihexadecyl hydrogen phosphate, 174

E
Electroactive polymers, 55
Electrochemical, 12, 52, 65, 67, 92, 139, 157, 169
Electrochemical processes, 2
Electrochemical reactions, 4
Electrochemistry, 108, 158
Electrode, 159, 160
Electrodeposition, 7
Electrolyte, 110
Electronic, 147
Electron Microscopy (EM), 12
Electrons, 5
Environment, 52
Environmental, 52, 127, 146, 189

F
Fabricated, 69
Film-modified, 52
Functionalized, 70

G
Glassy, 52
Graphene, 70
Graphene-based, 138
Graphene-based nanocomposites, 16

H
Hydrazine hydrate, 172
Hydrothermal, 143
Hydroxyethyl imidazoline, 94

I
Immobilization, 197
Impedance, 14
Inorganic, 65
Inorganic ions, 66
Ions detected, 56

O. M. Ama et al. (eds.), *Modified Nanomaterials for Environmental Applications*, Engineering Materials, https://doi.org/10.1007/978-3-030-85555-0

M
Material science, 1
Mesoporous, 96
Microorganisms, 201
Microscopic, 168
Morphological, 113, 163
Multiwalled carbon nanotubes, 159

N
Nanocomposite, 52, 56, 93, 162
Nanocomposite-based, 18
Nanocomposite coatings, 100
Nanocrystalline, 92
Nanofillers, 21
Nanomaterials, 8, 93, 126, 157, 158, 173
Nano-molecularly, 159
Nanoparticle, 11, 94, 160
Nanoparticle-modified, 21
Nanotechnology, 110
Non-diamond, 163
Nyquist, 15

O
Organic nanomaterials, 69
Oxidation, 97

P
Pharmaceutical, 155, 172
Pharmaceutical determination, 13
Photo-active, 116
Photoanode, 114, 116, 137
Photocatalysts, 110, 116
Photocatalytic, 138
Photoelectrochemical, 107, 124
Photosensitizer, 128
Photovoltaics, 148
Physical contamination, 55
Physicochemical, 69
Physiochemical, 189
Polymer-based, 54
Potentiometric technique, 12

S
Sensing, 67
Sensors, 189, 197
Sodium, 172
Solar, 142
Solar-to-hydrogen, 148
Surface area, 176

T
Toxic, 26
Toxicity, 51

V
Voltammetry, 56

W
Wastewater pollutant, 115

Z
Zinc, 52
Zirconium oxide, 170